14TH INTERNATIONAL CONFERENCE ON TURBOCHARGERS AND TURBOCHARGING

T0271971

PROCEEDINGS OF THE INTERNATIONAL CONFERENCE ON
TURBOCHARGERS AND TURBOCHARGING (LONDON, UK, 2021)

14TH INTERNATIONAL CONFERENCE ON

TURBOCHARGERS AND TURBOCHARGING

Edited by

Institution of Mechanical Engineers

CRC Press
Taylor & Francis Group
Boca Raton London New York Leiden

CRC Press is an imprint of the
Taylor & Francis Group, an **informa** business

A BALKEMA BOOK

CRC Press/Balkema is an imprint of the Taylor & Francis Group, an informa business

Typeset by Integra Software Services Pvt. Ltd., Pondicherry, India

Published by: CRC Press/Balkema
Schipholweg 107C, 2316XC Leiden, The Netherlands
e-mail: Pub.NL@taylorandfrancis.com
www.routledge.com – www.taylorandfrancis.com

ISBN: 978-0-367-67645-2 (Hbk)
ISBN: 978-1-003-13217-2 (eBook)
DOI: 10.1201/9781003132172
https://doi.org/10.1201/9781003132172

Table of Contents

ADDITIONAL PAPERS

Organising Committee

Powertrains Systems and Fuels Group

The Institution of Mechanical Engineers

Member Credits

Ricardo Martinez-Botas (Chair)	Imperial College London
Peter Davies (Vice-Chair)	Garett – Advancing Motion
Jamie Archer	Cummins Turbo Technologies
Lutz Aurahs	MAN Energy Solutions
Alan Baker	Jaguar Land Rover
Kian Banisoleiman	Lloyd's Register EMEA
Andrew Banks	Ricardo
Dr Elias Chebli	Porsche
Colin Copeland	University of Bath
Michael Dolton	Cummins Turbo Technologies
Dietmar Filsinger	IHI Charging System International GmbH
Dino Imhof	ABB Turbocharging
Steve Johnson	Ford Motor Company
Rogier Lammers	Mitsubishi Turbocharger and Engine Europe B.V.
Per-Inge Larson	SCANIA Power Train Sweden
Nathan McArdle	BorgWarner
Takashi Mori	IHI Corporation

Takashi Otobe   Honda R&D

Prabhu Ramasamy   Caterpillar

Joerg Seume   Leibniz Universitaet Hannover

Stephen Spence   Queen's University Belfast

Bernd Wietholt   Volkswagen AG

Michael Willmann   ABB Switzerland Ltd

Design and evaluation of an active inlet swirl control device for automotive turbocharger compressors

C. Stuart, S.W. Spence

School of Mechanical and Aerospace Engineering, Queen's University Belfast, UK

S. Teichel, A. Starke

IHI Charging Systems International GmbH, Heidelberg, Germany

ABSTRACT

It is widely accepted that large swirl angles (>60°) are required to deliver meaningful surge margin extension at the relatively low pressure ratios typical of automotive turbocharger compressors. However, in order to maintain performance towards choke (and hence the rated power point of the engine), the requirement is for delivery of zero pre-swirl. These constraints cannot be met using traditional variable inlet guide vane systems, as whether flat plate or cambered vanes are chosen, significant losses are to be expected at low and high mass flow rates respectively.

Taking the above into account, the primary intention of this study was to develop a device capable of efficiently generating large swirl angles for surge margin extension and compressor efficiency enhancement at low mass flow operating points, without adversely affecting performance at other areas of the compressor map.

The chosen concept involved placing an axial fan upstream of a standard automotive turbocharger compressor stage. The fan was designed to act as a variable pre-swirl device, which due to being driven by an electric motor, was capable of operating completely independently of the centrifugal impeller. The chosen concept was progressed through 1-D and 3-D design phases to understand the feasibility of the system, before committing to hardware manufacture and an extensive experimental test campaign to validate the numerical findings and test the surge margin extension potential of the device.

1   INTRODUCTION

In order to ensure compliance with the increasingly stringent emissions targets being levied by legislators around the world, the uptake of the concept of engine downsizing (as well as additional complementary technologies such as Miller valve timing) in the automotive sector is becoming increasingly prevalent. From the perspective of the turbocharger, these strategies place additional demands upon enhancing compressor performance and stability at low mass flow operating conditions.

One approach to achieve this goal that is common in larger scale applications is the application of positive pre-swirl (introduction of a non-zero circumferential velocity component, V_{u1}, in the same direction as impeller rotation) at compressor inlet through the use of inlet guide vanes. For automotive turbocharging applications, the expected improvements can be summarised as follows. Firstly, improvements in compressor efficiency can be expected at low mass flow rates due to the alleviation of incidence based losses. One study utilising an automotive turbocharger compressor [1] cited up to a 3.0%pt improvement in efficiency at a pressure ratio of 2.5 through the application of pre-swirl. Secondly, the possibility to extend the compressor surge margin is a particularly attractive benefit. However, due to the relatively modest pressure ratios of automotive turbocharger compressors, it has been recognised that high levels of swirl are required to achieve a tangible improvement (60°-70° @ PR<3.0) [2]. Finally, through effectively unloading the compressor, application of positive pre-swirl has been demonstrated to offer improved transient response from the turbocharger, and hence reduced engine time-to-torque figures [3].

In spite of the above mentioned benefits, the uptake of inlet guide vane (IGV) systems (either fixed or variable) in automotive turbocharging applications has been very limited. The nature of the requirements imposed on the compressor stage necessitate a very wide operating range, which creates significant issues for the application of pre-swirl. At high mass flow rates, intolerably large reductions in compressor efficiency (arising from increased incidence losses) and choking mass flow rate [1,4] are to be expected. Ultimately then, an ideal IGV system should be capable of delivering high swirl angles for low mass flow operating conditions, and zero swirl towards choke. While this is clearly an impossibility from a fixed geometry system, unfortunately, even moving to a variable inlet guide vane (VIGV) configuration does not entirely overcome this limitation. A comparatively recent study [5] evaluated the feasibility of a number of advanced VIGV concepts for use with an automotive turbocharger compressor (including variable vane camber), but ultimately could not deliver meaningful improvements in performance.

As a consequence of these requirements, the current study sought to develop a device capable of efficiently generating large swirl angles at low mass flow rates, while also being capable of generating zero pre-swirl for operation close to choke. The chosen solution involved positioning an electrically driven axial fan upstream of a standard automotive turbocharger compressor stage. This concept was first presented in a patent document by Daimler [6] however despite apparently showing potential, the authors are not aware of it being investigated beyond an initial feasibility study. This architecture is advantageous from the perspective that the fan speed can be controlled completely independently of the turbocharger shaft speed, allowing the upstream axial fan to act as an active inlet swirl control device with the potential to utilise the work input to overcome a proportion of the losses typical of IGV systems. Furthermore, it was envisioned that the capability to actively interact with the impeller inlet recirculation would yield the potential to suppress its extent and hence further contribute to enhancing compressor stability [2].

It is also worth addressing the reasons behind choosing to use an electric motor to generate pre-swirl rather than direct boosting in this context. Primarily, targeting swirl generation allows the use of a much lower power motor than would be required to achieve meaningful pressure rise, making the system less expensive, more robust and easier to integrate with both the base turbocharger and the vehicle itself. Furthermore, utilising a two-stage architecture does not improve operating range, which represents one of the key targets of this study.

2   METHODOLOGY

Prior to engaging in a detailed description of the design methodology employed, it is worth first outlining the constraints involved. The primary aim from the commencement of the investigation was to arrive at a configuration that could be practically tested and feasibly implemented (with reasonably limited further design effort) into a real world application. As a part of this, and to control costs at the initial prototyping stage, it was deemed that the electrical motor had to be a readily available "off-the-shelf" unit. Consequently, due to the relatively large dimensions of suitable production electric motors, packaging of the system was not considered an overarching limitation. However, the use of a very large, slower rotating fan to generate the swirl was not deemed as being feasible in this case.

Regarding the choice of base compressor stage, in keeping with the desire to utilise production components where possible, a typical automotive turbocharger was chosen from a gasoline engine application. The compressor impeller comprised of six main and six splitter blades, with D_2=51mm and β_{2b} = -44.0°. In order to focus the design effort on the most important regions of compressor operation, and to streamline to presentation of results and subsequent analysis, six key operating points of interest were defined, as illustrated in Figure 1. The designation describing each operating point can be decomposed in the following fashion; "120k" represents a turbocharger shaft speed of 120,000rpm, with "NS," "DP," and "NC" representing near surge, design point and near choke operation respectively.

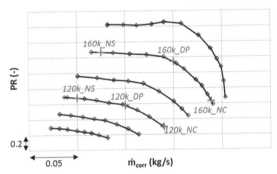

Figure 1. Base compressor map highlighting key operating points of interest.

2.1   Modelling
The aerodynamic modelling work undertaken comprised of 1-D fan design, as well as 3-D CFD design and performance prediction of the coupled centrifugal stage and axial fan. The 1-D design phase permitted selection of appropriate initial fan design parameters, and highlighted the need for a part span configuration to allow generation of the required swirl levels while minimising the impact on the choke flow capacity of the complete compression system. This point will be clarified in the coming section, however the unavoidable (and significant) interaction between the impeller inlet recirculation and the fan quickly dictated that CFD analysis was the only meaningful way to proceed.

Having settled on a geometry specification, the design was progressed through FEA modelling of the fan and spinner to ensure structural integrity at the operating points derived from the aerodynamic study. In addition, a simplified brushless direct current (BLDC) motor performance model was developed and utilised to ensure that the chosen operating points were capable of being attained by the electric motor. The wide ranging constraints involved dictated something of an iterative approach to the above modelling work, however it is presented herein in terms of the logical, high level, global work flow.

2.1.1 *CFD methodology*
The approach used for the baseline CFD configuration was that as presented by Stuart *et al.* [7], representing a typical single passage analysis comprising of two stationary domains (inlet and vaneless diffuser) and one rotating domain (impeller). Consequently, for the sake of brevity, the current publication will not enter into a detailed description of this method. However, a schematic representation of the setup

3

employed has been included in Figure 2 for completeness sake. In each case, stage performance was defined from measurement planes coincident with the inlet and vaneless diffuser exit (MP0 to MP3).

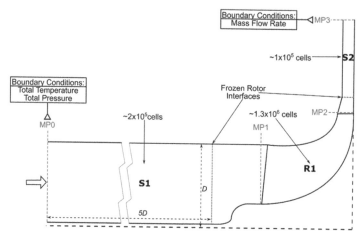

Figure 2. Baseline CFD setup [7].

The first piece of analysis undertaken using the CFD setup was to evaluate the potential benefits of inlet swirl at the previously defined operating points of interest. In order to define a "best case scenario" for this evaluation, a uniformly distributed non-zero theta component was added to define the flow direction at the inlet boundary, thus signifying a condition whereby the pre-swirl could be generated on an entirely loss free basis. From this it was determined that the optimum swirl levels to maximise compressor efficiency were 60° at the NS points, 20° at the DP points and 0° at choke. The overall impact of the inclusion of swirl is presented and discussed towards the conclusion of this section.

Having understood the magnitude of the theoretically achievable benefits available, the next step was to evaluate how closely it would be possible to replicate these results using the proposed concept. In order to facilitate this, the stationary inlet domain (represented by "S1" in Figure 2) was replaced with a single passage rotating domain containing the fan geometry. All other aspects of the model remained unchanged in comparison to the baseline configuration. The new inlet domain was generated in ANSYS BladeGen and meshed in ANSYS TurboGrid (in both cases utilising version 17.2 of the software), requiring approximately three million cells to satisfy mesh quality and grid independence requirements. The resulting domain setup is illustrated in Figure 3, which highlights the different measurement planes used to define performance of both the fan and the overall system, as well as the treatment of the different walls to render certain regions effectively stationary as necessary.

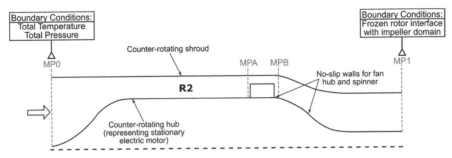

Figure 3. CFD setup for fan domain and measurement plane naming conventions.

Figure 3 illustrates the "unconventional" inlet geometry brought about by the large diameter of the chosen electric motor (36mm). While this posed challenges in ensuring that any total pressure losses arising from the change in section were minimised, it did facilitate the use of slightly lower fan speeds to generate a given level of swirl at compressor inlet (due to conservation of angular momentum considerations). The profile of the resulting "S-duct" connecting the fan to the impeller was optimised to minimise total pressure losses, while the extended fan spinner was identified as an important inclusion early in the CFD design process. Upon removal of this feature, stage efficiency was witnessed to reduce by up to 4%pts due to reduced flow guidance.

Figure 3 also demonstrates the relatively large axial separation between the fan and centrifugal impeller. This decision was taken during the CFD design process for two reasons. Firstly, in a more closely coupled arrangement, the interaction between the fan and the shroud side impeller inlet recirculating flow proved to be beyond what the steady single passage CFD methodology could rationalise. Consequently, in order to allow the solver to achieve converged solutions (particularly at lower mass flow operating conditions), the axial separation was increased as shown. Secondly, reducing (but not eliminating) the direct interaction between the fan and impeller inlet recirculation made it feasible to evaluate fan performance in isolation.

It is worth explicitly noting at this stage that the intention for the CFD study was not to gain direct insight into the potential for surge margin extension; one cannot expect a steady single passage analysis to accurately predict the onset of a fundamentally unsteady, system dependant phenomenon such as surge [4]. However, the ability to generate the required levels of pre-swirl (determined from the literature review) was theorised to be sufficiently indicative for this stage of the design process. Consequently, the primary objective for the CFD was to optimise fan performance within the existing confines of the compressor map (while maintaining the ability to generate large swirl angles for surge margin extension, and also satisfying the operational limits of the chosen electric motor).

The complete CFD setup, which encompassed details illustrated by both Figure 2 and Figure 3, was then utilised to evaluate the fan speed that maximised overall compressor efficiency (MP0 to MP3) for each of the six operating points of interest. A summary of the final fan dimensions are presented in Table 1, with the corresponding operating points outlined in Table 2. It is worth emphasising again that the fan was a part span design, with the blades occupying the lower 60% of the inlet section span.

Table 1. Final fan design parameters.

D_{Ah} (mm)	36
D_{As} (mm)	52
Z (-)	11
Blade Height (mm)	4.8
Axial Length (mm)	9
$\beta_{A30\%\_span}$ (°)	59.0
$\beta_{B30\%\_span}$ (°)	44.1

Table 2. Chosen fan operating points.

N (rpm)	n (rpm)	Description (-)
120000	50000	NS
120000	60000	DP
120000	70000	NC
160000	60000	NS
160000	90000	DP
160000	90000	DP

2.2 Results

The results of the aforementioned comparisons are presented in Figure 4 for the predictions of compressor pressure ratio and efficiency. In order to maximise the clarity of the results, the data for all six operating points have been presented as a change relative to the baseline configuration depicted in Figure 2. It is worth noting that the apparent absence of "Baseline with Inlet Swirl" data for the choke points is representative of the fact that the ideal swirl value at these operating points was equal to zero.

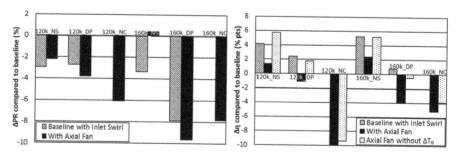

Figure 4. CFD predicted change in stage pressure ratio and efficiency relative to baseline at six operating points of interest.

Examining the pressure ratio prediction within Figure 4 first of all, it is readily apparent that the application of inlet pre-swirl, even when generated on a loss free basis, results in a decrement in total pressure ratio across the stage. This is of course not an unexpected result; the Euler turbomachinery equation clearly dictates that positive pre-swirl at inlet results in a reduction in the specific work input of the stage, and hence the expected total pressure rise.

Looking at performance with the fan in particular, it is clear that while the fan does add work to the flow, it is a very modest contribution, and with the exception of the 160k_NS point, it was not sufficient to overcome the swirl based decrement in impeller work input. Unfortunately, the modest amount of pressure rise across the fan was nullified by losses through the S-duct, which were an unavoidable consequence of the relatively large electric motor casing diameter (and correspondingly large fan hub diameter). In spite of the concession to utilise a part span fan design, this issue clearly increases in prominence as mass flow rate increases, with the near choke points in particular exhibiting significant drops in stage total pressure ratio (6.1% and 7.8% for 120k_NC and 160k_NC respectively). Ultimately, it was found that while increasing fan speed could help to overcome the pressure losses through S-duct at higher mass flow rates, the intolerance of the centrifugal stage to the additional pre-swirl generated resulted in an even more significant stage performance decrement, and hence the operating points illustrated in Figure 4 represent the best compromise. Away from the near choke points however, it was promising to witness that the reductions in stage total pressure ratio were comparable to what was witnessed in the baseline case with swirl applied as an inlet boundary condition.

Moving onto the efficiency comparison, it is worth firstly clarifying the meaning of the additional data field presented in Figure 4. The "Axial Fan without ΔT_0" field represents the stage efficiency evaluated without including the total temperature rise (work input) across the fan, i.e. in reference to Figure 2 and Figure 3:

$$\eta_{\text{Axial Fan without } \Delta T_0} = \frac{\left(\frac{p_{03}}{p_{00}}\right)^{\frac{\gamma-1}{\gamma}} - 1}{\frac{T_{03}}{T_{0B}} - 1} \tag{1}$$

This efficiency definition relates to a situation whereby the energy required to drive the fan is available "for free." From a practical perspective, this is not unreasonable with the excess of electrical energy available on modern automotive installations (48V mild hybrid systems etc) and the relatively low power values required in this application (<600W). The results are therefore comparable to the idealised "Baseline with Inlet Swirl" case. However, while the work input from the fan was excluded in this definition, it was not feasible to accurately isolate the total pressure loss through the S-duct in CFD. While the work input could be isolated by evaluating total temperature at Measurement Plane (MP) B, completion of the same exercise with total pressure necessitated evaluation at MP1. Unfortunately, with the large amounts of inlet recirculation present, it was not possible to extract a representative total pressure value at MP1. As a result, while the "Axial Fan without ΔT_0" results should approach those depicted by the idealised baseline with inlet swirl case, they do not replicate the idealised case exactly. While not particularly relevant at the lower mass flow operating points (which exhibited a relatively minor S-duct derived total pressure decrement over the idealised case), this becomes a point that is necessary to bear in mind at higher mass flow operating points where the large motor diameter (and resulting S-duct) caused significant total pressure losses.

7

Focusing on the efficiency results presented in Figure 4 then, the benefit of inclusion of the fan at the near surge points is readily apparent. At the low mass flow rate operating points, improvements in efficiency of 1.4%pts and 2.4%pts for 120k_NS and 160k_NS were witnessed, respectively. Removal of fan work saw these increase to 5.7%pts and 5.2%pts respectively, successfully matching (or exceeding in the case of 120k_NS) the idealised baseline with inlet swirl case. Consequently, it was clear that the fan was successfully bringing about improvements close to the surge line.

In spite of the demonstrated limitations with the system towards the choke side of the operating range (which are recognised to originate predominantly from the large diameter motor required for the prototype system), it was hypothesised at this point that the predicted efficiency improvements at low mass flow rates were sufficient to justify proceeding through to experimental validation. Furthermore, while the CFD results indicated a tangible interaction between the fan and the impeller inlet recirculation region (demonstrating reduced inlet recirculation extent when operating with the fan, which has been correlated to improved compressor stability [2]), experimental testing provides the only means of truly understanding the potential of the device to extend the compressor surge margin.

3   DEVELOPMENT OF EXPERIMENTAL HARDWARE

Having demonstrated sufficient potential from the system during the numerical modelling section to warrant undertaking experimental validation, the first step was to consider the most appropriate means of representing the simulated configuration during physical testing.

The intention from the outset was to mimic the configuration used in the CFD study as closely as possible. Consequently, all of the main geometric parameters were maintained, including the meridional profile of the inlet duct and the axial distance between the axial and radial stages. The design of the axial fan and spinner targeted minimising mass; the chosen electric motor (Fusion Exceed 3.5T) originated from a remote control car application, and hence the standard bearing system was never intended to support a large overhung mass. A static structural analysis was conducted in ANSYS Mechanical 17.2 to evaluate the structural integrity of the chosen fan and spinner designs. The simulation was conducted at the maximum rotational speed of 90krpm with no applied aerodynamic loading (very low pressure ratio design), and using 7075-T6 as the chosen material. The results indicated no issues with the design, with adequate safety factors and insignificant deflection values throughout.

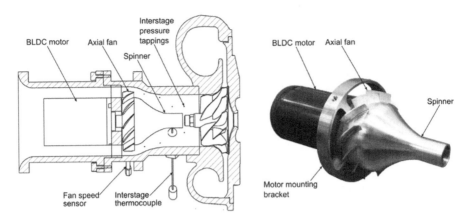

Figure 5. Illustration of experimentally tested hardware configuration.

4   EXPERIMENTAL TESTING RESULTS

The experimental test campaign was conducted using the standard methodology and instrumentation used for compressor testing in Queen's University Belfast (QUB), as outlined in [7]. The results are presented subsequently in two distinct sections; firstly, evaluation of stage performance with and without the axial fan for the six previously defined operating points of interest. This is then followed by an evaluation of the potential of the device to extend the surge margin of the baseline compressor stage. The data in each case was non-dimensionalised using the maximum value of a given parameter encountered during testing of the baseline configuration.

4.1   Stage performance

The results from the experimental test campaign for both pressure ratio and efficiency are presented in Figure 6, depicting the cases for the baseline and with the axial fan operating at the speed values defined in Table 2.

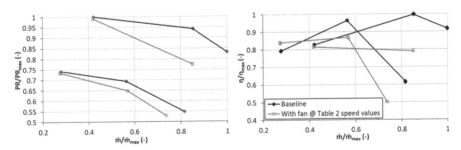

Figure 6. Comparison of measured compressor stage performance parameters at 120krpm and 160kpm.

Upon examining Figure 6, one obvious omission from the data is readily apparent. It was not possible to conduct testing at the 160k_NC point due to a fan overspeed issue. It transpired that, even with motor turned off (at windmill conditions), the fan speed looked set to exceed 100krpm. This was beyond the operating limits set during the design process and those of the electric motor, so to ensure safety of the hardware (and considering it is not one of the more critical operating points), the decision was taken to neglect this operating point.

Returning to the results, comparison of Figure 6 with the CFD predictions depicted in Figure 4 illustrates trends in the experimental data that are effectively exaggerated versions of what was witnessed in the numerical study. While the increase in stage efficiency of 3.0%pts at 120k_NS exceeded expectations, the efficiency decrement present at 160k_NS was unexpected, and lead to a question about whether the CFD had in fact highlighted the most appropriate fan speed values to maximise stage efficiency. Variations in fan speed were tested for the near surge points, where it was shown for the 160k_NS operating points that it was possible to recover from a 1.0%pt efficiency drop at a fan speed of 60krpm to a 0.4%pt increase by increasing the fan speed to 75krpm. By comparison, at 120k_NS the CFD simulations succeeded in predicting the most efficient speed to operate the fan at.

Regarding the higher mass flow operating points, the significant pressure ratio and efficiency reductions can have the same explanation attributed to them as during the CFD study (albeit the reductions in performance, and mass flow capacity for 120k_NC, proved to be more pronounced). While it would ideally be desirable to increase fan speed at the DP and NC operating points to overcome the increasing proportion of losses in the inlet duct, this was not possible. Increasing fan speed would indeed increase the pressure ratio generated by the fan, but also increases the swirl generated. For the near choke points in particular, this was an intolerable scenario, as the resulting increase in losses in the centrifugal stage were shown to far outweigh the benefits in reducing losses upstream of the impeller.

4.2   Impact on surge line

Following evaluation of stage performance within the existing stable operating range of the compressor, the next step was to understand the potential of the fan to improve compressor stability at low mass flow rate operating conditions. Testing was conducted at turbocharger shaft speeds of between 100krpm and 180krpm (56% to 100% speed) to ensure the key on-engine operating points were covered. While the rationale for choosing fan operating points within the existing stable operating range of the compressor stage centred on maximising fan efficiency, the target for the range extension tests was to maximise fan speed to deliver as much swirl as possible to the impeller. The resulting comparison between the surge line for the baseline configuration, and that with the fan are presented in Figure 7.

10

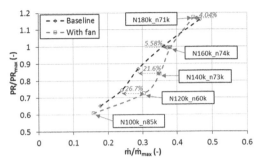

Figure 7. Comparison of baseline surge line and surge line with axial fan.

Upon scrutinising the results depicted in Figure 7, it is clear that operation with the fan triggered reduced stability at low flows almost universally, with the exception being the 100% speedline. While the improvement at 100% speed could be theorized as being related to the fact that as pressure ratio increases, the impeller becomes more receptive to the benefits brought about by inlet swirl (as discussed in the Introduction), it does not explain the reduced stability elsewhere. Unfortunately, it is notable that the only point where a benefit was obtained in terms of compressor stability yields no tangible advantage when considering on-engine operation.

Similarly, measured performance at 100krpm also proved to be somewhat unexpected, whereby an apparent reduction in surge mass flow rate was nullified by the compressor operating at significantly lower pressure ratio and a 9.1%pt efficiency decrement. As a consequence, the reduced surge mass flow rate at 100krpm cannot be viewed as a useful stability enhancement in comparison to the baseline case.

Further analysis of the experimental test results alongside additional CFD simulation work indicated that operation of the fan on the positive slope of its pressure rise characteristic was at the root of the problem. The aerodynamically unstable nature of the fan when operating at high rotational speeds and very low flow conditions acted to destabilise the complete compression system, leading to the early onset of surge. However, without operating at high rotational speeds, the fan could not generate the high (>60°) swirl angles known to be required for surge margin extension at the relatively low pressure ratios typical of automotive turbochargers [2]. Owing to the design compromises imposed by the chosen electric motor and the required wide range of operation, this was unfortunately something of an unavoidable scenario.

5   CONCLUSIONS

An active inlet swirl control device concept was successfully progressed through design, manufacture and testing phases. 1-D, 3-D CFD and FEA analyses were used to develop a functional prototype utilising off-the-shelf electrical components and hardware manufactured in QUB. The prototype system successfully completed the test campaign and performed in line with what was required following the initial CFD study. Regarding the CFD methodology, the approach employed was ultimately shown to result in an adequate design tool, with the overall trends having been well represented, but ultimately more pronounced in the test data.

Inclusion of the axial fan brought about both stage efficiency and pressure ratio benefits for the near surge operating conditions, typified by a 3.0%pt increase in efficiency for the 120k_NS operating point in comparison to the baseline configuration. However, similarly to what was predicted in the CFD study, the fan was not capable of achieving the desired performance/neutral influence at the higher mass flow rate operating points. This was largely the result of the constraints on inlet geometry imposed by the relatively large diameter electric motor, with the resulting total pressure losses being incapable of being overcome by the action of the fan. The natural inclination to increase fan speed to help offset the increasing total pressure drop in the inlet section as mass flow rate increased was curtailed either by mechanical considerations for the motor/fan, or through the intolerance of the centrifugal stage to the additional swirl generated (manifesting itself through further reductions in compressor efficiency and choke flow capacity).

Similarly, while the potential to generate high swirl angles with the fan was demonstrated, these did not translate into improvements in surge margin for the compressor stage. The high pre-swirl angles required to bring about benefits in the relatively low pressure ratio compressor typical of automotive turbochargers, combined with the design compromises required to enable a wide range of operation, resulted in a fan that was aerodynamically unstable at the required high rotational speed, low flow conditions. This instability ultimately triggered the onset of surge at higher mass flow rates than were witnessed for the baseline configuration.

In terms of an outlook, there is little doubt that a smaller diameter, faster rotating electric motor could significantly negate the performance decrements witnessed in this study at the higher mass flow rate operating conditions. Coupling removal of the requirement for the S-duct inlet geometry with a higher speed motor would be highly beneficial in improving choke side performance and would also open up additional design freedom to overcome the surge margin issues. However, specification of such a motor, suitable control electronics and the fan itself may yield a system that is too expensive to be feasible for a product destined for series production. Furthermore, combining such a system with a specific design/selection of base compressor stage to work with inlet swirl would be worthy of investigation.

ACKNOWLEDGEMENTS

The authors would like to sincerely thank IHI Charging Systems International GmbH for provision of the necessary compressor hardware for testing, as well as for their continuing technical support. The authors would also like to extend their thanks to ANSYS Inc. for the use of their CFD software and their technical support during this program of research.

NOMENCLATURE

D	Diameter (m)
n	Electric motor speed (rpm)
N	Turbocharger shaft speed (rpm)
$\dot{m}$	Mass flow rate (kg/s)
p	Static pressure (Pa)
p_0	Total pressure (Pa)
PR	Total pressure ratio (-)
QUB	Queen's University Belfast
T	Temperature (K)
T_0	Total temperature (K)
U	Blade speed (m/s)
V	Absolute velocity (m/s)
Z	Blade number (-)
β	Relative flow angle measured from meridional direction (°)
γ	Ratio of specific heats (-)
η	Isentropic total-to-total efficiency (-)
rotn.	Rotating
stn.	Stationary
BLDC	Brushless Direct Current
DP	Design point
MP	Measurement Plane
NC	Near choke
NS	Near surge
VIGV	Variable inlet guide vane
Subscripts:	
a	Axial direction
A	Fan leading edge
b	Blade
B	Fan trailing edge
corr	Corrected
h	Hub location
s	Shroud location
u	Tangential direction
0	Stagnation/Total conditions

(*Continued*)

13

(*Continued*)

1	Impeller inlet
2	Impeller exit/vaneless diffuser inlet
3	Vaneless diffuser exit
max	Maximum measured value for a given parameter

REFERENCES

[1] Ishino, M., et al. 1999, "Effects of Variable Inlet Guide Vanes on Small Centrifugal Compressor Performance," *Proceedings of ASME International Gas Turbine & Aero-engine Congress*, Indianapolis, IN, USA, 99-GT–157.
[2] Whitfield, A., & Abdullah, A.H., 1998, "The Performance of a Centrifugal Compressor with High Inlet Prewhirl," Trans. ASME Journal of Turbomachinery, vol.102(3), pp 487–493.
[3] Fraser, N., et al., 2007, "Development of a Fully Variable Compressor Map Enhancer for Automotive Application," *Proceedings of SAE World Congress*, Detroit, MI, USA, SAE 2007-01–1558.
[4] Lüdtke, K. H., 2004, Process Centrifugal Compressors: Basics, Function, Operation, Design, Application, Springer-Verlag, Berlin.
[5] Kleine Sextro, T., Steglich, T., Seume J.R., 2015, "Variable Inlet Guide Vane Device for a Turbocharger Compressor," *Proceedings of International Gas Turbine Congress*, Tokyo, Japan, IGTC2015–0161.
[6] Fledersbacher, P., Löffler, P., Sumser, S., & Wirbeleit, F., 2002, United States Patent No. US 6,481,205 B2.
[7] Stuart, C., Spence, SW., Filsinger, D., Starke, A., & Kim, S.I., 2018, "Characterising the Influence of Impeller Exit Recirculation on Centrifugal Compressor Work Input," *ASME Journal of Turbomachinery*, Vol.140 (1), doi: 10.1115/14038120.

Steady-state CFD calculation of a complete turbocharger radial turbine performance map: Mass flow rate and efficiency

G. SALAMEH, P. CHESSE, D. CHALET, P. MARTY

Ecole Centrale de Nantes, France

ABSTRACT

Turbochargers performance maps need to be extrapolated by automotive manufacturers: experimental techniques could be difficult and extrapolation models could be too simplified. A CFD model is presented with a reduced calculation time and power required for a radial turbine complete performance map: mass flow rate and efficiency. The turbine is studied as an entire system without separating the different components. The calculation is carried with a non-moving rotor mesh and extended to the negative mass flow rate region. The tendency of the efficiency curves is studied as a function of the turbine expansion ratio and the blade speed ratio.

Keywords: Turbocharger, radial turbine, simulation, CFD, performance map, extrapolation, mass flow rate, efficiency.

1   INTRODUCTION

Diesel engines require compressed air at engine intake. Turbocharging is used to recover the energy from the exhaust gases and use it for engine intake air compression. The exhaust gases expansion takes place in the turbine that transforms the fluid energy into power on the turbocharger shaft that turns the compressor impeller and compresses air going into the cylinders. Turbocharging is also used in spark ignition engines since its necessity for downsizing. Downsizing is reducing the engine volume and mass and preserving the power produced: this requires compressed air at the engine intake to have the same mass of air as in a smaller volume. Downsizing is used to reduce fuel consumption and engine emissions. Higher density air is going from the compressor to the intake, which allows a higher power production. Radial turbines are used in automotive applications due to space restrictions: they are cheap, compact, and robust and correspond to the application requirements since they allow high expansion ratios.

The turbocharger performance data is given by the supplier as compressor and turbine data maps. For the turbine, there is a flow rate performance map where the mass flow rate is represented as a function of the expansion ratio and an efficiency performance map where the turbine isentropic efficiency is represented as a function of the expansion ratio or the blade speed ratio. The data is given for constant rotational speed curves. The supplier turbine flow rate and efficiency performance maps shown in Figure 1 and Figure 2 are called SAE performance maps and are measured according to the *Turbocharger Gas Stand Test Code, J1826 SAE* [1] on the turbocharger test bench described in the work of Salameh *et al* [2]. The measurement area of the SAE performance maps is narrow compared to the operating area and this is caused by the compressor. On one side, there is the surge limit in the compressor and the measurements cannot go further on this side to avoid surge. On the other side, there is the choke limit where the air density at the compressor inlet is not enough to get to higher mass flow rates.

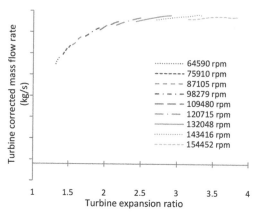

Figure 1. Supplier radial turbine mass flow rate performance map: corrected mass flow rate vs. expansion ratio.

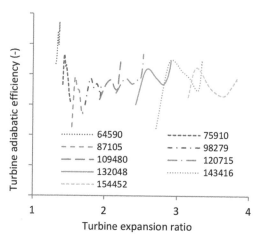

Figure 2. Supplier radial turbine efficiency performance map: turbine adiabatic efficiency vs. expansion ratio.

Due to narrow measurements area on the supplier performance maps, there are many studies to extend the turbine mass flow rate performance maps experimentally on modified test benches. SAE performance maps are measured in hot conditions at the turbine inlet (around 873 K for Diesel engine studies and 1173 K for gasoline engine studies). In the work of M. Frelin [3] and Salameh *et al* [2] [4], measurements on the test bench are done with cold air at the turbine inlet: a different inlet air temperature gives a different density and therefore a different area on the performance maps. Measurements are also done at intermediate temperatures to extend the curves in the area between the hot and the cold measurements. To extend the curves to lower power regions, the measurements can be done in an opened loop as presented by Venson and Barros [5]. To get to higher power and rotational speeds, the measurements can be done in a closed loop: this is the case in the work of Scharf *et al* [6]. In the closed loop configuration, the same air going into the compressor goes into the turbine. To extend the curves on the surge limit side, surge can be avoided by creating a swirl effect at the compressor inlet using a guide vane; this technique is presented by Serrano et al [7]. Measurements can also be extended through the surge limit by blowing compressed air in the compressor outlet, and can be extended through the choke limit by blowing compressed air in the compressor inlet: this technique extends the curves on both sides, high and low mass flow rates. This method is presented in detail in the work of Salameh et al [4][2] and the result is shown in Figure 3. The turbine is driven by a reversed compressor to measure the negative mass flow rate values.

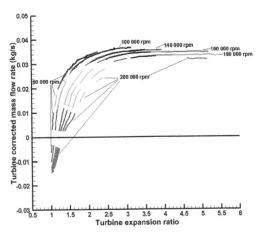

Figure 3. Turbine mass flow rate performance map experimental data: corrected mass flow rate vs. turbine expansion ratio (Salameh et al [2]).

All the techniques presented above are applied on a complete turbocharger without major modifications or dismantling of the turbocharger. It is possible to remove the compressor and replace it with a bladeless disk, a transmission system or an electric motor as is the case in the work of Scharf et al [6]. This technique allows the extension of the curves at low mass flow rates and low rotational speeds: that is to say low power region. Dismantling a turbocharger is a very delicate matter because it jeopardizes the axial and radial balance between the turbine, the compressor and the central housing.

The experimental data with the extended turbine performance maps is used to calibrate and validate extrapolation models used to calculate a complete performance map using the data given by the supplier, the geometrical data of the turbocharger and the fitting coefficients. Extrapolation models are used by automotive manufacturers in numerical simulations and engine calibration (test new architectures, ECU, HIL). This means that an accurate and reliable extrapolation model is needed to generate an extended performance map.

The experimental modifications presented above to extend the turbine performance maps are difficult to execute. It is not possible to do the experimental measurements for every turbocharger to validate the model. Different models are developed to extrapolate the turbine performance maps with different levels of precision and complexity. Simpler models are empirical models based on polynomial or exponential or other mathematical equation with coefficients to calibrate using the supplier map. This is the case with the models of Sieros et al [8], Orkisz and Stawarz [9], and Fang et al [10][11]. These models are easy to calibrate but it is difficult to be sure that it is reliable outside of the narrow measurement area. More complicated models are semi empirical models: they are based on physical thermodynamics and turbomachinery equations with assumptions and hypothesis applied to simplify the calculation and reduce the number of variables. Most of the semi empirical models represent the fluid expansion in the turbine as an expansion in one or more nozzles: the expansion can be isentropic or not. Semi empirical models are presented by Jensen *et al* [12], Canova *et al* [13][14], Payri *et al* [15][16], Mseddi *et al* [17], Serrano *et al* [18]and Salameh *et al* [19].

Empirical and semi empirical models are used because of the complexity and computer power needed for three dimensional calculations. CFD calculations can be used to study unsteady flow behavior, thermodynamics and aerodynamics at the stator-rotor interface and the blade to blade gap (L. Jawad [20]). CFD calculations require also a detailed knowledge of the turbocharger's geometric details [21] which is not always available for automotive users who need complete performance maps for engine simulations. In this study, a CFD calculation is presented with the minimum required calculation power and time to produce a complete turbine mass flow rate performance map without going into the details of the flow in every part of the turbine for every operating point. Nevertheless, it is possible to compare the flow characteristics at different parts of the turbine for different regimes and expansion ratios to understand why the curves take the shape of an isentropic nozzle expansion curve.

The objective of this work is to provide a complete performance map for the turbine mas flow rate and efficiency. This can be done experimentally with a complicated test bench and measurement techniques that are not always available for the user. These tests could take weeks or even months as well as the preparation of the test bench. Extrapolation models can be fast and accurate but require calibration and validation for every turbocharger type and require also experimental data. The purpose here is to be able to extrapolate the turbine performance maps using only the turbine CAD and the supplier performance map with a moderate preparation and calculation time since the objective is not to provide a detailed study of the flow inside the different components of the turbine.

2   TURBOCHARGER PRESENTATION

The calculation is done on a small diesel engine turbocharger radial turbine. The turbine geometry is divided into different parts: the inlet pipe, the volute, the vaneless distributor, the impeller, the outlet stator, and the outlet pipe shown in Figure 6. The inlet and outlet pipes are not parts of the turbocharger but they are added to the model to represent a turbine mounted on a turbocharger test bench. They allow the boundary layer to develop and the flow to become uniform at the turbine inlet and outlet sections. The vaneless distributor is considered as a continuity of the volute and the fluid characteristics are calculated at the turbine inlet (volute inlet) and the distributor outlet (volute outlet). The impeller inlet is radial and its outlet is axial. The impeller has 12 blades shown in Figure 4 and Figure 5. The turbine dimensions are presented in Table 1. The different index numbers are given in Table 2 and in Figure 6 and Figure 7. The turbine walls are modeled as adiabatic: the heat transfer from the turbine is neglected because of the insulation of the turbine case.

Table 1. Radial turbine features.

Turbine feature	Dimension
Turbine inlet pipe length	100 mm
Turbine outlet pipe length	100 mm
Turbine inlet section diameter	35 mm
Turbine outlet section diameter	46 mm
Impeller inlet diameter	35 mm
Impeller blades inlet width	4.4 mm
Impeller outlet internal diameter	11 mm
Impeller outlet external diameter	29 mm
Impeller blades inlet angle	90°
Impeller blades outlet angle	24°
Impeller blades number	12

Table 2. Cross section index and position.

Cross section number	Description
0	Domain inlet
1	Turbine inlet (volute inlet)
2	Volute outlet-Rotor inlet
3	Rotor outlet
4	Turbine outlet
5	Domain outlet

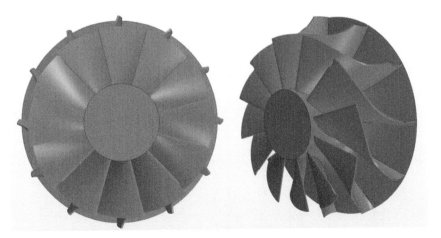

Figure 4. Three dimensional view of the radial turbine impeller.

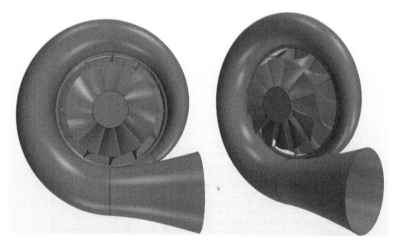

Figure 5. Three dimensional views of the radial turbine volute and impeller.

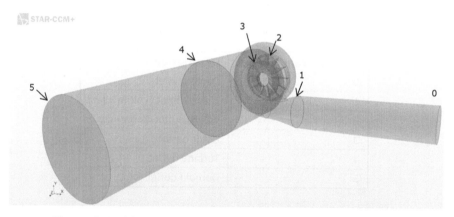

Figure 6. Turbine overall view with the inlet and outlet pipes.

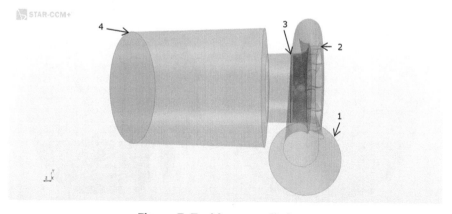

Figure 7. Turbine overall view.

3   CFD METHODOLOGY

The StarCCM+ 12.02.011 code is used to calculate the turbine performance maps. The turbine CAD is introduced into StarCCM+ and a surface repair is necessary to fix the surface tessellation errors such as parts intersections or gaps. The surface extrusions are then executed at the turbine inlet and outlet sections to create the inlet and outlet pipes. The fluid chosen is an ideal gas going through a steady-state flow.

The solver is a coupled implicit solver. The implicit approach solves all cells at the same time while the explicit approach solves for all variables cell by cell. The implicit approach requires more memory but the solution of a steady-state flow converges faster than in the case of an explicit solver (Galindo *et al* [22]).

The turbulence model chosen for this study is the *k-ω* SST RANS model. It presented good results in the literature. The RANS approach is less accurate than the DES model but it is chosen for its affordable computational cost. The SST model has been used in turbomachinery simulations and showed good agreement between the simulations and the experiments (Menter *et al* [23]). It was also used in turbomachinery applications by Pecnik *et al* [24] and Simpson *et al* [25].

3.1   Mesh

The mesh settings are chosen based on previous mesh sensitivity studies from similar work on turbochargers in the literature as in the work of Jawad LH [20] and De Souza and Filho [26]. The mesh is more coarse downstream the impeller and is refined in the volute and the impeller. Downstream the turbine, the mesh is a cylindrical extrusion and it is not a part of the turbine: it is used to represent the pipes at the turbine inlet and outlet to let the flow develop. The boundary layer is generated with a prism layer mesh model. The mesh quality is tested and validated. The mesh is topologically valid and has no negative volume cells. The face validity, volume change and skewness angle criteria are validated. The Y+ value is adapted to the prism layer model for most of the parts taken into consideration in the turbine calculation.

The meshing is polyhedral mesh with a base size of 1 mm, a surface growth rate of 1.3, a surface remesher and a prism layer mesher as shown in Figure 9 and Figure 10. The domain is divided into two regions: stator and rotor. The rotor region is the impeller mesh. The stator contains the inlet pipe (extension), the volute, the turbine outlet part, and the outlet pipe (extension). In the rotor, the momentum interchange happens, which is not the case in the rest of the turbine; it is then preferable to have a smaller cell size in the rotor region [22]. A cell refinement is then conducted in the rotor region where the base size of the cells is reduced. The mesh is also adjusted in the volute and stator parts close to the rotor to avoid a too high surface growth rate. The total number of cells for the turbine alone is 852 687 cells. The mesh in the turbine stator has 635 756 cells: 308 974 cells for the turbine stator (volute and outlet part) and 326 782 cells for the extensions at the turbine inlet and outlet. The number of elements distribution in the turbine and the extensions is presented in Table 3.

Table 3. Number of elements in mesh of different turbine components.

Region	Part	Number of cells
Stator	Volute	148 650
	Outlet part	160 324
	Inlet extension	78 397
	Outlet extension	248 385
Rotor	Impeller	543 713
Total		1 179 469

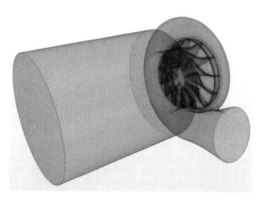

Figure 8. Turbine overall mesh view (without the extensions).

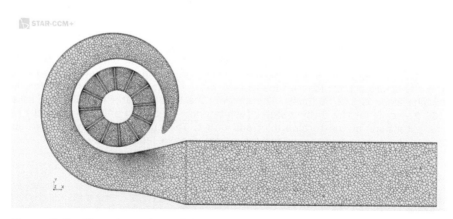

Figure 9. Section plane view of the mesh: normal to z (turbo shaft direction).

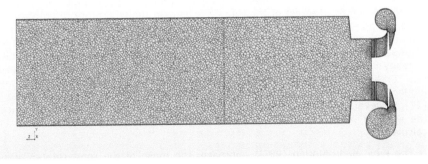

Figure 10. Section plane view of the mesh: normal to x (inlet pipe direction).

3.2   Boundary conditions
The turbocharger is tested under steady-flow conditions. It is the standard measurement procedure for manufacturers. The boundary conditions in this study can be divided into two sets: positive and negative mass flow rate. In both cases, the operating corrected rotational speeds are 0; 30 000; 60 000; 90 000; 120 000; 150 000; 170 000; 190 000; 210 000; 250 000; 300 000 rpm.

3.2.1 *Positive mass flow rate*
In this part of the study, the pressure at the turbine inlet is high enough to assure the flow of the fluid from the inlet to the outlet. The boundary conditions in this part are inlet total pressure and temperature and outlet static pressure. The outlet static pressure is fixed to atmospheric pressure of 101 325 Pa. The inlet total temperature is fixed to 823 K, and the total inlet pressure varies between 101 325 and 607 950 Pa to cover an expansion ratio area between 1 and 6. The choice of the imposed mass flow rate at the inlet is abandoned because the combination of total pressure at the inlet and static pressure at the outlet gives good numerical stability and convergence rates in radial turbine simulations [25].

3.2.2 *Negative mass flow rate*
At low expansion ratios (close to 1), the head losses in the turbine can cause the flow to go from the outlet to the inlet. In this case, the fluid is driven by the rotor and the turbine acts as a compressor with a reverse flow direction as shown by Salameh *et al* [2]. The flow rate measured is then negative and the turbine map is extended to negative mass flow rate values at low expansion ratios. For these points, the boundary conditions are inlet total pressure and temperature applied to the turbine outlet and static outlet pressure applied to the turbine inlet. The total temperature and pressure applied to the turbine outlet (flow inlet) are 300 K and 101 325 Pa, and the static pressure applied to the turbine inlet (flow outlet) varies but remains under 111 325 Pa.

3.3   Impeller rotation strategies
There are two strategies possible to simulate the rotor motion: the Moving Reference Frame (MRF) and the Rigid Body Motion (RBM).

23

In the MRF approach, the mesh does not move. The reference frame (coordinate system) can move with the rotor and the Navier-Stokes equations can be solved in this moving reference frame where the Coriolis and centrifugal forces are included. The calculation for the rest of the turbine is done in the laboratory reference frame which is a non-rotational coordinate system. MRF models assume that the angular velocity of the body is constant and the mesh is rigid. The main advantage of the MRF approach is the low computational cost and being able to do the calculation in a steady-flow condition. It is used by De Souza and Filho [26] to calculate a steady-flow turbine map. It is not possible to study the unsteady flow effects. The mesh motion is then essential to study the stator-rotor interaction as shown by Liu and Hill [27]. However, it is still a controversial points in turbomachinery simulations, since in the work of Aymanns et al [28], the MRF approach is used for pulsating flow calculations.

In the Rigid Body Motion (RBM) approach, the mesh of the rotor rotates around the turbo shaft and the stator is fixed and the two regions are connected with sliding interfaces. The cells connectivity changes on each side at every time step. The RBM is used for unsteady-flow calculations and the flow will not converge to a value; it will oscillate. The stopping criterion is a periodic solution. Pressure waves can generate from this moving mesh upstream and downstream of the rotating part and can have an effect on the solution but the results can be satisfactory (Galindo et al [22]).

In this study, we are resolving a system-level flow behavior: the objective is to get a complete mass flow rate performance map for the turbine with the hypothesis of a steady-state flow. The Moving Reference Frame (MRF) is then chosen to calculate the data needed to draw the map. The regions of the stator (Lab reference frame) and the rotor (Moving Reference Frame MRF) are connected with interfaces.

3.4   Initial conditions and convergence criteria
The initial conditions are fixed for the first calculation point: atmospheric pressure and temperature, axial velocity, viscosity ratio... Once the calculation is finished with this file, it is used as an initial condition where the inlet and outlet conditions are changed and the calculations is done once again but with a better distribution of initial conditions that can help reduce the calculation time and let the file converge much faster. For this study, there are different convergence criteria: the residuals and the continuity principle. The calculation is considered converged when the 5 residuals (continuity, Tke, Tdr, X-momentum, Y-momentum, Z-momentum) converge to low values (10^{-8} for continuity, 10^{-6} for X-momentum, Y-momentum, Z-momentum and Tke, 10^{-3} for Energy and Sdr) and the mass flow rate at the inlet and the outlet converge to the same value. Three asymptotic convergence criteria are added: inlet mass flow rate, outlet mass flow rate, inlet-outlet mass flow rate difference.

The calculations are executed on a two 3.2 GHz processors computer with 64 GB of memory. Eight calculations are launched in parallel and each point takes around 10 hours for the solution to converge. With 190 points in total, there is around 15 days of calculations

4   TURBINE PERFORMANCE MAPS PLOTS

Data recovered from the calculation file is used to calculate the characteristics needed to draw the performance maps: mass flow rate, inlet total temperature, inlet total pressure, outlet total temperature, outlet total pressure, inlet total enthalpy, outlet total enthalpy, regime. These data are calculated using a mass flow rate average applied on the cells on the inlet and outlet sections. The mass flow rate and the rotational speed are then transformed into corrected mass flow rate and corrected rotational speed to eliminate the effect of the measuring conditions. The turbine adiabatic efficiency and the turbine expansion ratio are dimensionless values and do not need correction. Corrected values are calculated using the measured mass flow rate, rotational speed, inlet conditions and reference conditions as shown in equations (1) and (2):

$$\dot{m}_{corr} = \dot{m} \frac{\sqrt{\frac{T_{i,0}}{T_{ref}}}}{\frac{P_{i,0}}{P_{ref}}} \tag{1}$$

$$N_{corr} = N \sqrt{\frac{T_{ref}}{T_{i,0}}} \tag{2}$$

Where P_{ref}, and T_{ref}, are the ambient pressure and temperature.

For negative flow rates, the corrected values are calculated using the fluid inlet conditions (turbine outlet conditions) and the equations become:

$$\dot{m}_{corr} = \dot{m} \frac{\sqrt{\frac{T_{i,5}}{T_{ref}}}}{\frac{P_{i,5}}{P_{ref}}} \tag{3}$$

$$N_{corr} = N \sqrt{\frac{T_{ref}}{T_{i,5}}} \tag{4}$$

The turbine adiabatic efficiency is the ratio of the the turbine real power to the power produced by an isentropic expansion. Applying the first law of thermodynamics on a steady state unidimensional flow, it becomes the ratio between the total enthalpy difference to the total enthalpy difference in the isentropic expansion case. In this study, the efficieny calculated is the total-to-static adiabatic efficiency: it is calculated using the total enthalpy at the turbine inlet and outlet and the static isentropic outlet enthalpy. The efficiency is calculated using equation (5).

$$\eta_{t-s} = \frac{Pow_{real}}{Pow_{isentr}} = \frac{h_{i,0} - h_{i,5}}{h_{i,0} - h_{isentr,5}} \tag{5}$$

The turbine efficiency map can be drawn as a function of the expansion ratio (Figure 12) or the blade speed ratio (Figure 13). The turbine expansion ratio is the ratio between the total inlet pressure and the static outlet pressure. The blade speed ratio is the rotor tip linear velocity (U) divided by the velocity (C_{isentr}) achieved if the gas has an isentropic expansion, as shown in equation (6):

$$SR = \frac{U}{C_{isentr}} = \frac{\pi DN/60}{\sqrt{2c_p T_{i,0}\left(1 - \left(P_5/P_{i,0}\right)^{\frac{\gamma-1}{\gamma}}\right)}} \tag{6}$$

5   RESULTS AND ANALYSIS

The results of the previous calculations are 190 operating points at 10 different rotational speeds (except 0 rpm). These points are plotted on a mass flow rate performance map shown in Figure 11 and an adiabatic efficiency map shown in Figure 12 and Figure 13. In the mass flow rate map, there is the data calculated at 0 rpm which is not used in the calculation of the efficiency. Experimental validation is not available for this turbocharger. The CAD used in this calculation is close to the turbocharger previously tested in the experimental work of Salameh et al. [2] but not exactly the same so the shape of the curves can be compared but not the values.

5.1 Mass flow rate map

The mass flow rate performance map shown in Figure 11 represents the corrected mass flow rate as a function of the turbine expansion ratio for different rotational speeds. The shape of the curve looks like the shape of a leaned Barré de Saint Venant curve for an isentropic expansion in an ideal nozzle. The choke limit is lower than the case of a nozzle since there is a rotational part. The experimental data for a complete mass flow rate performance map is presented in the work of Salameh et al [2] in Figure 3. The shape of the curves is close to the results from the CFD model. The curves at low rotational speeds intersect and the 0 rpm curve goes beneath some curves with non-zero rotational speeds which can be explained by the losses due non-adaptation occurring in the turbine. From a certain value of the regime, the values of the mass flow rate get lower with higher rotational speeds for the same expansion ratio. The choking value also decreases with the increasing rotational speed due to the increasing Mach number. The negative values are calculated when the turbine rotor is driven by the turbo shaft and not the fluid as is the case in Figure 3.

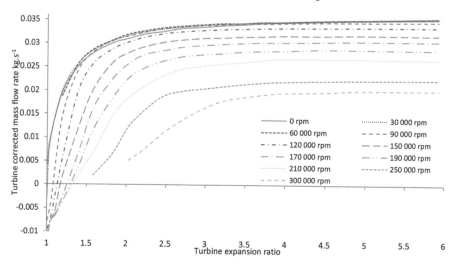

Figure 11. Turbine mass flow rate CFD performance map: corrected mass flow rate vs. Turbine expansion ratio.

5.2 Efficiency map

As mentioned above, the turbine adiabatic efficiency map can be plotted as a function of the expansion ratio (Figure 12) or the blade speed ratio (Figure 13). Experimental measurements were carried out to plot the turbine adiabatic efficiency curves as in the work of Marelli et al [29] and Zimmerman et al [30]. Models were developed to extrapolate the turbine efficiency map like the models of Jensen et al [12] and G. Martin [31] and the results from these models correspond to the results from the CFD model presented in Figure 12. The efficiency curves shape changes from a rotational speed to another. For lower and intermediate rotational speeds, the efficiency increases with the expansion ratio and then decreases after going through the optimum adapted operating point at this rotational speed. For higher rotational speeds, the efficiency increases with the expansion ratio since an adapted point is not reached yet at such high speeds: it belongs at higher expansion ratios.

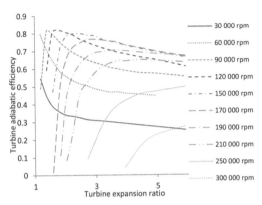

Figure 12. Turbine efficiency CFD performance map: Turbine adiabatic efficiency vs. Turbine expansion ratio.

Another representation of the results is possible with the turbine blade speed ratio (Figure 13). In this representation of the efficiency, the shape of the curve corresponds to the map given by Watson and Janota [32]. In the turbine efficiency map with the blade speed ratio (Figure 13), the efficiency follows a certain bell shape: there is an adapted point where the efficiency is at its maximum and above and beneath this point the efficiency drops. The efficiency drops steeply above a certain value of SR: it is a characteristic of the radial turbine. At low mass flow rates and high rotational speeds (high SR) the flow rate is not enough to overcome the centrifugal forces [32]. The efficiency decreases at high SR until it reaches zero; above this point, the turbine acts as a compressor. This is shown in the mass flow rate map (Figure 11) with the negative flow rate values: at low expansion ratios, the turbine acts as a compressor and the pressure difference is not enough to compensate the centrifugal forces and the losses and the flow goes backwards and the turbine is acting as a compressor.

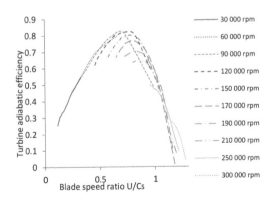

Figure 13. Turbine efficiency CFD performance map: Turbine adiabatic efficiency vs. Blade speed ratio.

6   CONCLUSION

This paper presents a CFD calculation for a radial turbine performance map. Many experimental studies are presented to show the difficulty of a wide turbine map measurement. Extrapolation models are also presented: they are not very accurate in extrapolating a turbine map outside the supplier given region without having any experimental data. CFD calculations usually require a lot of time and computational power which is not the case in this study. The purpose of this calculation is not to provide a precise knowledge of the fluid properties at each point of the volume, but rather to provide an overall study of the turbine as a whole unit in order to plot a mass flow rate map and an efficiency map on a wider range than the one given by the supplier. For this reason, the calculations are executed in steady state conditions with a non moving rotor mesh to reduce the calculation time and power needed. 190 operating points are calculated using this code and the results give the turbine mass flow rate and adiabatic efficiency performance maps. The mass flow rate is plotted as a function of the turbine expansion ratio for values going from 1 to 6. At expansion ratios close to 1, the mass flow rate is backwards because of the centrifugal forces and the losses in the turbine. The efficiency is also plotted as function of the expansion ratio and the blade speed ratio. In this last representation, the maximum efficiency corresponds to an adpated operating point and the efficiency degrades for higher or lower values than the optimal blade speed ratio. These maps can be compared to the ones measured experimentally for similar turbines in the literature and the shape is the same. This model could be tested on multiple turbines to obtain a complete performance map. The next step could be to do the calculation for the compressor, the central housing and the entire turbocharger. In this case the coupling effect could be taken into consideration as well as the heat transfers.

NOTATIONS

C [m/s]	gas velocity
c_p [$J.K^{-1}.kg^{-1}$]	specific heat capacity
D [m]	rotor diameter
h [J/Kg]	enthalpy
$\dot{m}$ [kg/s]	mass flow rate
N [r/min]	turbine speed
P [Pa]	pressure
P_i [Pa]	total (stagnation) pressure
T [K]	temperature
T_i [K]	total (stagnation) temperature
U [m/s]	blade speed
γ [-]	specific heat ratio
η [-]	isentropic efficiency
CAD	Computer Aid Design
CFD	Computational Fluid Dynamics

(Continued)

(Continued)

ECU	Engine Control Unit
HIL	Hardware In the Loop
LES	Large-Eddy Simulation
MRF	Moving Reference Frame
Pow	Power [Watt]
RANS	Reynolds-Averaged Navier-Stokes
RBM	Rigid Body Motion
SAE	Society of Automotive Engineers
SST	Shear-Stress-Transport
SR []	blade speed ratio

SUBSCRIPTS

corr	corrected
i	total, stagnation
i,0	turbine inlet state or stagnation condition
isentr	isentropic
ref	reference
t-s	total-to-static
0-5	state points

REFERENCES

[1] SAE. SAE J1826 Turbocharger Gas Stand Test Code. Surf Veh Recomm Pract 1995:1–12. doi: 10.4271/J1826_199503.

[2] Salameh G, Chesse P, Chalet D. Different Measurement Techniques for Wider Small Radial Turbine Performance Maps. Exp Tech 2016;40:1511–25. doi: 10.1007/s40799-016-0107-85.

[3] Frelin M. Prévision des caractéristiques d'une turbine radiale à partir des données géométriques, PhD Thesis. Paris 6, 1991.

[4] Salameh G, Chesse P, Chalet D, Talon V. Experimental Study of Automotive Turbocharger Turbine Performance Maps Extrapolation. SAE Tech. Pap. 2016-01-1034, vol. 2016- April, 2016. doi: 10.4271/2016-01-1034.

[5] Venson GG, Barros JEM. TURBOCHARGER PERFORMANCE MAPS BUILDING USING A HOT GAS TEST STAND. Proc ASME Turbo Expo 2008 Power L Sea Air GT2008 June 913 2008 Berlin Ger 2008;GT2008-509: 1–9. doi: 10.1115/GT2008-50994.

[6] J Scharf, Schorn N, Smiljanovski V, Uhlmann T, Aymanns R. Methods for extended turbocharger mapping and turbocharger assessment. ATK - 15. Aufladetechnische Konf., Dokumentation Kraftfahrtwesen e.V.; 2010, p. 443–65.

[7] Serrano JR, Tiseira A, García-Cuevas LM, Inhestern LB, Tartoussi H. Radial turbine performance measurement under extreme off-design conditions. Energy 2017;125:72–84. doi: 10.1016/j.energy.2017.02.118.

[8] Sieros G, Stamatis A, Mathioudakis K. Jet engine component maps for perform-ance modeling and diagnosis. J Propuls Power 1997;13:665–74.

[9] Orkisz M, Stawarz S. Modeling of turbine engine axial-flow compressor and tur-bine characteristics. J Propuls Power 2000;16:336–9.

[10] Fang X, Dai Q, Yin Y, Xu Y. A compact and accurate empirical model for turbine mass flow characteristics. Energy 2010;35:4819–23. doi: 10.1016/j. energy.2010.09.006.

[11] Fang X, Dai Q. Modeling of turbine mass flow rate performances using the Taylor expansion. Appl Therm Eng 2010;30:1824–31. doi: 10.1016/j. applthermaleng.2010.04.016.

[12] Jensen J-P, Kristensen AF, Sorenson SC, Houbak N, Hendricks E. Mean Value Modeling of a Small Turbocharged Diesel Engine. 1991. doi: 10.4271/910070.

[13] Canova M, Midlam-Mohler S, Guezennec Y, Rizzoni G. Mean Value Modeling and Analysis of HCCI Diesel Engines With External Mixture Formation. J Dyn Syst Meas Control 2009;131:011002. doi: 10.1115/1.2977465.

[14] Canova M. Development and validation of a control-oriented library for the simu-lation of automotive engines. Int J Engine Res 2004;5:219–28. doi: 10.1243/ 1468087041549625.

[15] Payri F, Benajes J, Reyes M. Modelling of supercharger turbines in internal-combustion engines. Int J Mech Sci 1996;38:853–69. doi: 10.1016/ 0020-7403%2895%2900105-0.

[16] Payri F, Serrano JR, Fajardo P, Reyes-Belmonte MA, Gozalbo-Belles R. A physically based methodology to extrapolate performance maps of radial turbines. Energy Convers Manag 2012;55:149–63. doi: 10.1016/j. enconman.2011.11.003.

[17] Mseddi M, Baccar M, Kchaou H, Abid MS. Modélisation des turbines radiales de suralimentation. Mec Ind 2002;3:35–44. doi: 10.1016/S1296-2139%2801% 2901131-9.

[18] Serrano JR, Arnau FJ, Dolz V, Tiseira A, Cervelló C. A model of turbocharger radial turbines appropriate to be used in zero- and one-dimensional gas dynam-ics codes for internal combustion engines modelling. Energy Convers Manag 2008;49:3729–45. doi: 10.1016/j.enconman.2008.06.031.

[19] Salameh G, Chesse P, Chalet D. Mass flow extrapolation model for automotive turbine and confrontation to experiments. Energy 2018.

[20] Jawad LH. Numerical Prediction of a Radial TurbinePerformanceDesignedfor Automotive engines Turbocharger. J Univ Babylon 2018;26:132–46.

[21] Palfreyman D, Martinez-Botas RF. The Pulsating Flow Field in a Mixed Flow Turbo-charger Turbine: An Experimental and Computational Study. J Turbomach 2005;127:144. doi: 10.1115/1.1812322.

[22] Galindo J, Hoyas S, Fajardo P, Navarro R. Set-up analysis and optimization of CFD simulations for radial turbines. Eng Appl Comput Fluid Mech 2013;7:441–60.

[23] Menter FR, Langtry R, Hansen T. CFD simulation of turbomachinery flows-verification, validation and modelling. Eur. Congr. Comput. Methods Appl. Sci. Eng. ECCOMAS, 2004.

[24] Pecnik R, Witteveen J, Iaccarino G. Uncertainty quantification for laminar-turbulent transition prediction in RANS turbomachinery applications. 49th AIAA Aerosp. Sci. Meet. Incl. New Horizons Forum Aerosp. Expo., 2011, p. 660.

[25] Simpson AT, Spence SW, Watterson JK. A comparison of the flow structures and losses within vaned and vaneless stators for radial turbines. J Turbomach 2009;131:31010.

[26] de Souza RC, Krieger Filho GC. Automotive turbocharger radial turbine CFD and comparison to gas stand data. 2011.

[27] Liu Z, Hill DL. Issues surrounding multiple frames of reference models for turbo compressor applications 2000.

[28] Aymanns R, Scharf J, Uhlmann T, Lückmann D. A revision of quasi steady modelling of turbocharger turbines in the simulation of pulse charged engines. 16th Supercharging Conf. Dresden, Ger. Sept, 2011, p. 29–30.

[29] Marelli S, Marmorato G, Capobianco M, Boulanger J-M. Towards the Direct Evaluation of Turbine Isentropic Efficiency in Turbocharger Testing. 2016.

[30] Zimmermann R, Baar R, Biet C. Determination of the isentropic turbine efficiency due to adiabatic measurements and the validation of the conditions via a new criterion. Proc Inst Mech Eng Part C J Mech Eng Sci 2018;232:4485–94.

[31] Martin G.Modèlisation 0D-1D de la chaîne d'air des moteurs á combustion interne dédiée au contrôle. Université d'Orléans, 2010.

[32] Watson N, Janota MS. Turbocharging the Internal Combustion Engine. 2015. doi: 10.1007/978-1-349-04024-7.

Aerodynamic design of a fuel cell compressor for passenger car application

H. Chen[1], L. Huang[2], K. Guo[2], K. Kramer[2], Z.Y. Zhang[2]

[1]Naval Architecture & Ocean Engineering College, Dalian Maritime University, China
[2]Turbocharging Department, Great Wall Motors Ltd., China

ABSTRACT

Hydrogen powered fuel cell is considered a clean energy source for passenger cars. The cell uses the oxygen of air and hydrogen fuel to generate electricity to power the cars. A compressor is required to rise the air pressure to provide sufficient oxygen to the cell. This paper describes the aerodynamic design of a hydrogen fuel cell compressor for passenger car application by Great Wall Motors (GWM). The requirements for such a compressor are first given and discussed. One challenging requirement is that only single stage compressor is allowed for packaging reason. Because of the speed limit of the driving electric motor, compressor diameter must be increased and the compressor becomes a low specific speed one and works away from the ideal condition in the Cordier diagram. Aerodynamic implications of this and strategy and techniques employed in the design to overcome the difficulty are discussed, and CFD results are deliberated. Test results of the first prototype compressor are also shown. The compressor reaches required pressure ratio and has a good flow range, but its efficiency needs further improvements.

Keywords: Passenger cars, hydrogen fuel cell, compressor aerodynamics

1 INTRODUCTION

Fuel cell that uses pressurised hydrogen and the oxygen in air to generate electricity is considered a relatively clean energy source for transportation. As the exhaust from the cell is water, it has the same benefit of zero 'tailpipe' emission as rechargeable batteries but also has the advantage of higher energy density and capacity. Hydrogen powered passenger cars, such Toyota's Mirai, Honda's Clarity and Hyundai's Nexo, all have a driving range more than 500km, and Nexo released in 2018 has a range of 800km [1]. The second generation of Mirai for 2020 from Toyota claims to have 30% more range than previous model [2]. They can also be refuelled in a similar time as petrol- or diesel-powered cars, which current battery powered electric cars cannot.

One of the key components of fuel cell power systems is the air compressor that provides compressed air to the cell. Any type of compressors maybe used, for passenger cars with limited onboard space however, centrifugal compressor is a good choice with its compact size and quiet operation [3]. Honeywell Turbo Technologies now Garrett, has developed a two-stage, centrifugal compressors for Honda's fuel cell car Clarity [3, 4]. The company showed the compressor and did a presentation on this compressor at IMechE's Turbochargers and Turbocharging conference in 2018 [5]. In the presentation, many aspects of the compressor were discussed although information related to its aerodynamic design is limited. It took Honeywell several years to develop the compressor, and a portion of that time and efforts were spent on compressor aerodynamics. Indeed, fuel cell compressors differ considerably from turbocharger compressors for internal combustion engines. They are driven by high-speed electric motors and are difficult to design, at least for peoples used to work on high speed turbomachinery such as turbochargers.

GWM is committed to green transportation, and has invested heavily in hydrogen fuel cell technology. It has the largest hydrogen fuel cell test laboratory in China. The company is looking into building a hydrogen powered car in near future. An air compressor for the fuel cell of this car is required and is currently under development [6, 7]. In this paper, the aerodynamic aspect of this compressor and the lessons learnt in its development are described.

2   FUEL CELL REQUIREMENTS

2.1   Aerodynamic Requirement

Figure 1 shows predicted compressor operating line superimposed on the map of a state-of-art, 44mm, high pressure ratio turbocharger compressor. The line is slightly simplified to emphasis its two end points. It can be seen from the figure that the flow capacity of the compressor is too large, and some reduction of the capacity is required. This reduction first looks easy: a slight reduction of impeller size would accomplish the job. The reality is, however, more difficult: turbocharger compressors can run at very high rpms, but compressors driven by electric motors cannot. In this case, to have a machine Mach number of 1.45 required to meet the needs of PR and flow of the fuel cell at 140,000rpm, the compressor size is calculated to be about 66.5mm, which is significantly larger than 44mm. To pass the same mass flow in such a large impeller, the trim or blade height of the impeller will have to be small, and be well outside the optimum trim values (45-55) of turbocharger compressors at this diameter. On the other hand, unlike turbocharging internal combustion engine at high flows when compressor efficiency may not be critical, compressor efficiency for the fuel cell system needs to be as high as possible to maximise system efficiency, so this is going to be a challenging task.

At such a size, compressor flow range will be reduced when compressor trim is reduced. The flow range requirement, while may not be a problem with the map in Figure 1, may become an issue. Map width enhancement measures, such as ported shroud casing treatment and variable geometry may then be necessary.

2.2   Other constraints to the compressor

The overall dimension of the compressor is restricted to that of a turbocharger compressor for similar power rating passenger cars. An implication of this size restriction is that with a wheel size much larger than that of a turbocharger compressor, the outlet-to-inlet diameter ratio of diffuser and the volute size of the fuel cell compressor will be reduced. This will have negative impact on compressor efficiency, adding to the difficulty of compressor aerodynamic design.

Second constraint of the compressor is its axial force: the force must be small. Air foil bearings, not oil bearings nor ball bearings, are to be used to support the compressor and the high-speed motor driver, to avoid contamination by lubricant oil or grease vapour to the fuel cell. The air bearings have very limited thrust load capability, and without a turbine or a second compressor to balance the thrust force, the axial force generated by the compressor needs to be reduced. The force is pointed toward the compressor inlet direction, and can be reduced by properly design of the cavity and seal behind the impeller. Axial force reduction of the compressor, however, will not be discussed in this paper.

3   COMPRESOSR AERODYNAMIC DESIGN

3.1   Specific speed

Specific speed Ω_s and specific diameter D_s are commonly used to select the type of turbomachines and to judge if a design is likely to be efficient, they are defined as

$$\Omega_s = \phi^{1/2}/\Psi^{3/4}, D_s = \Psi^{1/4}/\phi^{1/2}, \qquad (1),(2)$$

where ϕ and Ψ are flow coefficient and loading coefficient of the turbomachine respectively. Using the 1000m altitude, higher flow point in Figure 1 and assume a compressor efficiency of 70% at this operating point, following values are obtained: $\phi = 0.250$, $\Psi = 0.648$, $\Omega_s = 0.693$ and $D_s = 1.79$. Looking at the Cordier diagram for machine selection in Figure 2, the compressor is inside centrifugal compressor region, and is a low specific speed machine. The compressor is, however, some distance away from the Cordier line, meaning that it is away from high efficiency locations.

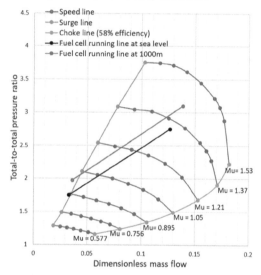

Figure 1. Fuel cell running lines on a 44mm turbocharger compressor map.

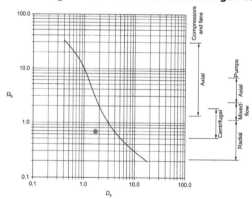

Figure 2. Cordier diagram for machine selection. Taken from [8]. Turbomachines of good efficiency should be on or near the Cordier curve. Red dot represents the fuel cell compressor to be designed.

A low specific speed centrifugal compressor has small blade height, and its flow passage is narrow and long (which is used to increase blade loading). The friction loss of the passage's boundary-layer and tip leakage loss will be large and affect efficiency. Many industrial compressors are low specific speed types, so there are experiences in designing them. Compared with the fuel cell compressor considered here however, these compressors are usually large. This means that their Re, which is proportional to the size of machines, is large, and their tip clearance relative to machines' size is small, so the effects of low specific speed are generally less severe. In recent years, there are some publications on fuel cell compressor design such as one mentioned earlier [5], and Zheng et al [9] described the aerodynamics of development of a low specific speed compressor for fuel cell application. There are also publications on electrically driven compressors for supercharging internal combustion engines [10-13], and these compressors are also of low specific speed type, and some interesting design features such as the use of forward swept of impeller and vaned diffuser were proposed. All these compressors, however, operate at much lower pressure ratios than current one: their pressure ratio is about 2:1 or less, compared with over 3:1 of the current compressor. The high pressure requirement restricts the tip width of the impeller for stability reason, and this, as will be seen, has a significant effect on compressor efficiency.

3.2   CFD method employed

A design optimisation was performed to find the best geometric configurations for the compressor. Some of the results from this optimisation are reported here, to highlight the influences of the low specific speed on design decisions and the lessons learnt. CFD was used as a design tool. Unless otherwise stated, commercial CFD code Fineturbo™ was employed, and single impeller and vanless diffuser flow passage was simulated. The CFD setup follows the practice established at GWM which has been calibrated with experiment. The code solves Reynolds averaged Navier-Stokes equations with finite-volume method. One-equation, Spalart-Allmaras model was used for turbulence closure. CFD was performed at 140,000rpm. Computational grid for the impeller of the compressor may be seen in Figures 10 and 11. When CFD code CFX was employed, two-equation, SST turbulence mode was used.

3.3   Optimisation of meridional geometry

3.3.1 *Effect of trim*

Figure 3 shows the two trims studied by CFD, wheel and vaneless diffuser were included. Figure 4 shows the results of CFD. When trim is reduced from 23 to 20 (12%), the peak efficiency is reduced by 2 points, peak pressure ratio by 3.4% and choke flow by 13%. The negative effect of the low specific speed on compressor efficiency is obvious. Calculated peak efficiency of the larger trim compressor is less than 76%, and a normal turbocharger compressor at this size would score a higher than 80% value.

35

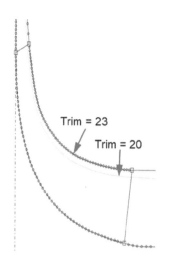

Trim = 23

Trim = 20

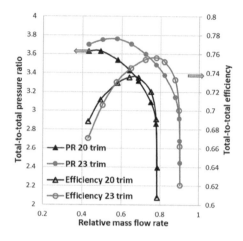

Figure 3. Two trims studied by Fineturbo™.

Figure 4. Effects of trim on perform-ance by Fineturbo™.

3.3.2 *EI and effect of tip clearances*

Ratio of clearance-to-blade height, which is the largest at impeller tip, has large nega-tive effects on compressor efficiency and pressure ratio. This ratio becomes large when impeller blade tip width decreases due to the reduction of compressor specific speed. A relatively large tip width or EI value, at least 10% larger than the typical values used by turbocharger compressors, see Figure 3, was found necessary and useful. Higher EI values would have been possible for lower pressure ratio compres-sors, such as similar sized electrically driven compressors for charging internal com-bustion engines, and highly beneficial to compressor efficiency; but high EI values make compressors less stable by reducing the radial velocity at diffuser inlet, and the stability requirement at high pressure ratios sets an upper limit of EI value of this compressor.

The effects of the clearance between the wheel and the housing on compressor per-formance were investigated by CFD with two configurations: one used typical clear-ance values of turbochargers, and the other reduced the clearance by 14% at the leading edge (LE) and 43% at the trailing edge (TE). These reduced clearances were considered the minimum clearances that the machine could safely operate. The simu-lation of the entire compressor stage was carried out by CFX, and the results are given in Figure 5. They show that the clearance reduction increases the peak efficiency of the compressor by nearly 2 points, and augments compressor pressure ratio.

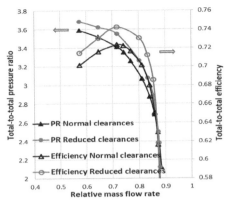

Figure 5. Effects of reducing shroud clearances by 14% and 43% at LE and TE respectively, CFX stage simulation.

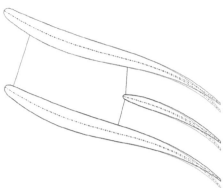

Figure 6. Hub of the two wheels with different backswept angles, grey one has ten degrees more backswept. TE is located at right-hand side.

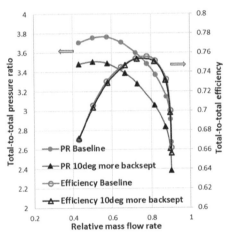

Figure 7. Effects of backswept angle on total-to-total performance by Fineturbo™

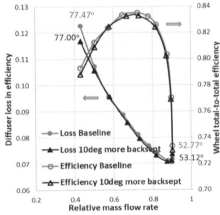

Figure 8. Effects of backswept angle on wheel & diffuser performance by Fineturbo™. Numbers labelled are diffuser inlet flow angle α_2 at flows of 0.4309 & 0.9014.

3.4   Effects of exit blade angles

The compressor wheel is of backswept type, two values of the backswept angle (the blade angle at exit) were studied. Figure 6 shows the blade-to-blade view of the hub of the two impellers. The grey and lighter colour one has ten degrees more backswept angle, the angle was varied by the same amount at the shroud as well. Calculated performance is given in Figure 7. When the backswept angle is increased, the pressure ratio is reduced, and the stability of the speed line enhanced as expected. However, the efficiency is hardly improved, and this differs from the behaviour of turbocharger compressors. Usually, the loss inside impeller will increase with the backswept angle because of the higher flow velocity inside the exducer and a longer exducer passage, but the loss inside diffuser will be reduced because of decreased diffuser inlet velocity and more radially outward flow at diffuser inlet, and the combination of the two effects increase compressor efficiency at reduced mass flow condition. So, what happen here?

Figure 8 compares *impeller* total-to-total efficiency and the efficiency loss caused by the *vaneless diffuser* in the two backswept cases. Both the efficiency and the loss are only slightly affected by the backswept angle except at the smallest mass flow, and the improvement in the diffuser performance is void by the worsening of impeller performance. The impeller outlet absolute flow angle, labelled in Figure 8 at two extreme mass flows, also differs little. From impeller exit velocity triangle, diffuser inlet flow angle α_2 can be expressed as

$$\tan\alpha_2 = V_{2u}/V_{2r} = (W_{2u} + U_2)/W_{2r} = \tan\beta_2 + 1/\phi_2, \tag{3}$$

so, when flow coefficient ϕ_2 is small, as is in this case, α_2 is less affected by the changes of relative flow angle β_2. Present compressor employs a diffuser with a smaller exit-to-inlet diameter ratio than normal turbocharger compressors, and the flow passage length and the loss inside the diffuser is less affected by the flow angle at diffuser inlet. These findings suggest that for low specific speed compressors, backswept angle may not be an efficiency critical parameter, and it may be manipulated for the benefits of pressure ratio of the compressors.

Figure 9 compares the static-to-total performance of the two compressors. The higher backswept impeller shows about 1-point higher static-to-total efficiency than the lower backswept one. The diffuser exit velocity at the two extreme mass flows is labelled in the figure, and the numbers show that the higher backswept angle reduces the velocity. This will decrease the loss in downstream components such as volute, and benefit stage efficiency. This advantage is however accompanied by a significant drop of compressor pressure ratio.

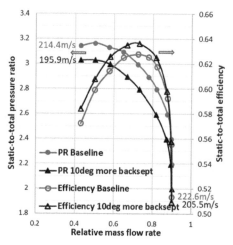

Figure 9. Effects of backswept angle on static-to-total performance by Fineturbo™. Diffuser outlet velocity is also shown at flows of 0.4309 & 0.9014.

38

The diffuser exit velocity is high, suggesting that the housing will be critical to stage efficiency. (This was found true in a housing study for a different fuel cell compressor.) It also signifies insufficient diffusions in the impellers and diffuser. Impeller diffusion factor, defined as the ratio of outlet-to-inlet relative velocities W_2/W_1, is 0.909 at the peak efficiency point for the baseline, which is considerably larger than the common value of 0.7 for centrifugal compressors. This happens because the small inlet blade diameters in this compressor (see Figure 3) reduce the inlet relative velocity W_1 compared with larger trim turbocharger compressors, and decrease impeller diffusion (increase the diffusion factor). One can raise the inlet diameters to augment the diffusion in the impeller with a negative effect on stability by reducing the centrifugal force contribution to impeller's pressure rise, and increasing inducer tip clearance-to-blade height ratio due to a reduction in inducer blade height.

3.5 Effects of blade fillet

In small blade heights, certain minor geometric features which usually have limited effects on compressor performance, may become influential. One of such features is blade root fillet necessary for the mechanical integrity of the impeller. Fillet's effect on the fuel cell compressor was studied: a uniform fillet of 1mm radius was added, see Figure 10. The number of spanwise gridlines in Figure 10(a) is 53, and 61 in (b) both including 17 gridlines for the shroud clearance. For the without fillet case, a grid sensitivity study was carried out further with 69 and 81 spanwise gridlines and little performance differences were found. Numerical results are given in Figure 11. It shows that the fillet can significantly affect compressor performance: it reduces the peak efficiency by about 2 points and decreases the choke flow and pressure ratio of the compressor. It also makes the compressor more stable. Such large influences have not been observed in normal turbocharger compressors.

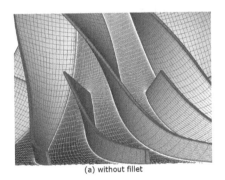
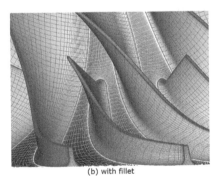

(a) without fillet   (b) with fillet

Figure 10. Computational grids with and without fillet.

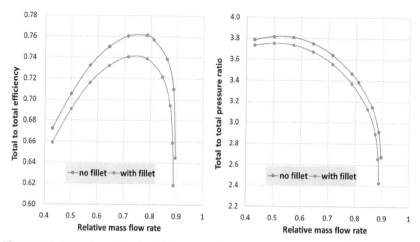

**Figure 11. Effects of blade fillet on performance by Fineturbo™. Left: effi-
ciency; Right: pressure ratio.**

The origin of the large loss associated with the fillet was investigated. One contributing factor is the intensified TE wake because TE thickness is increased at the hub substantially and the fillet now occupies a large portion of the TE. This can also explain the improved stability when the fillet was added: the stronger wake energises diffuser hub boundary-layer, making the vaneless diffuser more stable. A similar phenomenon was observed when blunt TE replaced elliptical one in turbocharger compressors [14]. Figure 12 compares the pitchwise averaged entropy with and without the fillet. There is a marked increase of entropy in the diffuser starting from the impeller near TE when the fillet is used. The inducer hub of the impeller is also affected. The blockage by the fillet reduces flow area and pushes flow toward shroud region, this decreases compressor choke flow. This happens to turbocharger compressors as well, but with reduced blade height, the effect is more significant here. The blockage also makes the diffuser inlet flow more tangential in the hub region and less so in other spans, Figure 13. This increases the passage length for the flow in the diffuser hub region and escalates the friction loss in the region.

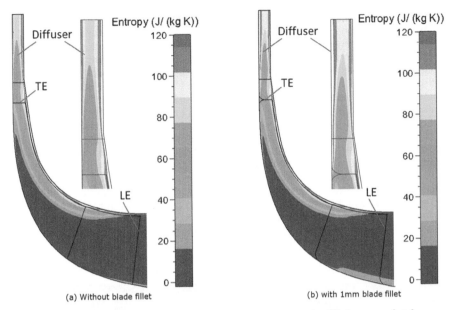

Figure 12. Pitchwise averaged entropy near peak efficiency point by Fineturbo™ with enlarged view of diffuser.

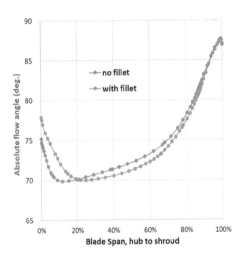

Figure 13. Pitchwise averaged diffuser inlet flow angle near peak efficiency point by Fineturbo™.

3.6   Use of splitters

The impeller is designed with splitter. This configuration was found better than using only full blades in meeting the flow range requirement of the application. The passage squareness of the impeller, that is, blade height-to-pitch ratio, however, is still very poor at impeller exit due to the small exit blade height. This increases individual exducer blade loading and the secondary flow at exducer and is considered bad for efficiency. An attempt was made to improve the squareness by introducing secondary splitters, a common practice used for highly loaded centrifugal compressors. It was thought that the additional splitter blades could also raise compressor pressure ratio, and by increasing blade number, individual blade loading could be reduced, and so would the overtip leakage. The baseline is a usual full blade + splitter design. Two different splitter designs were studied, and they are (1) full blade + short splitter + long splitter,) and (2) full blade + short splitter + long splitter + short splitter, Figure 14(b). In (1), all the blades are equally spaced in the circumferential direction, and in (2), the two short splitters are placed in the middle of the full blade and the long splitter. The shorter splitters were made by cutting the longer splitter. Figure 14(c) compares the predicted performance of three configurations. Compressor efficiency is reduced at all mass flows with the two new splitter configurations. In addition, Splitter (2) does not increase compressor pressure ratio, and Splitter (1) does so only at reduced mass flow region. Although the new splitters are not optimised, and some improvement to compressor efficiency may be expected if they are, the results are still disappointing. It is therefore worthwhile to investigate. Only Splitter (1) case will be discussed, Splitter (2) case is similar.

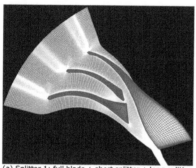 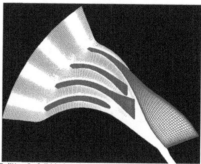

(a) Splitter 1: full blade + short splitter + long splitter (b) Splitter 2: full blade+short splitter+long splitter+short splitter

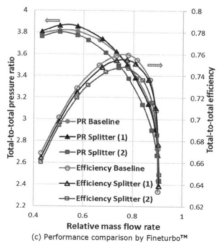

(c) Performance comparison by Fineturbo™

Figure 14. Splitter configuration study.

Figure 15 compares the entropy at 90% span of the baseline and splitter (1) at relative mass flow of 0.675. There is an increase of entropy when a shorter splitter is added. The streamlines at the same span are also plotted in the figure. The overtip leakage flow, shown as pitchwise flow across the blades, is particularly strong at the exducer (shroud clearance-to-blade height is the largest at exducer blade). Some of this flow reverses inward and is mixed with outward flow along the pressure side of the main blade of baseline or of the longer splitter of Splitter (1), generating substantial entropy. In Splitter (1) case, the flow reversing is stronger and the mixing loss higher.

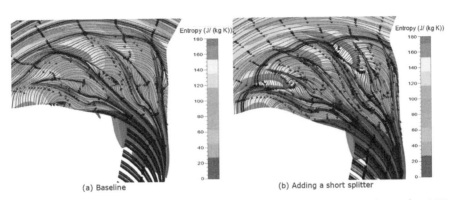

(a) Baseline   (b) Adding a short splitter

Figure 15. Entropy & streamlines at 90% span at relative mass flow of 0.675 (near peak efficiency).

Figure 16 compares the blade loading of the baseline and splitter (1) at 85% span and relative mass flow of 0.675. The exducer loading of the main blade is reduced as expected when a shorter splitter is added. The long splitter loading is reduced in its exducer too. The decreased loadings should weaken the overtip leakage of the main blade and the long splitter. The loading generated by the additional short splitter however creates its own overtip leakage, and this contributes to the entropy increase in the passage formed between the two splitters, Figure 15.

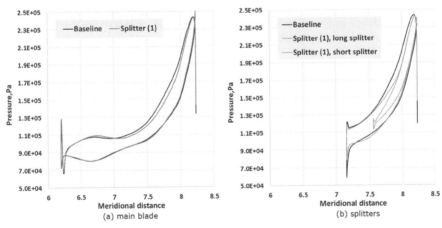

(a) main blade   (b) splitters

Figure 16. Blade loading at 85% span and relative mass flow of 0.675 by Fineturbo™.

43

Table 1 gives the overtip leakage mass flow for the baseline and Splitter (1). It can be seen that the reduced loading of main or full blade and splitter blade does decrease the leakage flow through these two blades, but the additional splitter increases the total overtip leakage flow by about 20%. The percentage of the leakage flow to compressor mass flow is 5.38% and 6.57% in the two cases respectively. This design exercise shows the importance of overtip leakage in such a small size, low specific speed impeller.

Table 1. Overtip leakage flow at relative mass flow of 0.675.

Leakage flow	Baseline	Splitter (1)
Main blade	0.0209	0.0199
Splitter	0.0154	0.0142
Short splitter	-	0.0103
Total leakage flow	0.0363	0.0444
% of comp. flow	5.38	6.57

3.7   Use of vane diffuser

Because of the modest pressure ratios, a vaned diffuser was not considered useful. A CFD study was however carried out with a low-solidity vane diffuser and no advantage was found, and the idea was abandoned.

4   RESULTS

A prototype compressor was made and tested at GWM. A photo of the impeller is given in Figure 17. A volute housing with vaneless diffuser was employed. The severer size constraint to the compressor means that a small diffuser and a small housing had to be used. The test was carried out on an automatic gas-stand of GWM and the compressor was driven by a turbine fed with hot gas (this may decrease measured compressor efficiency at low pressure ratios due to the heat transfer from hot turbine to compressor and the way compressor efficiency is calculated.). Figure 18 displays the measured compressor performance. The compressor satisfies the requirements of the fuel cell at sea level in both flow range and efficiency. It also has sufficient altitude margin. However, the compressor efficiency at high speeds is not as high as wanted, and some improvement to compressor stability is needed to cover the low-end point at the 1000m altitude.

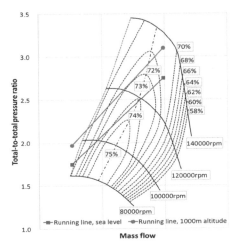

Figure 17. Prototype impeller.

Figure 18. Measured compressor map, fuel cell running lines imposed.

5   CONCLUSIONS AND DISCUSSION

A fuel cell compressor is under development at GWM, and aerodynamics of the compressor is discussed. Limited by the speed of driving electric motor and single stage configuration, the compressor is of low specific speed, and departs from the Cordier line. A stringent size constraint limits the size of diffuser and housing. These makes it difficult to achieve high stage efficiency that is required for the application. Several lessons were learnt during the development, and they could be useful for compressor designers who face similar challenges. These are:

(1) Backswept angle has less effect on stage efficiency than in higher specific speed compressors such as turbocharger compressors. Both impeller efficiency and vaneless diffuser loss are not greatly affected by the angle.

(2) When inducer blade height is small and the inducer is placed at small radii, the diffusion factor of impeller can be large, or the diffusion of relative velocity by the impeller can be quite small. While this may be good for compressor stability, it is detrimental to compressor efficiency.

(3) From points (1) and (2), a conclusion can be drawn that the housing will have a great influence on stage efficiency and be an important component. While not reported here, this has been confirmed by a separated housing study. In additional to geometry configuration, the surface smoothness of the housing passage will be important, and die-cast or surface grinding techniques can be beneficial.

Different design strategies from the one adopted here may also worth trying. A large EI value up to 1 or greater will produce a small tip clearance-to-blade height ratio and improve compressor efficiency, and ported shroud [15] or low solidity vane diffuser [16] may be applied to stabilise the flow for good flow range. Optimisation of Splitter (1) case may then make sense and yield better results. Large Blade swept, lean or bowing may be considered to decrease shroud loading to reduce the leakage and supress secondary flows, now that the mechanical design is less an issue than turbocharger compressors due to lower tip speed. Compressor housing needs a further investigation for this

45

particular application because of its unusually high inlet velocity. Surface finish of the housing is important so grinding of housing surface will be beneficial. A proper matching of housing to diffuser can minimise housing loss [17], and should be practiced.

NOMENCLATURE

A	Area
D	Diameter
D_s	Specific diameter
EI	Exit-to-inlet annulus area ratio of impeller
GWA	Great Wall Auto
LE	Leading edge of impeller
Mu	Machine Mach number
PR	Total-to-total pressure ratio
Re	Reynolds number
TE	Trailing edge of impeller
trim	$(D_{1s}/D_{2h})2 \times 100$
U	Blade tip speed
V	Absolute flow velocity
W	Relative flow velocity
α	Absolute flow angle
β	Relative flow angle
ϕ	Flow coefficient
Ω_s	Specific speed
Ψ	Loading coefficient
Subscript	
0	Total state
1	Impeller inlet
2	Impeller outlet
h	Hub
r	Radial
s	Shroud, specific
u	Tangential

ACKNOWLEDGEMENTS

The authors are grateful to the management of GWM for their permission to publish this paper. Several people at GWM contributed to the drawings, hardware procurement and tests of the compressor, and this is acknowledged. The support from the Project of the Scientific Research Leaders of Dalian Maritime University (002530) is greatly appreciated. The authors thank the unknown reviewers for their useful and constructive comments and suggestions, many of which are implemented.

REFERENCES

[1] https://en.wikipedia.org/wiki/Fuel_cell_vehicle.

[2] https://global.toyota/en/newsroom/toyota/29933463.html? padid=ag478_from_kv.

[3] T. Sugawara, T. Kanazawa, N. Imai, and Y. Tachibana, Development of motorized turbo compressor for Clarity fuel cell. SAE Technical Paper 2017-01-1187, 2017, doi: 10.4271/2017-01-1187.

[4] T. Sugawara, Air supply system for Honda fuel cell vehicle. Int. Conf. Turbochargers & Turbocharging Asia Pacific, 8–9 May 2019, Singapore.

[5] Honeywell's presentation at 13th International Conference on, IMechE, 16–17 May 2018, London.

[6] K. Kramer, The Challenge of Fuel Cell Compressor Development. Int. Conf. Turbochargers & Turbocharging Asia Pacific, 8–9 May 2019, Singapore.

[7] K. Kramer and P. M. Wu, Fuel Cell Compressor development, a Great Wall Motors perspective. Turbocharging Seminar 2019, 10–11 Sept. 2019, Harbin, China.

[8] S. L. Dixon and C. A. Hall, Fluid mechanics and thermodynamics of turbomachinery. 7th edition, Elsevier Inc., 2014. ISBN: 978-0-12-415954-9.

[9] X. Q. Zheng et al, Design of a centrifugal compressor with low specific speed for automotive fuel cell. Proc. ASME Turbo Expo 2008, GT2008-50468, June 9–13, 2008, Berlin, Germany.

[10] K. R. Pullen, S. Etemad and R. Cattelland, Experimental investigation of the Turboclaw® low specific speed turbocompressor. Proc. ASME Turbo Expo 2012, GT2012-69074, June 11–15, 2012, Copenhagen, Denmark.

[11] T. Lefevre, The S-turbo, a Low Speed turbocharger for small gasoline engine. Turbocharging Seminar 2015, 23–24 Sept. 2015, Tianjin, China.

[12] B Richards, et al, A high-performance electric supercharger to improve low-end torque and transient response in a heavily downsized engine. Int. Conf. Turbochargers ; Turbocharging, 17–18 May 2016, London.

[13] I. Ghazaly and M. Zangeneh, Design of compressor for electrically decoupled turbocharger in downsized gasoline engine by 3D inverse design. Int. Conf. Turbochargers ; Turbocharging, 17–18 May 2016, London.

[14] H. Chen and J. Yin, Turbocharger compressor development for diesel passenger car applications. 8th International Conference on Turbochargers and Turbocharging, IMechE, May 2006, London.

[15] H. Chen and V. M. Lee, Casing treatment & inlet swirl of centrifugal compressors. ASME Turbomachinery Journ. July 2013, Vol. 135/041010-1~8.

[16] H. Tamaki, Experimental study on matching between centrifugal compressor impeller and low solidity diffuser. Turbocharging Seminar 2017, 19–20 September 2017, Dalian, China.

[17] T. Ceyrowsky, A. Hildebrandt, M. Heinrich and R. Schwarze, Assessment of the loss map of a centrifugal compressor's external volute. ASME Turbo Expo 2020, GT2020-14291. June 2020.

Design and validation of a pulse generator for the pulse shaping of the turbine inlet pressure at the hot-gas test bench

P. Nachtigal, H. Rätz, J. Seume

Leibniz University Hannover, Institute of Turbomachinery and Fluid Dynamics (TFD), Germany

H. Mai

Kratzer Automation AG, Germany

ABSTRACT

In order to reproduce the unsteady turbocharger-turbine inlet conditions generated by an engine at a hot-gas test bench, a novel pulse generator unit (PGU) has been developed. This PGU modulates pressure-oscillations to the turbocharger intake-flow and is designed to work continuously in tough industrial environment. It consists of a circular throttle, which is rotated with non-uniform velocity in order to modify the shape and the frequency of the pressure oscillation. In this paper, the design and first experimental results of the PGU are presented. The focus lays on generating close-to-engine operating conditions for the turbine in an engine operating range below 2000 min^{-1}. Therefore, the shapes of the pressure curves and their dependence of the pulse frequency are investigated in detail.

1   INTRODUCTION

Nowadays internal combustion engine turbocharging is common practice and is used to increase engine performance and process efficiency. From the various options for turbocharging, exhaust-gas turbocharging has established itself in passenger cars. Here, the power required to increase the boost pressure is obtained from the residual enthalpy of the exhaust gas. The turbocharger (TC) became an integral part of the internal combustion (IC)-engine and its transient performance is essential for the agile power output of the whole system. It is therefore necessary to pay great attention to the behaviour of the TC, especially in part load operation where the system efficiency is important for reducing fuel consumption.

The enthalpy transfer from the exhaust-gas to the turbine of the turbocharger happens under idealized, adiabatic conditions according to the following equation:

$$H_T(t) = \dot{m}_T(t) \cdot \eta_{is,T}(t) \cdot c_p(t) \cdot T_3(t) \cdot \left[1 - \left(\frac{p_4(t)}{p_3(t)} \right)^{k-1/k} \right] \tag{1}$$

It is important to note, that all quantities are time dependent and when the TC is operated on an IC engine. To be able to compare the unsteady behaviour of the TC on the engine with the steady performance at the hot-gas test bench, all those variables need to be assessed in a time-resolved fashion.

The different behaviour of those quantities between the operation of the TC at the IC engine and at the hot-gas test bench are described in the following two sections.

1.1 Operating behaviour of the TC on the hot-gas test bench

The flow conditions on the hot-gas test benches are usually uniform. Thus, the energy transfer of the turbine also takes place under steady conditions and at constant efficiencies. This is advantageous because operating conditions can be reliably reproduced, and characteristic maps can be easily compared between different test benches and turbochargers. However, the steady characteristic maps of the turbocharger (which were generated on the hot-gas test bench) cannot directly be used to derive the operating behaviour of the TC at the engine. Sens et al. (1) carried out a comparison of the characteristic maps between transient operation (on the engine) and steady operation (on the hot-gas test bench); the result is shown in Figure 1. The figure illustrates that the achievable compressor pressure ratio in engine operation is considerably lower than measured on the hot-gas test bench. There are several reasons for these deviations, such as the measurement position of the fluid properties or the size of the volumes behind and in front of the rotor. However, even if all those geometrical parameters are kept similar, the TC performance is not the same because the flow characteristics through the impeller deviate between steady and transient boundary conditions. This is described in the next section.

1.2 Operating behaviour of the turbocharger on the engine

Compared to the operation of the turbocharger under steady conditions at the hot-gas test bench, the fluid properties fluctuate strongly when the TC is operated on an IC-engine. The frequency at which the inlet conditions are fluctuating (which is referred to as *pulse frequency* in the following chapters) can be calculated according to the following equation.

$$f_{pulse}\left[s^{-1}\right] = \frac{n_{ICE}\left[min^{-1}\right]}{60\,\frac{s}{min}} \cdot \frac{number\ of\ cylinders}{i} \qquad (2)$$

with i = 2 for 4-stroke-engines.

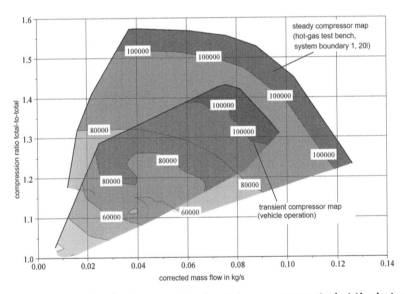

Figure 1. Comparison between a steady-state map generated at the hot-gas test bench and a transient map generated at the engine test bench (translated from: (1)).

49

With a 4-stroke, 4-cylinder engine, the pulse frequency corresponds to twice the engine speed, as one cylinder exhaust opens every half revolution of the crankshaft. According to (2), pulsations in turbines can have effects in three areas: they can influence the aerodynamic properties, the heat transfer and the friction in the bearings and seals. According to Equation (eq.1), the performance of the TC mainly depends on the expansion ratio p_3/p_4, the mass flow $\dot{m}_T$ and the turbine entry temperature T_3, which are all time variable quantities. Their influence is described below.

1.2.1 *The mass flow*
Due to the intermittent gas exchange of IC-engines, the flow to the turbine is not uniform but pulsating depending on the engine operating point. Whenever the exhaust valve of the engine is opening, a mass quantity is released to the exhaust system and consequently to the TC. In combination with the volute upstream of the turbine, the mass flow coming from the engine accumulates in the volume between exhaust valve and turbine rotor inlet, resulting in a rising pressure ratio across the turbine. As soon as the exhaust valve closes, the accumulated mass is discharged through the rotor and the pressure in the volute is falling again. This unsteady behaviour is shown in Figure 2 for a six-cylinder engine (3).

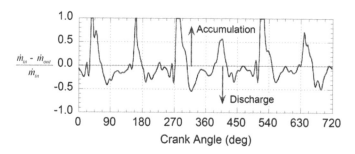

Figure 2. Unsteady mass storage in a turbine through one engine cycle (3).

1.2.2 *The expansion ratio*
The expansion ratio is directly depended from the accumulated mass quantity in the volute. When the turbocharger is operated on the combustion engine, the pressure ratio depends on the following variables in particular:

- the engine operating point (defined by engine speed and torque),
- the valve opening timings and
- the flow resistance of the turbine volute and impeller.

Figure 3 shows an example of the turbine inlet pressure for two different speeds at different engine loads (25%, 50% and full load). The figures illustrate how the peak pressure and the amplitude of the pressure pulsation increase with increasing engine load. Furthermore, it can be seen how the individual pressure pulses merge with increasing engine speed, such that the amplitudes effectively become smaller again. The maximum pressure amplitudes are therefore to be expected in the range of high loads at low speeds.

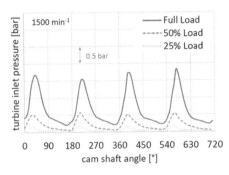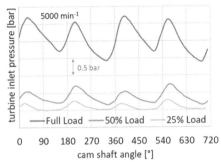

Figure 3. Load dependent pressure shape at the turbine entry for two cam shaft revolutions at 1500 min⁻¹ (left) und 5000 min⁻¹ (right) (4).

The pulsation of the turbine inlet pressure and mass flow has a significant influence on the operating behaviour and efficiency of the turbine and thus on the entire turbocharger. The basic mechanisms for this are summarized as follows: For a given geometry of the turbine volute, the absolute angle of inflow to the turbine impeller remains almost constant for a given operating point. During a pressure peak caused by opening the engine-exhaust valve, the mass flow through the turbine volute and consequently the absolute inflow velocity to the impeller increases. Since the circumferential speed of the rotor remains almost constant due to its inertia, the relative flow angle changes, resulting in an oscillating incidence angle to the impeller. Due to the misaligned flow, the transfer of work through the turbine and thus its efficiency is reduced, with the result that there is less power available for the compressor in comparison to steady inflow conditions at the same mean pressure.

1.2.3 *The turbine entry temperature*
The turbine entry temperature also strongly fluctuates during one engine cycle. One reason is the mixing of hot exhaust gas with cold intake air during load exchange with scavenging.

1.2.4 *Preliminary conclusion*
In order to be able to acquire the operating behaviour of the TC at the hot-gas test bench in the same way as on the IC engine, it is necessary to be able to map the pulsating flow conditions of the engine on the hot-gas test benches. This requires a pulse generator, which imposes a pressure and mass flow oscillation on the steady hot-gas flow. In order to obtain similar operating behaviour between engine and hot-gas test bench, it is important to:

- keep the enclosed volumes in both setups similar and to
- keep the characteristics of the pressure pulses as comparable as possible.

For measurements on the hot-gas test bench, it is therefore important to generate pressure pulses, which are similar in comparison to those on the IC engine. The influencing of the shape of the pressure pulse on the hot-gas test bench is in the following called *pulse shaping* and described in detail in chapter 4.

The temperature pulsation cannot be reproduced at the hot-gas test bench. Therefore, it must be adjusted to a constant value at which the same cycle averaged enthalpy can be transferred to the turbine.

1.3   Classification of pressure shapes

In order to describe the fluid mechanical characteristics of the pressure pulses, Zinner (5) introduced the parameters Alpha and Beta. The *quantity coefficient* Alpha describes the ratio between the actual mass flow $\dot{m}_{pulse}$, which flows through a nozzle of given cross-section during a pressure pulse, and the steady flow $\dot{m}_{steady}$, which would occur with an equivalent, constant, mean pressure and temperature curve.

$$\alpha = \frac{\dot{m}_{pulse}}{\dot{m}_{steady}} \qquad (3)$$

Under pulsating conditions, less fluid flows through a given turbine cross-section than what would be the case under equivalent, time-averaged flow conditions. This means that Alpha is always smaller than one under pulsating conditions. The so-called *energy coefficient* Beta describes the ratio of enthalpy in pulsating flow, to enthalpy in steady flow with the same average mass-flow per period.

$$\beta = \frac{H_{pulse}}{H_{steady}} = \frac{x^{-1} \cdot \sum_x h_{pulse,x} \cdot \dot{m}_{T,x}}{h_{steady} \cdot \dot{m}_T(t)} \qquad (4)$$

Under pulsating conditions, a relatively larger mass fraction flows through the turbine with high enthalpy (at high temperatures and pressures), compared to the corresponding time interval in average conditions. Therefore, Beta is always larger than one. The larger the pressure amplitudes at pulsating flow, the more Alpha and Beta deviate from one.

To have an additional parameter to characterize the shape of the pressure pulses in an easy way, the dimensionless parameter pos_{max} is introduced. This index describes the relative position pos_{max} of the pressure maximum p_{max} in relation to the period duration $T = f_{pulse}^{-1}$.

$$pos_{max} = t_{p,max} \cdot f_{pulse} \qquad (5)$$

The variable pos_{max} is used to describe the asymmetry of the pressure pulses. The analysis of the pressure curves from Figure 3 for the position of the peak pressure results in the following overview:

Table 1. Relative position of the pressure peak pos_{max}.

	1500 min^{-1}	5000 min^{-1}
25% load	0.22	0.36
50% load	0.19	0.33
Full load	0.31	0.33

Depending on the engine speed and load, the maximum pressure lies in the range from 0.2 to 0.4 times the period duration, with an average value of $pos_{max} \approx 0.3$. In order to be able to generate similar pressure curves on the hot-gas test bench to those on the engine, it is therefore necessary to shift the position of the pressure peak depending on the engine speed.

2   ALREADY EXISTING DESIGNS FOR PULSE GENERATORS

Investigations into the behaviour of turbochargers under pulsating boundary conditions have already been carried out in the past. The most important designs for pressure pulse generators on hot-gas test benches will be briefly presented below.

Already in 1990, Winterbone installed a pulse generator at his hot-gas test bench (6). This consists of a rotating cylinder with lateral recesses through which air could periodically escape into an enclosing, larger cylinder. Capobianco et al. (7) use two rotating ball valves, which are driven by electric motors, with which pulse frequencies of 10 to 200 Hz can be generated. In a second design, a modified cylinder head is used, which is mounted upstream the turbine inlet and in which pressure pulses with a variable profile can be set by a variable valve train. Karamanis et al. (8), on the other hand, use two rotating chopper discs as pulse generators, which rotate in opposite directions and which contain special cut-outs for generating the pulse profile. The advantage of this design is that the pulse shape can be varied by varying the shape of the cut-outs. Zimmermann et al. (2) also use a modified, electrically-driven cylinder head with variable valve train at their hot-gas test bench to generate the pressure pulsation. An overview of the advantages and disadvantages of the various pulsator types are provided by Reuter (9) in his doctoral thesis.

3   DESIGN OF THE NOVEL PULSE GENERATOR

In this section, the new generation of a Pulse-Generator-Unit (PGU) will be presented, which has been developed in cooperation between the *Institute of turbomachinery and Fluid Dynamics* of the *Leibniz University Hannover* together with the *Kratzer Automation AG*. The PGU is going to be installed at the hot-gas test bench directly in front of the turbocharger. When choosing the position, it is important to use a defined pipe length between the outlet of the PGU and the inlet of the turbine, which equals the volume on the IC engine. The newly developed PGU should be as robust as possible, and suitable for continuous use on the hot-gas test bench. It was therefore decided to generate the pressure pulsations by means of a rotating throttle valve, as this design is comparatively simple and can be adapted to any size and operating mode. However, in order to map the characteristics of the engine's pressure pulsation, the throttle valve is not rotated uniformly rather driven by a highly dynamic electric motor. This motor should be able to accelerate and decelerate the throttle valve several times per motor revolution. A possible, exemplary driving profile is shown in Figure 5 in section 4.

In order to achieve the highest possible dynamic response of the PGU, it is essential to keep the inertia of the entire rotating system as low as possible. In order to enable the *pulse-shaping* described above, a motor is required which can generate the maximum possible torque with minimum rotor inertia and thus the best possible acceleration capacity. The maximum possible acceleration of the motor can be calculated as a function of the motor speed using the following equation:

$$\dot{\omega}(n) = \frac{M_{EM}(n)}{J_{ges}} \tag{6}$$

It must be considered that the maximum torque which can be generated by the motor decreases with increasing speed. For the construction of the PGU, the water-cooled "*High Dynamic*" synchronous motor 1FT7065 from Siemens was chosen, as this motor has the best possible ratio of rotor inertia to torque of the motors in question. At speeds of $n < 2000\ min^{-1}$ the motor can generate a maximum torque of $M_{EM,max} = 45\ Nm$ with a rotor inertia of $J_{EM} = 6,4 \cdot 10^{-4}$ kg m^2. In addition to the inertia of the motor, the inertia of the throttle flap and the inertia of the installed clutch must be considered as well. For the overall assembly, a maximum acceleration capacity of the motor at speeds $n < 2000$ min^{-1} is estimated from equation (eq.6) as $\dot{\omega}_{max} = 44300 \frac{rad}{s^2}$.

For the control of the motor, the target value of the nominal angular velocity is set via an analogue voltage signal, which is transmitted to the control unit of the motor by the process control computer. The motor controller evaluates the signal with a cycle time of 32.25 µs and regulates the motor with a cycle time of 125 µs, which corresponds to a frequency of up to 8000 Hz. At this frequency the motor speed can be varied, which has huge potential for highly dynamic control.

Since the PGU is going to be used at hot-gas temperatures of up to 800 °C, the thermal expansion of the components used had to be considered in the design. With a total installation length of the PGU and the associated measuring tube for acquiring the turbine inlet conditions of approximately 1000 mm and a maximum turbine inlet temperature of 800 °C, there is a difference in length of max. 14 mm, which must be compensated in the test bench setup. For this purpose, the complete construction (PGU, motor and coupling), is mounted on rails, which allow a movement of the assembly in axial direction. The finished PGU including static pressure and temperature measuring points is shown in Figure 4 on the left. On the right side, a detailed view of the CAD model with the geometry of the throttle valve is shown.

Figure 4. Finished PGU and detail of the throttle valve (© *Kratzer Automation AG*).

4 CONTROL OF THE PGU

The requirement for the motor to be able to accelerate and decelerate the throttle valve per engine revolution requires complex control. To illustrate the planned operating behaviour, Figure 5 shows an exemplary course of the motor torque M_{EM} as a function of the throttle flap position. With the beginning of a new cycle at time t = 0, the flap is completely closed (φ = 0°) and rotates with the angular velocity ω_0. Within the time t_{acc}, the flap is accelerated by the motor to the velocity ω_{max} and rotates uniformly after the acceleration phase up to the 90° position. At this point, the motor decelerates the flap back to its initial angular velocity ω_0. As soon as the flap has reached its original speed ω_0 again, it rotates at this speed up to the 180°-position, from where the cycle starts again. In principle, all other arbitrary speed profiles can also be specified, as long as the maximum torque of the motor is sufficient to accelerate and decelerate the flap in the specified time.

In order to maintain the required pulse frequency and the position of the pressure maximum pos_{max} relative to the period duration, it is necessary to control the times listed in Table 2 accordingly.

Table 2. Boundary conditions for the control of the throttle flap (10).

Time	Description	Condition	Angle
0	Beginn of cylce	t = 0	$\varphi = 0°$
t_{acc}	Time of accelaration	Dependend from $M_{EM,max}$	
$t_{p,max}$	Time of maximum throttle opening	$t_{p,max} = pos_{peak} * t_{cycle}$	$\varphi = 90°$
t_{cycle}	Cylcle time	$t_{cycle} = 1/f_{pulse}$	$\Delta\varphi = 180°$

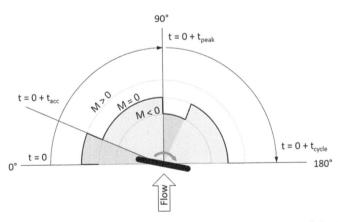

Figure 5. Torque of the motor in dependence of the position of the throttle flap (10).

4.1   Implementation of the controller

The control of the PGUs motor is carried out via the process control unit (PCU) of the test bench, in which a *Matlab Simulink* © model is executed in real time. In this model, the required course of the angular velocity is compared with the currently measured rotor position and the controlled variable (the required velocity of the motor) is calculated from it. The communication between PCU and PGU takes place via an EtherCAT coupler from *Beckhoff* using analogue voltage signals.

4.2   Simulated pressure curves downstream of the PGU

In order to be able to estimate the pressure curves which can be achieved with the PGU, a preliminary *Matlab* © script is written which contains the real inertia and bearing friction of the components used. The velocity profile used for the calculation was the angular velocity curve presented in the previous section. The pressure peak positions pos_{max}, achievable with the PGU are shown in Figure 6.

The pulse shape defined as *engine-like* with a peak position of $pos_{max} = 0.3$ can be achieved up to a pulse frequency of $f_{pulse} = 38$ Hz with the components used. At the cut-off frequency, the motor accelerates the complete first 30 % of the cycle time. A further increase is therefore no longer possible with the given components. If the driving profile of the motor is adapted (e.g. the motor starts accelerating, before the motor zero mark is reached), other pulse frequencies may also be realized. Naturally the generation of generic, symmetrical pulses ($pos_{max} = 0.5$) is possible up to the maximum motor speed of n = 6000 min^{-1} (results in $f_{pulse} = 200$ Hz).

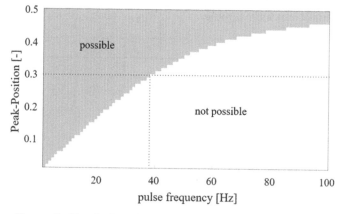

Figure 6. Simulation about achievable pressure shapes.

5   EXPERIMENTAL SETUP

The experimental measurements are performed with the test bench whose layout and instrumentation are shown in Figure 7.

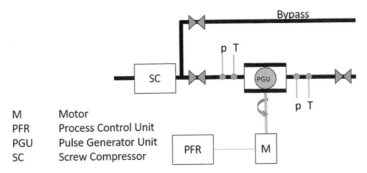

M   Motor
PFR   Process Control Unit
PGU   Pulse Generator Unit
SC   Screw Compressor

Figure 7. Experimental Setup for initial commissioning of the PGU.

For the commissioning, the PGU is connected directly to a compressed air line (without a hot-gas generator in-between). The pressure upstream of the PGU is regulated to $p_{in} \approx 2200$ hPa. The PGU is then rotated at different speeds and the pressure downstream the throttle is recorded with a data rate of 10 kHz. The results are presented below.

6   RESULTS

Figure 8 shows the chronological sequence of the normalised pressure downstream the PGU, at different uniform motor speeds (and therefore pulse frequencies). As expected from a circular throttle, the shape of the pressure pulses is sinusoidal. For the pressure measurements the combined measurement uncertainty of the used components is $u_p = 0.6$ % of the full-scale-output of the pressure transducer. This corresponds to an absolute uncertainty of $\Delta p = \pm 21$ hPa. The absolute uncertainty of the incremental encoder used to measure the rotor speed is $\Delta \varphi = \pm 18''$ ($\pm$ 0.005°). Because the error bars appear to be very small, they are not shown in the graphics below.

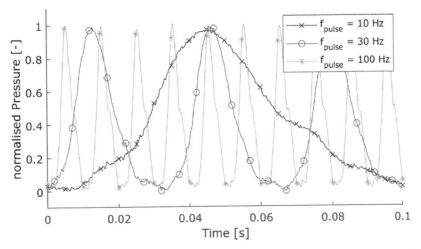

Figure 8. Comparison of the pressure curves at different pulse frequencies (uniform motor speed, $pos_{max} = 0.5$).

In Figure 9, the pressure curves at a pulse frequency of f_{pulse} = 1 Hz with two different pressure peak positions is shown. In parallel, the velocity profile of the motor is plotted for the two investigated peak-positions. For the uniform rotation (pos_{max} = 0.5), the angular velocity remains constant at $d\phi/dt$ = 180 deg/s. For the non-uniform rotation (pos_{max} = 0.2), the peak velocity of the rotor is around $d\phi/dt$ = 360 deg/s, whereas the minimum velocity is around $d\phi/dt$ = 126 deg/s. This results in the same time aver-aged velocity as for the uniform rotation.

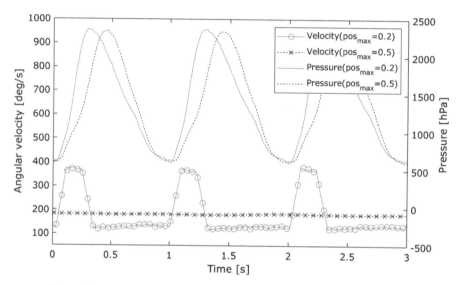

Figure 9. Example of the pressure pulse shaping for f_{pulse} = 1 Hz.

Figure 10 shows the pressure curves for different peak positions pos_{max} at a constant pulse frequency of f_{pulse} = 10 Hz. The earlier the pressure peak is located, the more the motor must accelerate the throttle and therefore the gradient of the rising edge is steeper. At constant period durations, this implies a lower gradient for the falling edge with smaller pos_{max}. With using different velocity profiles for the rotation of the throt-tle, this gradient can be adjusted.

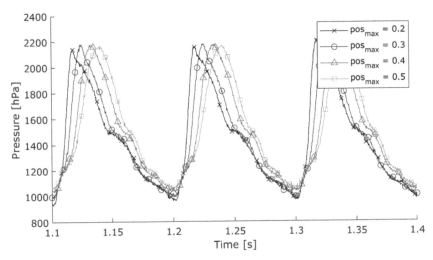

Figure 10. Comparison of different peak positions for f_{pulse} = 10 Hz.

In conclusion, Figure 11 shows the comparison of different pulse frequencies at a constant relative position of the pressure peak pos_{max} = 0.3. The time scale is normalised with the pulse frequency to get a comparable depiction of the pressure curves. At higher pulse frequencies, the deceleration time of the throttle becomes relatively longer, and thus conversely the throttle remains at a higher speed for a longer time.

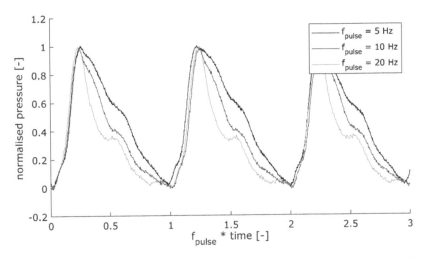

Figure 11. Comparison of different pulse frequencies with pos_{max} = 0.3.

59

7   CONCLUSIONS

The measurements show that it is possible to generate engine-like pressure oscillations using the newly designed PGU for low rotational speeds. The simple and robust design allows to easily adjust the shape of the pressure oscillation by choosing different control signals for the electric motor driving the throttle flap. This opens new possibilities to experimentally investigate the impact of the shape of the pressure oscillation on the performance of a turbocharger.

8   OUTLOOK

In the later application at the test bench of the *Leibniz University Hannover*, the PGU is used at inlet temperatures of up to 800 °C and up to six bar inlet pressure. Within the close future, the PGU will be used in various projects. In the FVV project 1311 *"Exhaust gas pulsation and turbocharger interaction"*, the PGU will be integrated into a hardware-in-the-loop system, where all operating parameters relevant for the turbocharger will be calculated with a real-time *GT-power* © engine model and then used in the test bench automation as set point variables for the turbocharger. The measurement data recorded in this way shall be compared with measurement data of the same TC from engine measurements in order to be able to quantify the different influences of the pressure pulsation on the operating behaviour.

NOMENCLATURE

c_p	Specific Heat Capacity	m	Mass Flow
f	Frequency	n	Rotational Speed
H	Enthalpy	p	Pressure
h	Specific Enthalpy	pos	Relative peak position
i	Engine Rotation per Work Cycle	T	Temperature
J	Polar Moment of Inertia	u	Relative uncertainty
κ	Isentropic Coefficient	η	Efficiency
M	Torque	φ	Angle of Throttle Flap

Indices/Subscripts

3	Turbine Stage Inlet	p	Pressure
4	Turbine Stage Exit	T	Turbine
EM	Electric Motor	t	Total Amount
is	isentropic	TC	Turbo Charger
max	Maximum, Peak	tot	total
min	Minimum		

Abbreviations

CFD	Computational Fluid Dynamics	PCU	Process Control Unit
DOF	Degree of Freedom	PGU	Pulse Generator Unit
ICE	Internal Combustion Engine	TC	Turbocharger

REFERENCES

[1] Sens, M.; Grigoriadis, P.; Nickel; Pucher, Helmut: *Untersuchungeen zum dynamischen Verhalten von Abgasturboladern im Fahrzeug*; 11. Aufladetechnische Konferenz; 2006.

[2] Zimmermann, R., Hakansson, S.: *Wirkungsgrad von Turbolader-Turbinen bei Beaufschlagung mit pulsierendem Abgasstrom*, FVV Vorhaben Nr. 1103; FVV Abschlussbericht; 2016.

[3] Winterbone, D. E.; Nikpour, B.; Frost, H.: *A contribution to the understanding of turbocharger turbine performance in pulsating flow*; IMechE paper C433/011; 1991.

[4] Iosfidis, G.; Walkingshaw, J.; Dreher, B.; Flisinger, D.; Ikeya, N.; Ehrhard, J.: *Tailor-Made Mixed Flow Turbocharger Turbines for best Steady-State and Transient Engine Performance*, Engine processes, ISBN: 9783816932222, 2013.

[5] Zinner, K.: *Aufladung von Verbrennungsmotoren*; ISBN: 978-3-540-07300-0; DOI: 10.1007/978-3-662-05914-2; 1975.

[6] Winterbone, D. E.; Nikpour, B.; Alexander, G. I.: *Measurement of the performance of a radial inflow turbine in conditional steady and unsteady flow*; 4th international conference IMechE; C405/015; 1990.

[7] Capobianco, M.; Marelli, S.: *Transient Performance of Automotive Turbochargers: Test Facility and Preliminary Experimental Analysis*; SAE paper 2005-24-066; DOI: 10.4271/2005-24-066

[8] Karamanis, N.; Martinez-Botas, R. F.; Su, C. C.: *Mixed Flow Turbines: Inlet and Exit Flow Under Steady and Pulsating Conditions*; ASME, Journal of Turbomachinery, Vol. 123; DOI: 10.1115/1.1354141; 2001.

[9] Reuter, S.: *Erweiterung des Turbinenkennfeldes von Pkw-Abgasturboladern durch Impulsbeaufschlagung*; Dissertation; 2010.

[10] Nachtigal, P.; Mai, H.; Rätz, H.; Seume, J. Willers, O.: *Auslegung und Regelung eines Pulsationsgenerators für Turbolader-Heißgasprüfstände*; Baar Gedenkschrift, TU Berlin; 2020.

1D gas exchange modelling of double scroll turbines: Experimental validation and improved modelling

Tetsu Suzuki[1], Georgios Iosifidis[1], Jörg Starzmann[1], Wataru Sato[2], Dietmar Filsinger[1], Takahiro Bamba[2]

[1]IHI Charging Systems International GmbH, Heidelberg, Germany
[2]IHI Corporation, Yokohama, Japan

ABSTRACT

Automotive turbocharger suppliers are continuing to refine their product portfolio to meet increasingly stringent emission regulations. IHI developed the double scroll turbine concept as one of the technologies suitable for the high-specific-power market segment of passenger vehicles. Despite the recent market introduction of this technology, its 1D gas exchange modelling methodology remains largely underrepresented in the relevant literature. In this paper, extensive experimental data are used to improve the steady-state turbine mapping. Furthermore, 1D modelling methodologies are validated against experiments under pulsating flow conditions. Finally, it is concluded that the turbine wheel can be treated as a quasi-steady component while any unsteadiness of the turbine stage can be considered to be due to mass storage effects in the turbine volute. On top of this, the effect of windage (ventilation) loss is not negligible in case of double scroll turbine under low frequency pulsating flow condition. By considering aforementioned points, an improvement of engine performance predictability is confirmed.

1   INTRODUCTION

For the purpose of turbine performance improvement, the ability to accurately predict turbine performance under engine operating conditions with lower dimension tools, such as 1D gas exchange models, is crucial. 1D engine modelling plays a central role in automotive engine development. Within a project cycle it serves as a tool to define targets for component development as well as to optimize and eventually evaluate the performance of the system. Novel turbocharger technologies like double scroll turbine designs call for further refinement of gas exchange simulation capabilities, particularly when they interfere with the degree of unsteadiness experienced by the turbine stage.

For mono scroll turbines Copeland et al. [1] have done extensive analysis of turbine performance, introducing parameters which can be used to quantify turbine unsteadiness. Based on these investigations Teng et al. [2] established unsteady parameters to judge whether each turbine component (volute, wheel) must be treated in a steady or unsteady manner. Costall et al. [3] compared representative frequencies and have shown that turbine wheels can be modelled as quasi steady. In parallel Aymanns et al. [4] suggest turbine one-dimensional modelling techniques for adequate turbine performance prediction, mainly by means of 3D-CFD. In a much more experimental approach Lüddecke et al. [5] investigated unsteady turbine behavior by using a novel method of contactless torque measurement. The authors concluded that for mono scroll turbines under pulsating flow conditions the turbine wheel behaves quasi-steadily whereas the volute must be treated unsteadily. For unsteadiness levels relevant to engine operating conditions this conclusion supports one-dimensional turbine modelling without the introduction of any special losses. For the double scroll turbine Newton et al. [6] have analyzed turbine performance under pulsating flow, showing that turbine performance of vaned double scroll turbines can be predicted reasonably well by means of 3D-CFD. The question however remains: to what extend can double scroll turbine performance be predicted with one-dimensional approaches? To address this point, one-dimensional predictions of unsteady turbine behaviour are compared with both experimental as well as 3D-CFD results.

Targets of the present investigation are to (a) clarify whether double scroll turbine performance can be accounted for by traditional 1D engine simulation tools such as GT-POWER, (b) provide insights about the challenges of double scroll turbine 1D simulation and (c) make proposals for further modifications of the modelling strategy.

2   ONE DIMENSIONAL DOUBLE SCROLL MODELLING

In order to establish a reference the basic one-dimensional modelling technique for double scroll turbines must be introduced. The commercial 1D gas exchange simulation tool GT-POWER has been used in this study. It is one of the solutions most extensively used in the automotive industry. There are several ways to model the double scroll turbine in GT-POWER [7]. The most rudimentary possibility is to deploy one turbine map (object) connected to two turbine inlets. In this case the turbine map flow capacity at each inlet can only be adjusted by a multiplier and a single common efficiency must be used as input. To model the cross flow rate through the gap between tongue and wheel a simple orifice can be used. This is, of course, a non-predictive approach and the effective area has to be calibrated based on either experimental or 3D-CFD data. In case of double scroll turbines it is well-established that the crossflow amount is significantly smaller compared to twin scroll turbines owing to the much smaller area between the two scrolls. Therefore, accurate prediction of the crossflow rate is much less important for double scroll turbines. On the other hand efficiency differences between full admission and single admission are much more pronounced for the double scroll turbine, since single admission effectively only utilizes half of the available turbine area. It should also be mentioned that in order to harvest the full potential of the double scroll turbine concept scroll connection valves are an almost-necessary add-on feature for passenger car applications [8]. Hence, turbine maps for both single and full admission conditions are required.

To address the inherent problems of this simple approach more detailed modelling strategies have recently appeared [9-11]. These consist largely of two turbine objects representing the turbine wheel (one for each scroll) and additionally secondary turbine objects which attempt to model the crossflow area depending on the turbine operation condition. In this case multiple maps can be used to more accurately describe various admission conditions of each side. Turbine flow capacity and efficiency can be looked up according to the turbine's admission condition, which is expressed by the Mass Flow Ratio (MFR) at each scroll inlet.

In the current investigation turbine performance under single, full and unequal admission condition is measured by using a dynamometer test bench enabling direct and wide-range torque measurement. These turbine maps are pre-processed using the IHI extrapolation method. Importantly, only single and full admission condition but no unequal admission maps are used as input for GT-POWER. This restriction is introduced so that the methodology does not deviate from daily practice, where the engineer rarely gets access to extended double scroll maps. (This type of mapping requires not only time and effort but also a dedicated double-burner gas stand [12].) Hence, an interpolation of turbine capacity and efficiency over the MFR between full and single admission is mandatory. This is done with the revised method based on [9] to improve the prediction at unequal admission condition during the engine's operation cycle. The duct volume in front of the turbine object is corresponding to the actual turbine volute volume of each scroll and is aiming to reproduce the mass storage effect in the 1D simulation. The inner surface friction is ignored as it is assumed that the measured efficiency already includes the related losses.

3   VALIDATION OF DOUBLE SCROLL MODELLING

3.1   Overview of measurement setup

The experimental investigation of the present double scroll turbine has been obtained at the well-established Imperial College London test facility. The tested turbine stage consists of a mixed flow wheel with 8 blades and an outer diameter of 86mm. In the following a brief introduction of the measurement setup is given but a more detailed description is given by Szymko et al. [13]. The turbine test rig operates with compressed air and typical turbine inlet temperatures are 40°C to 50°C. Figure 1 shows the mounted double scroll turbine and a schematic layout of the test bench [14, 15]. In the latter graph it is depicted that two V-cone flow meters are used to measure the mass flow rate of the inner and outer scroll of the double scroll turbine under steady-state operation. The pulse generator is located further downstream where a rotating plate with cut-offs is generating 180° phase-shifted flow pulses fed into each turbine scroll. In order to determine the mass flow rate under pulsating flow conditions at position 1 and 2 in Figure 1 a fine wire is kept at constant temperature. In the so-called constant temperature hot wire anemometry the voltage to maintain the wire temperature can be correlated to the mass flow rate. The required calibration is done using the V-cone flow meter data under steady operation.

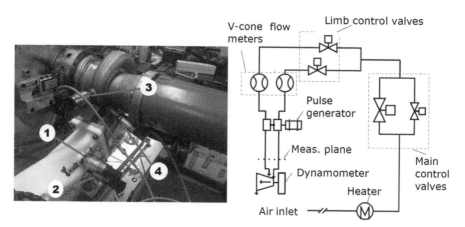

Figure 1. Turbine setup (left) and schematic test rig layout (right).

Beside the time-averaged pressures $\bar{p}$ and temperatures $\bar{T}$ also time-resolved pressure measurements were conducted and using the adiabatic relationship,

$$T = \bar{T}\left(\frac{p}{\bar{p}}\right)^{\frac{\gamma-1}{\gamma}}$$

(1)

the instantaneous inlet temperature is calculated for each scroll. This simplified method has been compared to a dual hot wire system and an uncertainty of $\pm 3°C$ was observed [6]. It is worth mentioning that stagnation conditions, which are used as simulation boundary condition, are derived by applying standard gas dynamic relations.

The Imperial College test facility allows a direct turbine power determination by recording the averaged torque τ via an eddy current dynamometer and the instantaneous speed ω via a 20-toothed optical encoder. Knowing the moment of inertia I the instantaneous torque τ follows,

$$\tau = \bar{\tau} + \tau' = \bar{\tau} + I\frac{d\omega}{dt}.$$

(2)

3.2   3D-CFD model description

Three dimensional steady and unsteady Reynolds-averaged numerical flow simulations were performed in Ansys CFX 19.2 and compared to the experimental and 1D simulation data. Unless otherwise stated the so-called 'frozen-rotor' approach has been applied and for turbulence closure the k-ε turbulence model with scalable wall function was selected. The present study focuses on global turbine performance rather than resolving near wall flow phenomenon and thus the overall y_+-values are in the range of 20 to 30 which, of course, also limits the computational costs. Mesh generation was carried out in ICEM CFD whereas for the volutes, the turbine diffuser and outlet pipe a structured, and for the rotor an unstructured mesh was generated. Features like blade fillets, scallops and the rotor back-disc cavity were included in the meshing process. Table 1 summarizes the mesh sizes for each of the computational flow domains.

Table 1. Summary of the CFD mesh statistic.

Domain	No. of mesh nodes	Type of mesh
Inner scroll and volute	267,620	Structured
Outer scroll and volute	432,800	Structured
Turbine wheel	1,842,920	Unstructured
Diffuser	228,850	Structured
Outlet pipe	149,280	Structured

In the course of a previous study [14] a thorough mesh independence study was conducted under steady operation conditions. The volute meshes were refined until the variation of Mach number at characteristic monitor points went below 5%. With about 1.8 million nodes (4.5million elements) the rotor mesh is already comparatively fine and regarding the computational time a further mesh refinement is not desired. In order to estimate the remaining mesh sensitivity an exemplary simulation with a 2.7 times finer rotor mesh revealed a negligible effect on the mass flow parameter (<0.5%) and an increase in efficiency by 0.5%-pts. This calculation has been conducted for an outer admission case at a pressure ratio of 1.5 and a turbine speed of 30krpm. The comparison of the 3D-CFD results with experimental data in chapter 3.4.2 confirms, that the mesh influence can be considered to be small compared to other effects.

3.3   1D results under steady-state conditions

As previously mentioned it is typically required to use double-burner testing facilities to measure unequal turbine inlet admission conditions which is a time- and cost-intensive activity. Such a double burner measurement can be avoided if the one-dimensional model can successfully interpolate turbine performance based solely on single and full admission maps. In order to establish whether the aforementioned 1D model can reliably predict steady-state turbine performance under unequal admission conditions while utilizing exclusively full and single admission maps, the latter were used in the simulation. The calculated performance at unequal admission was compared to the measurements. Here the total pressure at turbine housing inlet, the static pressure at turbine outlet and the turbine speed were defined as boundary conditions for the calculation. Thanks to the updated interpolation method of the mass flow parameter in the turbine object a satisfying forecast of mass flow rates at the turbine housing inlet for intermediate conditions was obtained. Since turbine isentropic power is dependent on turbine inlet temperature, mass flow rate and turbine pressure ratio the isentropic turbine power is also predicted well, see Figure 2. In terms of generated shaft power the error between measurement and one-dimensional calculation becomes bigger but is still less than ±5% at high power conditions where turbine power prediction under pulsating flow conditions is more critical, see Figure 3.

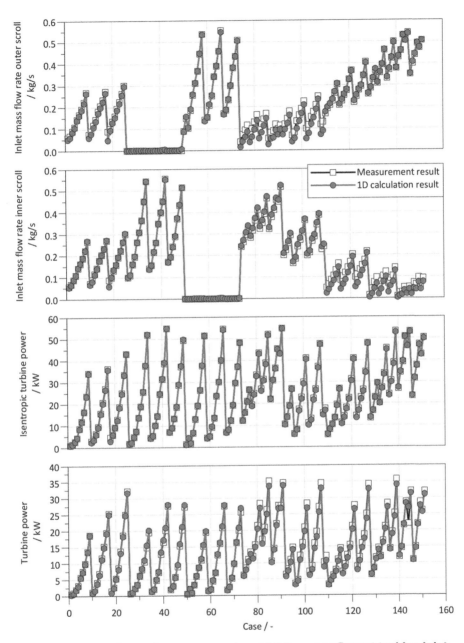

Figure 2. Comparison of measurement and 1D for mass flow at turbine inlet, shaft and isentropic turbine power under steady-state conditions.

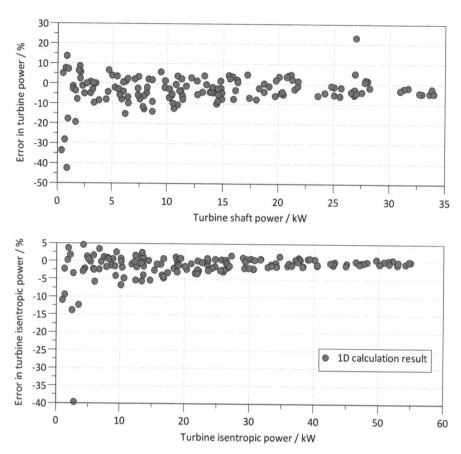

Figure 3. Percentage error of turbine shaft and isentropic power.

3.4   Results under pulsating conditions

3.4.1 *Evaluation of unsteadiness parameter*

As discussed earlier, in order to accurately predict turbocharged engine performance turbine operation must be considered under pulsating flow conditions. In case of mono scroll
turbine housings numerous authors argue that the volute flow should be considered
unsteady, due to filling and emptying of the volume, while the turbine wheel flow can be
treated quasi-steadily. For double scroll turbines the scroll separation is strong and the
positive pressure gradient is very high. Therefore, it is uncertain whether this conclusion
can be carried over from mono to double scroll designs. To study this issue measurement
results under pulsating inflow conditions were compared against one-dimensional double
scroll simulations where the turbine volute is treated unsteadily and the turbine wheel as
a quasi-steady element. Certain parameters to characterize engine operation conditions
in actual application were derived to ensure that a similar level of unsteadiness exists in
the experiment and in the simulation. In this investigation the unsteadiness criterion Λ
described by Copeland et al. [1] and the maximum turbocharger speed specific pressure
gradient Γ_p introduced by Aymanns et al. [4] have been considered. For a realistic evaluation of the 1D simulation model these turbine unsteadiness parameters were calculated
along the full load line of a 2.0L 4-cylinder gasoline engine with scroll connection valve as
a representative turbocharger configuration for this engine architecture. The controllable
scroll connection valve interconnects the two turbine volutes at higher engine speeds and

loads and facilitates the reduction of engine back pressure. This in turn enables a selection of smaller turbine stages which ultimately improves overall engine perform-ance [8, 16]. In Figure 4 the parameters Λ and Γ_p are shown for the pulse generator measurements and equivalent engine operating conditions.

$$\frac{2P}{P_0} \cdot \frac{L_0}{v_0 t_0} = \Pi \cdot \text{St} = \Lambda \qquad (3)$$

$$\Gamma_p = \frac{\mathrm{d}P}{\mathrm{d}t} \cdot \frac{1}{n_{\text{TC}}} \qquad (4)$$

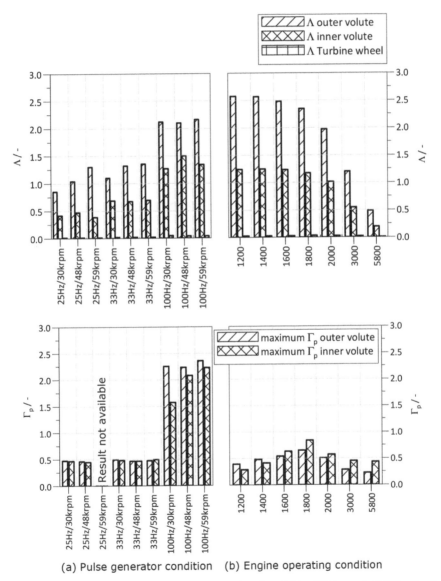

(a) Pulse generator condition   (b) Engine operating condition

Figure 4. Unsteadiness parameters Λ and Γ_p for 25Hz, 33Hz and 100Hz (test bench) and 1200rpm-5800rpm (exemplary 2.0L 4-cyl. gasoline engine).

Owing to the scroll connection valve the instantaneous pressure gradient at high engine speed is reduced so that the unsteadiness is smaller than that at low engine speeds which effectively leading to quasi-mono scroll operation. Therefore, turbine performance prediction for the scroll-connected turbochargers is more difficult at low engine speeds. The unsteadiness of such conditions (<2000rpm) is very similar to the 100Hz condition of the pulse generator when considering Λ, even if the multiplicator Π and the multiplicand St between pulse generator and the engine are different. Specifically, the pressure gradient Π of the pulse generator is almost twice that of the engine, while due to the frequency difference the Strouhal number St of the pulse generator is almost half that of the engine. Regarding Γ_p, the values of 25Hz and 33Hz from the pulse generator match the value of low engine speed conditions. This means that the pressure gradient for one turbine wheel rotation is similar between the pulse generator and the engine. This is critical regarding flow evaluation since it is known that the pressure gradient plays an important role when considering turbine stage unsteadiness. A turbocharger speed of 30krpm was selected because of the circumferential Mach number similarity between engine and pulse generator. Further discussion focuses on the 33Hz/30krpm operating condition.

Similar to the steady-state condition turbine housing inlet total and outlet static pressures are applied as boundary conditions in the one-dimensional model since these values are experimentally obtained and seem to be most accurate. The turbine braking torque is controlled so that the time-averaged shaft speed is same as that of the experimental condition. 3D-CFD was also conducted to confirm the flow condition.

3.4.2 *Comparison of measurement, 3D-CFD and 1D gas exchange*

Steady-state measured maps are used in the 1D simulation. A comparison with 3D-CFD simulation has been done for various speed lines. Figure 5 shows a representative result at 30krpm for the case of outer scroll admission with fully closed inner scroll. At pressure ratios below 1.5 the simulated and the measured mass flow parameter agree with each other but the discrepancy increases continuously in a way that at maximum pressure the turbine capacity is under-predicted by 7%. At peak efficiency the 3D-CFD turbine efficiency is about 6%-pts higher while this deviation is also increasing with pressure ratio. Such a behaviour can typically be explained by different types of leakage, e.g. across the bypass channel, but the tested design does not feature a wastegate. The test rig and measurement equipment are well-established and have been used in earlier studies. In addition to the mesh sensitivity study, the CFD model itself was carefully examined. Without any doubt, the 'frozen rotor' model leads to a severe simplification of the flow. Thus, a sliding mesh calculation with 15 time steps per blade passing event was conducted for a pressure ratio of 2.4. Figure 5 shows a negligible effect on the turbine capacity but the efficiency drops by 3.7%-pts. This reduces the deviation versus the experiment to an acceptable agreement but fails to fully explain the over-estimation of turbine performance. Double scroll turbine stage exhibit an extremely complex flow situation especially at the two tongue regions. Strong shear flows are present which makes the flow solution sensitive to turbulence modelling. A thorough study of these notoriously complex turbulence structures would not only come along with considerable simulation efforts but also requires a suitable experimental

validation both of are far beyond the scope of the current investigation. Moreover, the geometric 3D CFD model is generated based on nominal design data.
Deviation from casting tolerances can also serve to explain the mass flow deviation. It is worth mentioning that introducing surface roughness modelling into
the CFD model would tend to reduce the predicted flow capacity of the turbine
stage and is therefore not appropriate to resolve the discrepancy between simulation and experiment. In any case, it should be noted that the focus of the current work is on improving the 1D simulation capabilities rather than validating
the CFD model for this turbine geometry.

A comparison of the experimental data with the simulation results under pulsating inflow
conditions follows. Figure 6 shows the turbine inlet total pressure and temperature as well
as the turbine power and mass flow rate over the rotational angle of the pulse generator.
Since the total pressures and temperatures are used as boundary conditions there obviously exists no difference between simulation and experiment. Regarding the mass flow
rate at the turbine inlet, the 3D-CFD and the 1D-simulation predicting the pulse event in
good agreement to each other.

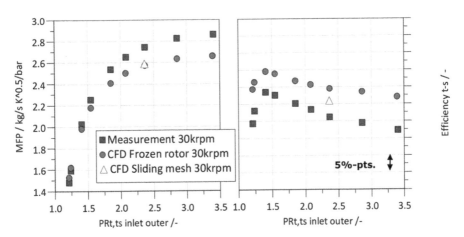

**Figure 5. MFP and efficiency comparison of measurements and 3D-CFD for
steady-state single admission flow condition.**

The predicted steady-state efficiency in 3D-CFD is higher than in the experiment
which causes higher predicted power in 3D-CFD compared to 1D (Figure 6).

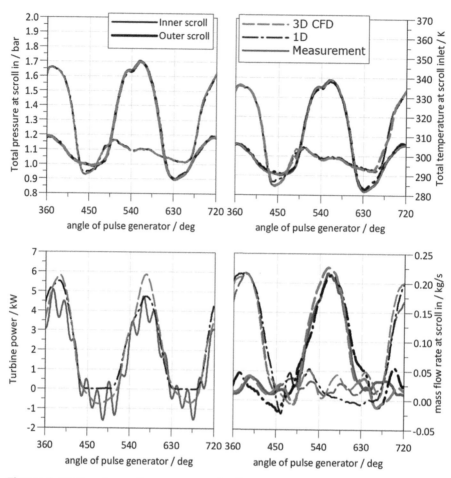

Figure 6. Comparison of measurements, 3D-CFD and 1D under pulsating flow conditions.

Remarkably enough, the turbine power achieves negative values both in the 3D-CFD calculation and the measurement. At these conditions Figure 6 indicates that the inlet total pressure is very low, and thus turbine pressure ratio is one or even smaller. It is obvious that no turbine work can be generated at such "no-flow" conditions. Furthermore the turbine wheel is spinning in a non-vacuum environment and is rather acting as a fan. These windage or ventilation losses leads to a reduction of available power [17].

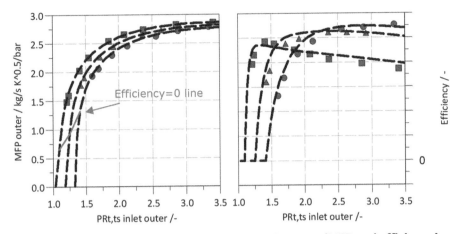

Figure 7. Extrapolated single admission turbine map (MFP and efficiency).

In GT-POWER efficiency is usually assumed to be zero under the so-called "run-away pressure ratio", see red line in Figure 7. In this condition, the turbine wheel is still rotating while generating windage losses. In case of single admission this effect is con-sidered in turbine efficiency but it should be additionally accounted for, when both vol-utes are operated under these conditions. Thus, a windage loss correction was implemented in the 1D model by employing of the following equation [17]:

$$\dot{W}_w = K_1 \left(\pi D_m h \frac{\rho U_m^{\;3}}{2} \right) (1 - \varepsilon) \tag{5}$$

K_1 is an empirical coefficient and was selected so that the absolute value of negative torque becomes similar to that of the CFD value. By considering windage losses it was possible to receive negative torque under conditions such that the pressure ratio of both turbine inlets is smaller than the run-away pressure ratio. Comparing the 3D-CFD, 1D and experimental results it was found that the 1D approach presented in Chapter 2, expanded with the windage loss model, has the capability to reproduce the 3D-CFD result reasonably well under the low engine speed conditions of the 2.0L 4-cyl gasoline engine.

After having compared the 3D-CFD and the 1D modelling results a comparison against the experimental results is certainly of interest. Although the flow capacity calculated by 3D-CFD is smaller than that of the measurement (see Figure 5), the predicted peak mass flow rate under pulsating conditions is particularly higher for inner scroll admis-sion (Figure 8). Such a difference can be mainly traced back to the storage effect of the scrolls; an in-depth analysis of this phenomenon is given by Aymanns et al. [4] but for a mono scroll turbine. It can be argued that this higher mass flow rate in the simulation leads to a 14% higher peak power at inner scroll admission.

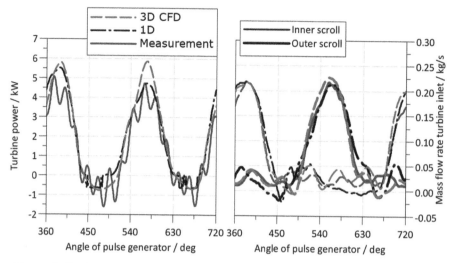

Figure 8. Instantaneous turbine power and mass flow rate with the revised 1D gas exchange model.

At outer scroll admission the turbine power between the 3D-CFD and the measurement differs by 19%. This over-estimation of peak power leads to a deviation of the cycle-averaged power by approximately 35%. In comparison to the investigation of Newton et al. [6], who studied a double scroll turbine with vaned nozzles and obtained a deviation of about 5%, this is a remarkable discrepancy. The over-prediction of instantaneous peak power and cycle-averaged power by 3D-CFD is somewhat surprising considering the very good agreement of the measured instantaneous static pressure in the outer volute, Figure 12. Further investigations are needed.

Regarding the flow direction within the experiment so far only the forward flow was is considered. Nevertheless, the current hot wire anemometry does not allow conclusions regarding the flow direction. Up to this point, all flow rates were treated as forward flow because no period of zero flow rate was detected – which was expected to be observed when flow direction changes. On the other hand, it is reasonable to assume flow reversal because one side of the pulse generator is closed when the opposite scroll side is fully open and flow is not admitted into both scroll sides simultaneously under this condition. Figure 9 shows the comparison before and after a modification of the flow direction. After this adjustment, 1D simulation was repeated with mass flow rate boundary condition. The result is shown in Figure 10. By changing the boundary condition the mass flow rate travelling through the turbine wheel becomes smaller and thus the estimated turbine power is also reduced. Both the instantaneous and the averaged power predicted by 1D are then well-matching with the experiment and the error is reduced from +34% to -7%. Thereby the error is calculated as follows:

$$error = \frac{Power_{1D} - Power_{exp}}{Power_{exp}} \cdot 100 \qquad (6)$$

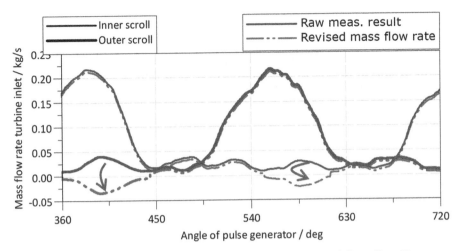

Figure 9. Revised mass flow rate to account for actual flow direction.

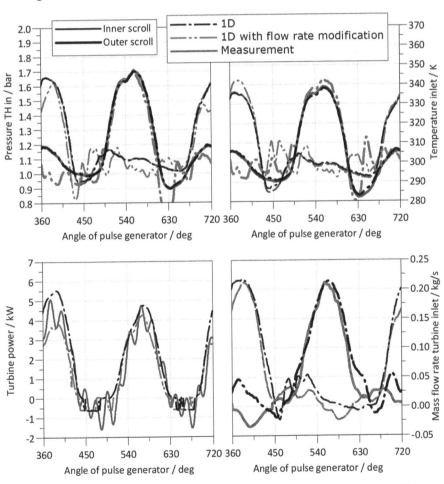

Figure 10. Modified turbine power and mass flow rate instantaneous behaviour with the revised flow direction.

3.4.3 *Further analysis*

In order to confirm in further detail whether the 1D approach is correctly capturing the actual flow condition, the instantaneous turbine operation was plotted on the characteristic map and compared to the steady-state result. The instantaneous operating line under pulsating conditions and the steady-state maps under full and single admission of the outer volute are depicted in Figure 11. The pressure delay is shown in Figure 12. The pressure delay was measured with the pressure sensor installed at the outer volute outlet. In 3D-CFD and in the 1D model the pressure at the same location was also extracted. The operating line for the turbine wheel and that of the turbine stage are plotted separately in order to more clearly visualize the differences in the turbine's components behaviour. It is visible that the turbine operating line hysteresis predicted by the 1D approach is well matching to the 3D-CFD as well as the measurement. For this confirmation the unknown representative characteristic length of the turbine volute was varied in GT-POWER while keeping the known total volute volume constant, whereby it was found that the definition of representative characteristic length is very important for the prediction of turbine performance. In other words, the turbine wheel can be considered as quasi-steady even in case of a double scroll housing design provided that the turbine volute is correctly modelled and therefore the filling and emptying effects are well captured.

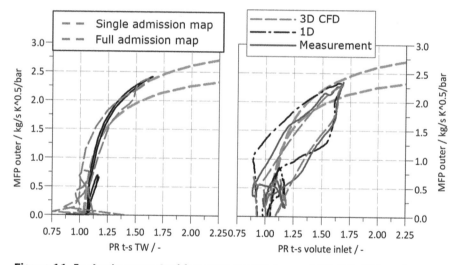

Figure 11. Instantaneous turbine map comparison of 1D, 3D-CFD and measurement results.

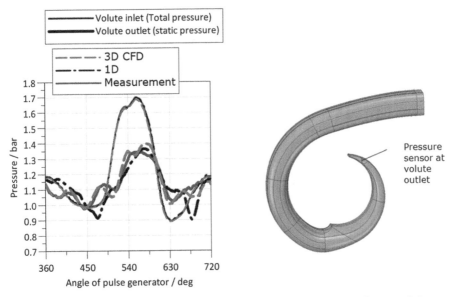

Figure 12. Pressure time delay (hysteresis) of the outer volute outlet.

4   IMPROVED MODELLING FOR ENGINE PROCESS SIMULATION

The objective of this paper is to improve the accuracy of engine performance prediction. The main findings from this investigation have been the significance of correct turbine volute modelling in one-dimensional models. Especially the representative characteristic length has to be defined properly. Also ventilation or windage losses have to be sufficiently considered. The latter phenomenon is important under pulse generator conditions but the influence on engine performance has to be confirmed since the on-engine shape of instantaneous pressure in front of the turbine is different. Figure 13 provides an instantaneous turbine inlet pressure of engine operating condition and a comparison of engine BMEP between engine performance measurement and 1D, both with and without consideration of windage at low engine condition of 1600rpm. Since the on-engine phase of pressure decrease is slower than on the pulse generator, the period during which windage loss must be considered on engine operation is very limited. During the engine cycle approximately 2.8% of its duration is effected from the windage effect. With this adjustment engine BMEP is reduced by 1.2%. In case of engines equipped with double scroll turbochargers with scroll connection valves, the exhaust valve duration is usually shorter because a larger flow capacity can be utilized at open scroll connection valve condition at higher engine speeds. In this case, the windage effect becomes more important as it does in the case of 2-cyl engines with longer pauses between consecutive pulses.

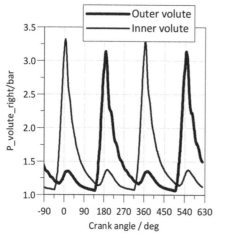
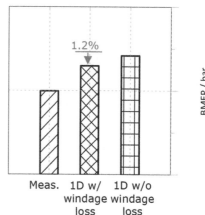

Figure 13. Instantaneous pressure traces for a scroll-connected double scroll 2.0L 4-cyl gasoline engine at 1600rpm and BMEP with and without consideration of windage losses.

5   CONCLUSIONS

In this paper double scroll turbine performance predictability under pulsating flow conditions by means of a 1D approach was investigated by utilising 3D-CFD as well as experimental results from a test rig equipped with a pulse generator. The instantaneous turbine operating line derived from the 1D approach showed good agreement with the experimental and also the 3D-CFD when the characteristic length was defined by the mean duct length of the volute. With this modelling detail the pressure delay (hysteresis) due to the volute's filling and emptying or in other words the mass storage effect in the 1D result also showed good agreement with the measurement and the 3D-CFD. During the comparison of instantaneous turbine behaviour it was found that the turbine wheel can be treated as a quasi-steady component while any unsteadiness of the turbine stage can be considered to be due to mass storage effects in the turbine volute. The modelling strategy that is explained in Chapter 2 provides good predictability, especially in terms of mass flow rate and if appropriate dimension have been specified. Another important finding is the windage loss effect during the pulsating cycle. In the examined pulse condition there exists a period where no flow is fed from any of the two volute sides. In the default GT-POWER setting the turbine efficiency is considered as zero while in reality the turbine wheel is acting as a fan and dissipates energy. The effect of this loss on engine BMEP at a relevant lower engine speed is approx. 1.2% even though the period where windage loss has to be taken into account is limited under engine operating conditions. By considering this effect, engine performance prediction at lower engine speed can be improved. On the other hand, the quality of 1D as well as the 3D-CFD predictability of turbine power generation requires further improvement.

NOMENCLATURE

T	Temperature
$\bar{T}$	Time averaged temperature
p	Pressure
$\bar{p}$	Time averaged pressure
γ	Ratio of specific heats
τ	Turbine torque
$\bar{\tau}$	Time averaged turbine torque
τ'	Turbine torque used to accelerate the shaft
I	Moment of inertia of turbine shaft
ω	Angular velocity
t	Time
Λ	Pressure weighted unsteady parameter
Γ_p	TC speed specific pressure gradient
L	Length
v	Fluid velocity
Π	Pressure amplitude weighting factor
S_t	Strouhal number
n_{TC}	Turbine speed
$\dot{W}_w$	Turbine windage power loss
K_1	Empirical constant
D_m	Mean diameter
h	Blade height
ρ	Density of fluid
U_m	Mean blade speed
ε	Arc of admission as a function of 360°

ACKNOWLEDGEMENTS

The authors would like to thank Imperial College London for their cooperation and collaboration as well as their continuous support and fruitful discussions. The authors also would like to express their gratitude to all IHI Co. colleagues who contributed to this project.

REFERENCES

[1] Copeland, C. D., Newton, P., Martinez-Botas, R. F., Seiler, M., 2012. "A comparison of timescales within a pulsed flow turbocharger turbine", IMECHE - 10th Int. Conf. on Turbochargers and Turbocharging, p.389–404.

[2] Teng, C., Liping, X., 2014. "A low order model for predicting turbocharger turbine unsteady performance", ASME Turbo Expo, GT2014-25913.

[3] Costall, A., Szymko, S., Martinez-Botas, R.F.; Filsinger, D., Nincovic, D., 2006. "Assessment of unsteady behaviour of turbocharger turbines", ASME Turbo Expo, GT2006-90348.

[4] Aymanns, R. Scharf, J., Vedder, R., Wedowski, S., Uhlmann, T., 2011. "A Revision of Quasi Steady Modelling of Turbocharger Turbines in the simulation of Pulse Charged Engines", Proc. 16th Supercharging Conference, Dresden.

[5] Lüddecke, B., Filsinger, D., Bargende, M.,2014. "Engine crank angle resolved turbocharger turbine performance measurements by contactless shaft torque detection", IMECHE - 11th Int. Conf. on Turbochargers and Turbocharging.

[6] Newton, P., Martinez-Botas, R. F., Seiler, M., 2014. "A 3-dimensional computational study of pusating flowinside a double entry turbine", ASME Turbo Expo, GT2014-26151.

[7] Gamma Technologies, 2017. GT-Suite: Flow Theory Manual. Gamma Technologies LLC.

[8] Walkingshaw, J., Iosifidis, G., Scheuermann, T., Filsinger, D., Ikeya, N., 2015. "A comparison of a Mono, Twin and Double Scroll" Turbine for Automotive Applications, Proceedings of ASME Turbo Expo, GT2015-43240.

[9] Lückmann, D., Stadermann, M., Aymanns, R. Pischinger, S., 2016. "Investigation of cross flow in double entry turbocharger turbines", ASME Turbo Expo, GT2016-57190.

[10] Uhlmann, T., Lückmann, D., Aymanns, R., Scharf, J., Höpke, B., Scassa, M., Schorn, N., Kindl, H., 2013. "Development and Matching of Double Entry Turbines for the Next Generation of Highly Boosted Gasoline Engines", 34. Internationales Wiener Motorensymposium 2013.

[11] Brinkert,N.,Sumser, S., Schulz, A., Weber, S.,Fieweger, K., Bauer, H., 2011, "Understanding the Twin Scroll Turbine - Flow Similarity,", ASME Turbo Expo, GT2011-46820.

[12] Lüddecke, B.,Mai, H., Spietzack, A., Walterscheid-Müller, P., Altvater, S., Weigl, S., 2011, "Hot gas test stand technology in the context of multi-scroll turbocharger turbine development", Proc. 21st Supercharging Conference, Dresden.

[13] Szymko, S., Martinez-Botas, R. F. and Pullen, K. R., 2005. "Experimental evalulation of turbocharger turbine performance under pulsating flow conditions", ASME Turbo Expo, GT2005-68878.

[14] Martinez-Botas, R. F. et al., 2018. "Investigation of steady and unsteady double entry turbine performance of automotive turbochargers", Internal technical report, Imperial College London, UK.

[15] Martinez-Botas, R. F., Srithar, R. and Newton, P., 2019. "Investigation of unsteady double scroll turbine performance of automotive turbocharger under pulsating inflow condition", Internal technical report, Imperial College London, UK.

[16] Rode, M., Suzuki, T., Iosifidis, G., Durbiano, D. Filsinger, D., Starke, A., Starzmann, J., Kasprzyk, N., Bamba, T., 2019. "Boosting the Future with IHI: a comparative evaluation of state-of-the-art TGDI turbo concepts", Proc. 24th Supercharging Conference, Dresden.

[17] Watson N and Janota M.S, 1982. "Turbocharging the Internal Combustion Engine" The MacMillan Press, Ltd.

14th International Conference on Turbochargers and Turbocharging
Institution of Mechanical Engineers, ISBN: 978-0-367-67645-2

Study on analytical method for thermal and flow field of a VG turbocharger

T. Kitamura, T. Suita, T. Yoshida

Mitsubishi Heavy Industries, LTD., Japan

Y. Danmoto, Y. Akiyama

Mitsubishi Heavy Industries Engine and Turbocharger, LTD., Japan

ABSTRACT

The analytical and experimental study on thermal and flow field of a VG (Variable Geometry) turbocharger has been conducted for attaining the high reliability thermal and structural design. CHT (Conjugate Heat Transfer) calculations, working for simulating heat transfer with mutual dependence between solid structures and fluid, are applied to the VG turbocharger. The computational domain includes the turbine section, the variable nozzle vane structure and the bearing housing to acquire the whole of thermal and flow field accurately. The thermal modelling around the variable nozzle vane structure has been developed.

In addition, the gas stand test demonstrated the VG turbocharger under high temperature condition to validate CHT calculations. Analytical results are evaluated against experimental data. Eventually, the proposed analytical method has been proved to have the advantage of thermal and structural design. The performance degradation induced by thermal energy loss is also discussed.

1   INTRODUCTION

VG (Variable Geometry) turbochargers have the variable nozzle vane structure to set the boost pressure optimum under various engine conditions. Eventually, they have widely been installed in diesel engine applications.

In recent years, the implement of VG turbochargers are strongly requested in also gasoline engines with Miller cycle, which realizes stable combustion and the improvement of thermodynamics efficiency. This trend obliges the variable nozzle vane structure to drive under the unlubricated and high temperature condition.

In VG turbochargers, thermal and flow field with thermal energy loss induces circumferential temperature distribution in the turbine section, which seriously influences the thermal field and durability of the variable nozzle vane structure.

Heat transfer and thermal energy loss from exhaust gas to various structures occur simultaneously. As a result, the CHT calculation with detailed modelling of a VG turbocharger is indispensable to simulate the aforementioned phenomena.

Several studies on CHT calculations of turbochargers have been conducted for the durability of turbines.

Bohn et al. [1] [2] carried out the steady CHT calculations for the whole of a turbocharger with the simplified geometry. Parametric study has been performed with variations of mass flow and turbine inlet temperature to investigate the performance characteristic of a compressor under diabatic condition.

Heuer et al. [3] applied steady and transient CHT calculations to a turbine housing with the integrated manifold. The stress analysis with the temperature field from the

CHT calculation was conducted and showed local peak stress at the locations of cracks by thermal shock tests. Additionally, the authors proposed the reasonable method of the transient CHT calculation, in which the mass, momentum and turbulence equations are frozen, and the only energy equation is solved on the basis of different time scales between fluid and solid states.

Numerical and experimental study on the thermo-mechanical load of a radial turbine wheel was reported by Heuer et al. [4]. Temperature of a turbine wheel was measured with telemetry system for the validation of CHT calculations, which supplied the temperature field to the stress analysis. As a result, the effect of thermal stress turned out to be low compared with that of centrifugal force.

Following the aforementioned study, Heuer et al. [5] investigated the analytical system surrounding the radial turbine wheel of CHT calculations. The effects of shaft specifications and adiabatic or diabatic condition applied to the walls of a turbine section have been confirmed.

Diefenthal et al. [6] modified the analytical method in Ref. [3] and measured transient temperature of a turbine wheel. The proposed analytical method, which means all transport equations are solved during the short period of fluid state varying largely, showed the improvement in evaluating the transient temperature variation of a turbine wheel through the comparison between analytical and experimental results.

Following the aforementioned study, Diefenthal et al. [7] developed the simplified method in which fluid and solid states are uncoupled. However, it is capable of guaranteeing the accuracy similar to the analytical method proposed in Ref. [3] with the reduction of calculation time.

Demelow [8] applied CHT calculations to the whole of a turbocharger to simulate transient temperature variation of the bearing housing under a hot shutdown event and thermo-mechanical fatigue life of the turbine housing under thermal cycle test condition. What should be recognized in the calculations is insufficient resolution of analytical grids in addition to no consideration of radiation heat transfer.

Kitamura et al. [9] developed the analytical method for thermal and flow field of a turbocharger with the catalyst unit. The gas stand test demonstrated the turbocharger under cold start-up condition to validate CHT calculations. The proposed method has been proved to have the advantage of designing for heating catalyst units.

This study aims to clarify thermal and flow characteristic of a VG turbocharger analytically and experimentally. This is the first actual study of CHT analysis on a VG turbocharger. The second chapter describes experimental program. The third chapter presents numerical analysis. Eventually, the experimental and analytical results are discussed in the last chapter. The useful information to attain the high reliability design for thermal and structural issues has been obtained.

2   EXPERIMENTAL PROGRAM

2.1   Experimental apparatus and method

Figure 1 shows the schematic drawing of an experimental apparatus for simulating cold start-up. The combustor generates the propane-burned gas of 800°C, which is bypassed to the exhaust chamber at the beginning. Shutting off the valve for the bypass-line and fully opening the valve toward the turbocharger simultaneously feed the hot combustion gas into the turbine section. The temperature in start-up condition was approximately 80°C in this experiment.

The test turbocharger is a VG turbocharger for diesel engines in small passenger vehicles. The opening of nozzle vanes, which are stationary, is approximately 80%.

2.2 Measurement

Local temperatures of the turbine housing and the variable nozzle vane structure were measured by sheathed thermocouples with a diameter of 1.0mm. They are embedded at the outer surfaces and the internal locations. The details of the measurement positions are shown in Figure 2.

The probes in Figure 3 were prepared for capturing the thermal field in the scroll passage of the turbine housing. They have a sheathed thermocouple with a diameter of 0.5mm in a SUS tube.

The rotational speed and the mass flow rate through the compressor and the turbine were measured by a magnet sensor and orifice plates respectively. The pressure and the temperature were also measured at the inlet and the outlet of the compressor and the turbine to determine the operating condition of the turbocharger.

The uncertainty of temperature measurements is at most ±3°C. Regarding pressure, uncertainty of measurements is at most ±0.5%. Figure 4 presents the measurement results about the operating condition. The measured operating condition has been implemented to the calculations in this study. Other measurement results will be examined in the later parts.

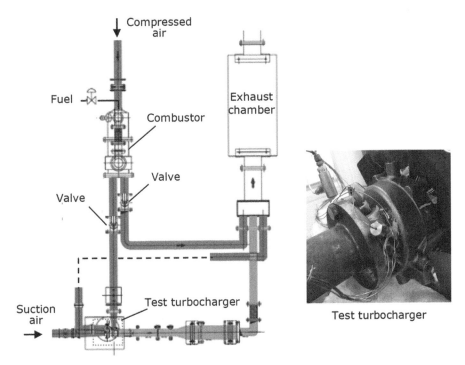

Figure 1. Experimental apparatus.

83

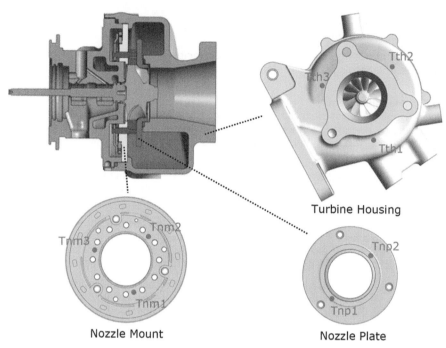

Turbine Housing

Nozzle Mount   Nozzle Plate

Figure 2. Measurement positions of metal temperature.

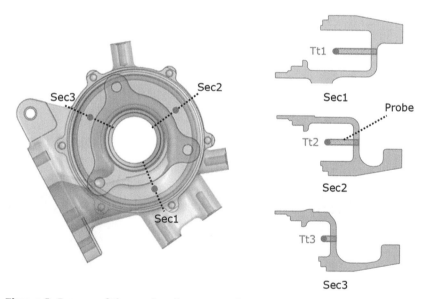

Figure 3. Image of the probes for measuring gas temperature in the scroll passage.

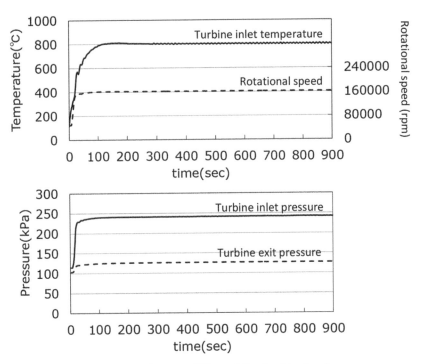

Figure 4. Operating condition of the test turbocharger.

3 NUMERICAL ANALYSIS

3.1 Analytical system and grid

The VG turbocharger stated in the previous section is the analytical system. The compu-tational domain, in the same geometry as Figure 2, consists of the turbine section and the bearing housing. The turbine wheel has been modeled as the full rotor passage. The interfaces between rotational and stationary system have been set at the inlet and exit of the turbine wheel. The analytical grid with tetrahedron elements has been generated. Multi-layered prisms have been placed at the regions adjacent to the walls. Conse-quently, it approximately has 19.9 million elements for fluid domains and 9.3 million elements for the solid domains. The non-dimensional wall distance y^+ is nearly 1.0 in fluid domains.

3.2 Numerical model and condition

The CHT calculation with the compressible CFD was performed with CFX14.5. Eddy-viscosity turbulence model according to SST k-ω model [10] has been used. The flow fields of the exhaust gas, water cooling passage and the cavity around the variable nozzle vane structure have been solved with the CFD. Stage interface has been applied at the boundaries between rotational and stationary system. Monte-Calro method takes consideration of radiation heat transfer at the cavity formed by the bearing housing and the variable nozzle vane structure. Additionally, the same method applied to the cavity formed by the back plate and the back surface of the tur-bine wheel. The emissivity is 0.8 at all of the walls.

Heat transfer at the oil gallery and the oil drain cavity is simulated with several sets of heat transfer coefficient and reference temperature. Similarly, thermal boundary conditions based on natural convection and radiation heat transfer with room temperature have been taken into consideration at the outer surfaces. In addition, heat conduction between components has been set with the appropriate thermal resistance as necessary. The summary of numerical scheme is illustrated in Figure 5.

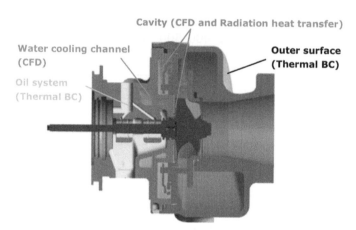

Figure 5. Summary of numerical scheme.

The propane-burned gas is regarded as ideal gas, and all thermal properties of the fluid and solid bodies are dependent on temperature. Table1 summarizes a set of the boundary condition for the steady state CHT calculation, which originates from the steady state data in Figure 4. The same method that Heuer et al. [3] proposed has been chosen in the transient CHT calculation. The mass, momentum and turbulence equations are frozen, and the only energy equation is solved with the time variation of turbine inlet temperature shown in Figure 4.

Table 1. A set of boundary condition.

Rotational speed	162200rpm
Inlet total pressure	242kPa
Outlet static pressure	127kPa
Inlet total temperature	798°C

4   RESULT AND DISCUSSION

4.1   Steady state CHT analysis

The total pressure and Mach number contours of the scroll and nozzle vane passage are presented in Figure 6. The flow field is circumferentially non-uniform under the large opening of nozzle vanes. The regions of low total pressure, which generate aerodynamics loss, are observed at the downstream side of nozzle supports. The flow field in the steady state CHT calculation is equivalent to that in CFD.

Subsequently, the streamlines and total temperature contours of the scroll and nozzle vane passage are shown in Figure 7. The total temperature field in CFD is generated by the aforementioned flow field. In addition, the thermal field in the CHT calculation has circumferential temperature distribution and reduction, which the heat loss toward various structures causes. At the outlet of the turbine housing, the mass flow-averaged temperature in the CHT calculation falls 18°C compared with that in CFD.

Figure 8 illustrates the temperature field in the meridional plane. The large thermal resistance between the nozzle mount and the bearing housing generates the temperature gap. The heat flow is also observed from the turbine housing to the bearing housing which is cooled with water cooling and oil system.

In addition to radiation heat transfer, natural convection induced heat transfer shown in Figure 8 occurs at the closed cavity formed by the bearing housing and the variable nozzle vane structure. The ratio of radiation to natural convection heat transfer is approximately 3 to 1.

Figure 9 reveals detailed temperature contours of the appearance. The temperature field of the turbine housing is determined by the complicated geometry, the heating from exhaust gas and external heat transfer. As described in 3.2, the natural convection and radiation heat transfer with room temperature are dominant at the outer surfaces in this study. Therefore, surrounding hot surface on an engine installation results in the higher temperature field of the turbine housing.

The comparison between the analytical and experimental temperatures of the turbine housing is also presented in Figure 9. The result of the steady state CHT calculation is in agreement with the experimental result qualitatively. The largest deviation is approximately +20°C.

Figure 10 shows the detailed temperature contour of the nozzle plate in which the highest temperature region is observed because of multi-surface heating based on turbine inlet temperature. The circumferential variation of temperature is mainly dependent on the total temperature field shown in Figure 7. Additionally, there are the good agreements between analytical and experimental temperatures of the nozzle plate.

Subsequently, Figure 10 also presents the detailed temperature contours of the nozzle mount and vanes. The temperature fields are affected by not only the total temperature field but also the turbine-side temperature of the bearing housing, which is also shown in Figure 10. In other words, the water-cooled bearing housing is thermally connected with the variable nozzle vane structure. Consequently, the temperature fields of the nozzle mount and vanes are complicated and non-uniform.

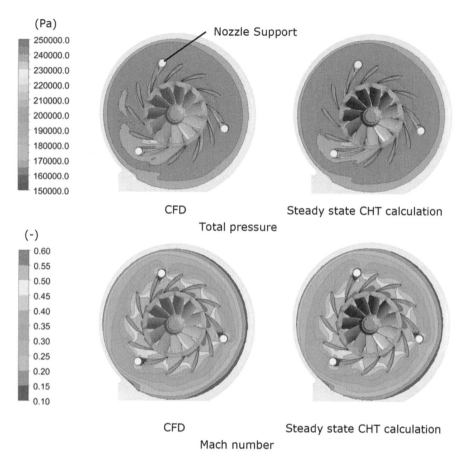

Figure 6. Total pressure and Mach number contours of the scroll and nozzle vane passage.

In addition, the comparison between the analytical and experimental temperatures of the nozzle mount is presented in Figure 10. The largest deviation, affected by the analytical accuracy of the thermal and flow field, is approximately +20°C.

The steady state CHT calculation with this analytical method is capable of predicting the thermal and flow field of a VG turbocharger.

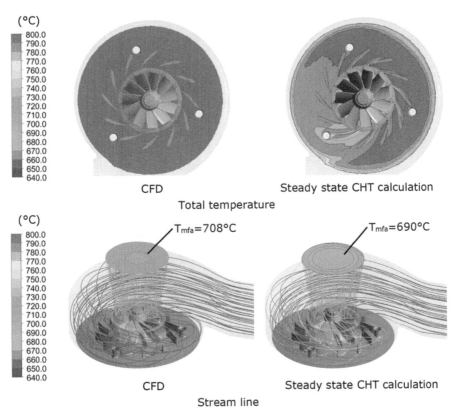

(°C)
800.0
790.0
780.0
770.0
760.0
750.0
740.0
730.0
720.0
710.0
700.0
690.0
680.0
670.0
660.0
650.0
640.0

CFD   Steady state CHT calculation

Total temperature

(°C)
800.0
790.0
780.0
770.0
760.0
750.0
740.0
730.0
720.0
710.0
700.0
690.0
680.0
670.0
660.0
650.0
640.0

$T_{mfa}=708°C$ $T_{mfa}=690°C$

CFD   Steady state CHT calculation

Stream line

Figure 7. Stream lines and total temperature contours of the scroll and nozzle vane passage.

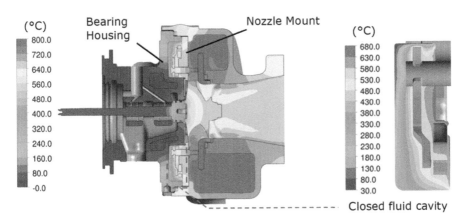

(°C)
800.0
720.0
640.0
560.0
480.0
400.0
320.0
240.0
160.0
80.0
-0.0

Bearing Housing   Nozzle Mount

(°C)
680.0
630.0
580.0
530.0
480.0
430.0
380.0
330.0
280.0
230.0
180.0
130.0
80.0
30.0

Closed fluid cavity

Figure 8. Temperature field in the meridional plane (Static temperature appears in the fluid region).

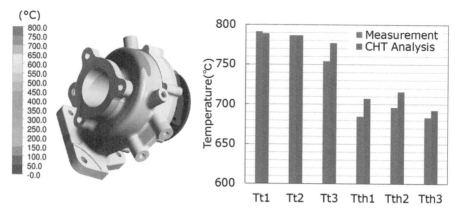

Figure 9. Temperature contour of appearance and temperatures at the measurement positions of the turbine housing.

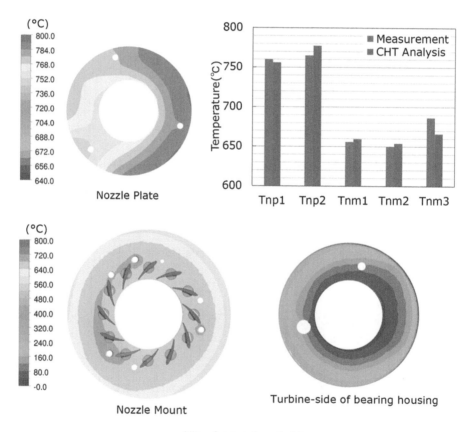

Figure 10. Temperature contour and temperatures at the measurement positions of the variable nozzle vane structure.

4.2　Transient CHT analysis

Figure 11 shows transient temperature contours of the turbine section including the static temperature field of fluid domains in the meridional plane. There is the temperature rise of the turbine section in receipt of the thermal energy from exhaust gas.

Figure 12 compares the time variations of analytical total temperatures with those of experimental ones. Regarding the temperatures of TT1 and TT2, there are the good agreements between analytical and experimental ones.

On the other hand, the analytical temperature of TT3 overestimates experimental one, which seems to be affected by the heat conduction to the wall of the scroll and tends to be lower value.

Subsequently, the transient temperatures at measurement positions of the turbine housing are presented in Figure 13. While the qualitative agreement in terms of temperature rise is observed, the transient responses of analytical temperatures overestimate those of experimental ones. The discrepancy between analytical and experimental temperatures soon after start-up seems to be attributed to the turbulence model which plays significant roles in simulating the heat transfer at the scroll region. Meanwhile, the reasonable overestimate of heat transfer can result in robust structural design.

Following the aforementioned results, Figure 14 and 15 show the time variations of analytical and experimental temperatures about the nozzle mount and the nozzle plate respectively. There are the good agreements between analytical and experimental ones although the deviation tends to enlarge under transient state.

Even though there is the partial issue about the analytical accuracy under transient condition, the proposed analytical method has been proved to have the advantage of thermal and structural design.

4.3　Thermal energy loss

Figure 16 shows the quantity of heat transfer toward individual components, which can be interpreted as thermal energy loss. The quantity of heat transfer Q is calculated as

$$Q = \sum q_i A_i \qquad (1)$$

The quantity of heat transfer in each element q_i and the heat transfer area in each element A_i are included in Equation (1).

The quantity of heat transfer is comparatively large under transient state condition, and gradually decreasing while reaching steady state. Regarding the turbine performance based on shaft power, the difference between the adiabatic CFD and the steady state CHT calculation is approximately 1%.

Furthermore, the scroll region of the turbine housing obviously dominates the thermal energy loss because it has the largest heat transfer area which is exposed to the forced convection heat transfer with the highest temperature. In any case, the reduction of thermal energy loss at the scroll region realizes in the prevention of performance degradation directly.

Additionally, what needs to be emphasized is the quantity of heat transfer at the diffuser. While the quantity of heat transfer is also comparatively large under transient state, it turns to be small under steady state condition. This phenomenon means that the heat flow from the scroll to the diffuser through the turbine housing itself occurs under steady state condition.

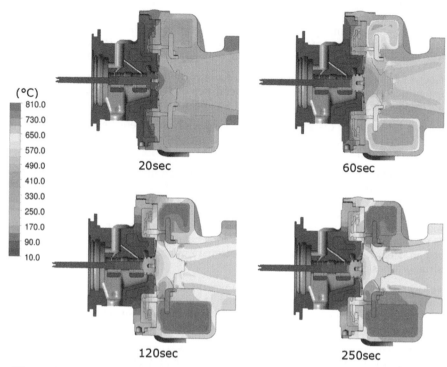

Figure 11. Transient temperature field in the meridional plane (Static temperature appears in the fluid region).

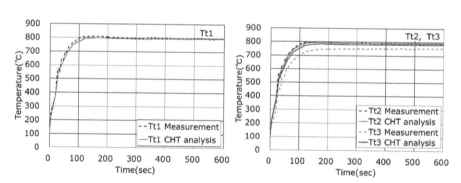

Figure 12. Time variations of total temperatures at measurement positions.

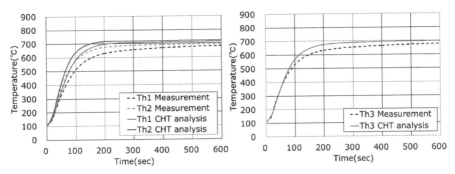

Figure 13. Time variations of temperatures at measurement positions of the turbine housing.

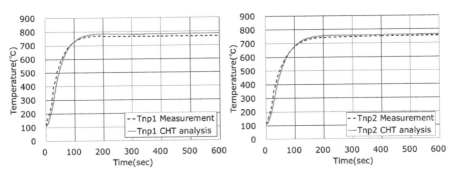

Figure 14. Time variations of temperatures at measurement positions of the nozzle plate.

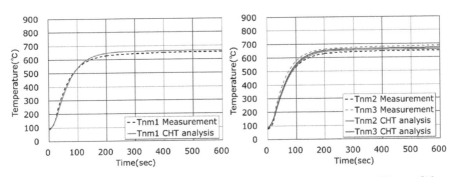

Figure 15. Time variations of temperatures at measurement positions of the nozzle mount.

93

4.4   Directions for improved design

The gas temperature drop at the inlet of the turbine wheel is one of the significant causes of turbine performance degradation. On the basis of the CHT calculations, there are 17°C and 45°C drop under steady and transient condition respectively. They are attributed to the quantity of heat transfer shown in Figure 16.

As mentioned in chapter 4.3, the reduction of thermal energy loss at the scroll region is effectively working for the improvement of turbine performance. As a specific solution, sheet metal components combined with insulators at the inside of the turbine housing are required for the purpose of reducing thermal energy loss. Additionally, they can contribute to making the temperature field uniform and preventing valuable nozzle vane structures from experiencing unfavourable thermal deformation.

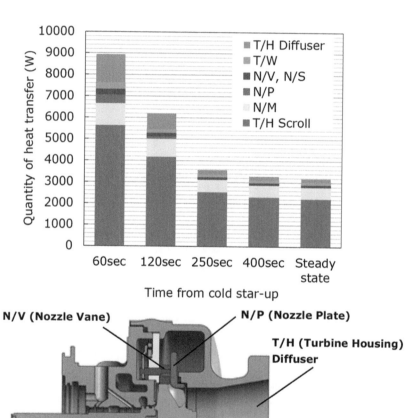

Figure 16. Quantity of heat transfer toward individual components.

5   CONCLUSIONS

The following conclusions were obtained through the analytical and experimental study on the thermal and flow field of a VG turbocharger.

- The analytical method for simulating thermal and flow characteristic of a VG turbocharger has been developed.
- The steady state CHT calculation with this analytical method is capable of predicting the thermal and flow field of a VG turbocharger.
- Even though there is the partial issue about the analytical accuracy under transient condition, the proposed analytical method has been proved to have the advantage of thermal and structural design.
- The quantity of heat transfer, which can be interpreted as thermal energy loss, has been clarified in a VG turbocharger. Regarding the turbine performance based on shaft power, the difference between the adiabatic CFD and the steady state CHT calculation is approximately 1%.
- The reduction of the thermal energy loss at the scroll region is effectively working for the improvement of turbine performance. It can also contribute to making the temperature field uniform and preventing valuable nozzle vane structures from experiencing unfavourable thermal deformation.

NOMENCLATURE

A_i	Heat transfer area in each element
BC	Boundary Condition
Q	Quantity of heat transfer
q_i	Quantity of heat transfer in each element
T_{mfa}	Mass flow-average temperature

ACKNOWLEDGEMENT

The authors wish to express their gratitude to Mitsubishi Heavy Industries, Ltd. and Mitsubishi Heavy Industries Engine & Turbocharger, Ltd. for permission to publish this paper.

REFERENCES

[1] Bohn, D., Heuer, T., and Kusterer, T, 2003,"Conjugate Flow and Heat Transfer Investigation of a Turbo Charger: Part I Numerical Results", ASME Turbo Expo, GT2003-38445, Atlanta-Georgia, USA.
[2] Bohn, D., Heuer, T., and Kusterer, T, 2005,"Conjugate Flow and Heat Transfer Investigation of a Turbo Charger", ASME J. Eng. Gas Turbine Power, 127(3), pp. 663–669
[3] Heuer, T., Engels, B., and Wollscheid, P, 2005, "Thermomechnical Analysis of a Turbocharger Based on Conjugate Heat Transfer", ASME Turbo Expo, GT2005-68059, Reno-Tahoe, USA.

[4] Heuer, T., Engels, B., Klein, A., and Heger, H., 2006, "Numerical and Experimental Analysis of the Thermo-mechanical Load on Turbine Wheels of Turbochargers," ASME Turbo Expo, GT2006-90526, Barcelona, Spain.

[5] Heuer, T., and Engels, B., 2007, "Numerical Analysis of the Heat Transfer in Radial Turbine Wheels of Turbochargers", ASME Turbo Expo, GT2007-27835, Montreal, Canada.

[6] Diefenthal, M., Tadesse, H., Rakut, C., Wirsum, M., and Heuer, T., 2014, "Experimental and Numerical Invetigation of Temperature Fields in a Radial Turbine Wheel", ASME Turbo Expo, GT2014-26323, Düsseldorf, Germany.

[7] Diefenthal, M., Luczynski, P., Rakut, C., Wirsum, M., and Heuer, T., 2017, "Thermomechanical Analysis of Transient temperatures in a Radial Turbine Wheel", ASME J. turbomach, 139(3), 091001–1–10

[8] Dimelow, A., 2018, "Steady and Transient Conjugate Heat Transfer Analysis of Turbochargers", IMechE The 13th International Conference on Turbochargers and Turbocharging, CON6516/202, London, UK.

[9] Kitamura, T., Ibaraki, S., Kihara, Y., Hoshi, T. and Ebisu, M, 2019, "Study on Analytical Method for Thermal and Flow Field of a Turbocharger with the Catalyst Unit", ASME Turbo Expo, GT2019-90451, Phoenix, Arizona, USA.

[10] Menter, F.R., 1993, "Zonal Two Equation k-ω Turbulence Models for Aerodynamic Flows", AIAA paper 93–2906

14th International Conference on Turbochargers and Turbocharging
Institution of Mechanical Engineers, ISBN: 978-0-367-67645-2

One dimensional modelling on twin-entry turbine: An application of TURBODYNA

B. Yang, R.F. Martinez-Botas, P. Newton

Department of Mechanical Engineering, Imperial College London

T. Hoshi, B. Gupta, S. Ibaraki

Fluid Dynamics Research Department, Research & Innovation Center, Mitsubishi Heavy Industries, Ltd

ABSTRACT

One-dimensional (1D) modelling is important for turbocharger unsteady performance prediction and system response assessment of internal combustion engine. Two limbs of twin-entry turbines deliver pulsating unsteady flows with 180° phase difference leading to significant mixing at the rotor inlet. Such effects cannot be reproduced in classic 1D modelling and hence the predictions by the latter are less satisfactory in twin-entry turbines. To solve this problem, the paper proposes a novel 1D modelling (TURBODYNA) and applies it to a twin-entry turbocharger turbine. Instead of applying constant pressure assumption at the limbs junction, TURBODYNA solves conservation equations during the mixing process. Unsteady source terms described by dynamic equations are added into Euler equation to simulation rotor unsteady performance. By comparing TURBODYNA with validated CFD, TURBODYNA not only provides a great agreement on turbine performances, but also accurately captures unsteady features with increased pulsating frequency.

1   INTRODUCTION

CO_2 emission control is a critical point for the development of internal combustion engines. In 2016, transport accounted for 29% of total CO_2 emission and the majority of transport emission is caused by internal combustion engines (1). An effective way of reducing transport CO_2 emission is engine downsizing. Among many technologies towards this aim, turbocharging is recognized as one of the most effective and applicable options for automobile and marine engines (2, 3).

Twin-entry turbines is superior to single-entry turbine in increasing turbocharger transient response and reducing engine exhaust overlap. Due to reciprocating motion of piston engines, separated two limbs of twin-entry turbine volute deliver high pressure, high frequency pulsating exhaust gas into the rotor with 180° phase difference. At the rotor inlet, two limbs merge into a junction and significant mixing happens. The high pulsating frequency and the mixing process take challenges to traditional 1D modelling, which is widely applied to assess turbine performance within an engine system. A typical traditional 1D modelling schematic diagram is shown in Figure 1. In limbs and exit pipe, Euler equation set is solved to capture pressure wave propagation and turbine steady performance maps (rotor) is used to construct boundary conditions between limbs and exit pipe. Two major assumptions are applied in traditional 1D modelling: The first assumption is that rotor's response to unsteady pulse is assumed to be quasi-steady, which suggests flow field timescale in the rotor is ignorable compared to pulsating frequency (4). The second assumption is that the mixing process is isobaric and incompressible (5, 6). For the first assumption, it has been approved that rotor flow field time scale is around 10% of pulsating frequency and a more accurate 1D prediction needs to take the time scale into consideration (9). For the second assumption, combination of iso-pressure and iso-density over-constrains the mixing problem and lacks of direct evidence that the assumption is valid once mixing happens under huge pressure difference (at partial admission condition, pressure in one limb can be around 4bar while 1bar in another limb). In addition, although mass conservation through mixing can be guaranteed by introducing an additional numerical iteration, it does not guarantee momentum and energy conservation. Due to these difficulties, traditional 1D modelling application on twin-entry turbine is limited. In fact, obvious differences between experimental and 1D results are observed (7, 8).

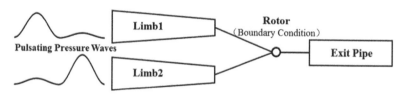

Figure 1. xTraditional 1D modelling schematic diagram.

This paper introduces an application of a noval developed 1D modelling - TURBODYNA (9). In TURBODYNA, rotor is taken into computational domain rather than used to construct boundary conditions among pipes. The benefits include more accurate prediction of pressure wave propagation by avoiding numerical waves due to boundary condition construction. Performance maps are not directly used. Instead, virtual 'wall friction', 'shaft work', 'virtual cascade' and 'mass migration' are introduced into the Euler equation set as 'source terms' to account for the effects of entropy generation, enthalpy variation, swirling flow strength and swallowing capacity difference in the rotor. Lag equation, with rotor flow field times scale as damping coefficient, is applied to calculate unsteady 'source terms'. The mixing process at the junction is assumed to happen within an infinite small length. Rather than adopt constant pressure assumption, Finite Volute Method (FVM) with Riemann-Solver is used to solver this spatial discontinuity problem (10). Mass, momentum and energy conservation are all guaranteed.

2   MODELLING

The schematic diagram of TURBODYNA is shown in Figure 2. TURBODYNA uses Finite Volume Method (FVM). Thus, geometry information is in the mesh rather than solved equations. The domain consists of three parts: volute, the rotor and exit pipe. Volute is symmetrically divided into two limbs. The area and length of the 1D domain are calculated from 3D geometry (Figure 3). Comparing 1D to 3D, limbs inlet areas and exit pipe outlet area are the same. As streamlines flow through the rotor inlet with a constant flow angle a, the effective rotor inlet area for 1D is $cos(a)$ times of 3D rotor inlet area. Volume of each part determines the 'filling and emptying' effect, which attributes to the main unsteadiness of a system. Thus, the volume of each part in 1D keeps the same as 3D and the length in 1D is given to satisfy this requirement.

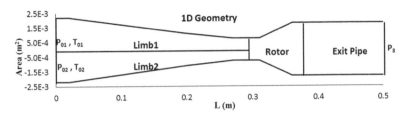

Figure 2. TURBODYNA schematic diagram.

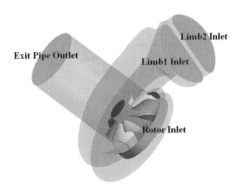

Figure 3. 3D geometry of a twin-entry turbine.

Once 1D geometry has been determined, mesh is generated uniformly along the length and the Euler equation set is numerically solved in the whole domain:

$$\frac{\partial E}{\partial t} + \frac{\partial F}{\partial x} = J \tag{1}$$

$$E = \begin{pmatrix} \rho \\ \rho u \\ \rho v \\ \rho \varepsilon \end{pmatrix}, \quad F = \begin{pmatrix} \rho u \\ \rho u \left(u + \frac{p}{\rho u} \right) \\ \rho u v \\ \rho u \left(\varepsilon + \frac{p}{\rho} \right) \end{pmatrix}, \quad J = \begin{pmatrix} \dot{S}_m \\ \dot{S}_{fx} \\ \dot{S}_{fy} \\ \dot{S}_E \end{pmatrix} \tag{2}$$

For radial turbine, swirling flows exists in exit pipe, to take swirling flow kinetic energy into consideration, velocity 'v' is also introduced in Eq. (2). 'J' in Eq. (2) is the turbo-machinery source terms, which returns to performance maps in steady cases and considers rotor unsteadiness (11) in unsteady cases. More details about source terms are given in the section 2.1. To accurately capture pressure wave propagation, Roe algorithm (Riemann solver) is applied to discretise and solve the Euler equation set (10). Two step Runge-Kutta is used to deal with time step. Characteristic boundary condition method (CBC) has been adopted to provide accurate boundary conditions (12, 13).

2.1 Turbomachinery Quasi-steady Source Terms

In TURBODYNA, performance maps are used to generate quasi-steady source terms in the 'rotor' in Figure 4. A detailed explanation on how to calculate source terms from performance maps can be found in reference (9). For single entry turbine, these sources terms represent for 'shaft work', 'entropy generation' and 'virtual cascade'. 'Shaft work' and 'entropy generation' determines turbine torque and efficiency, 'virtual cascade' controls total mass flow rate. For twin-entry turbine, as there is a swallowing capacity difference between two limbs, an addition mechanism refers to 'mass migration' is also needed. The positions where these source terms are placed is given in Figure 4. 'Mass migration' is placed at rotor inlet, entropy and enthalpy variations are placed within the rotor, 'virtual cascade' is placed at rotor outlet.

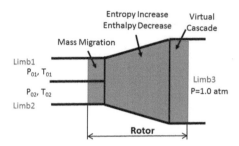

Figure 4. Positions of source terms.

Specifying the operation condition (giving P_{01}, P_{02}, T_{01}, T_{02}, p_3 and Ω), performance maps provide four equations: two MFPs, total-total efficiency and total-static efficiency. From these four equations, four variables $_1$, $_2$, T_{03}, and P_{03} can be calculated. Particularly 'shaft work' is:

$$W_{shaft} = C_p(\dot{m}_1 + \dot{m}_2)(T_{01}\frac{1}{1+2} + T_{02}\frac{2}{1+2} - T_{03}) \qquad (3)$$

or total temperature decrease:

$$\Delta T_0 = T_{01}\frac{1}{1+2} + T_{02}\frac{2}{1+2} - T_{03} \qquad (4)$$

total 'entropy increase' is:

$$\Delta S_{total} = S_3\frac{1}{-1+2} + S_1\frac{2}{1+2} - S_2 \qquad (5)$$

total entropy increase includes mixing entropy increase (ΔS_{mix}) at the junction and losses (ΔS_{rotor}) in blade passage:

$$\Delta S_{total} = \Delta S_{mix} + \Delta S_{rotor} \qquad (6)$$

ΔS_{mix} is calculated by applying Riemann Solver at the junction which is introduced in the section2.4. ΔS_{rotor} is added into Euler equation as source terms. 'Virtual cascade' is applied to control total mass flow rate, the cascade angle θ can be calculated from performance maps (9). 'Mass migration' (μ) is an unique feature for TURBODYNA. As one limb connects to the rotor hub while the other limb connects to the rotor shroud, there is a swallowing capacity difference. 'Mass migration' is applied to simulate this difference and reallocate mass flow rate in each limb. A diagram on 'mass migration' is illustrated in Figure 5a. Compared to Figure 5b which does not have 'mass migration', part of mass flow in Figure 5a is extracted from limb2 and injected into limb1. The migration process is assumed to be adiabatic and isentropic. Adiabatic assumption implies that the total temperature of injected mass flow into limb1 is T_{02}. Ignoring difference between T_{01} and T_{02},

100

isentropic assumption suggests the extraction and injection velocities are equals to local velocities in limb2 and limb1 respectively. 'Mass migration' is applied to correct mass flow rates in limbs. To calculate migrated mass flow rate $\Delta\dot{m}$, 1D steady calculations based on Figure 5b are first applied to calculate uncorrected mass flow rates $\dot{m}_{1-uncorrected}$, then the value of $\Delta\dot{m}$ is:

$$\Delta = \dot{m}_1 - \dot{m}_{1-uncorrected} \qquad (7)$$

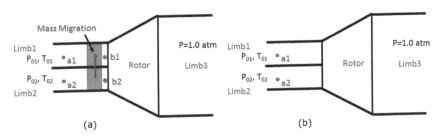

(a)   (b)

Figure 5. Profiles of (a) Twin-entry turbine with 'mass migration' and (b) Twin-entry turbine without 'mass migration'.

Figure 6 gives an example of ΔT_0, ΔS_{rotor}, θ and $\Delta\dot{m}$ for a twin-entry turbine at constant rotating speed (Ω=42000 rpm). Abscissa is mass flow rate at limb1 inlet and ordinate is mass flow rate at limb2 inlet. Mass flow rates are normalized by reference mass flow rate. In the paper, mass flow rate and efficiency are normalized for confidential issue. Seen from Figure 6, ΔT_0, ΔS_{rotor} and θ are almost symmetric to the line $y=x$. This implies that turbine global performance is almost symmetric to the limbs. However, the symmetry breaks down for $\Delta\dot{m}$, which reflects different swallowing capacities. Compared to $\dot{m}_1$ or $\dot{m}_2$, Δs a small value. Maximum ratio of $\Delta\dot{m}/(\dot{m}_1+\dot{m}_2)$ in this case is 4.6%.

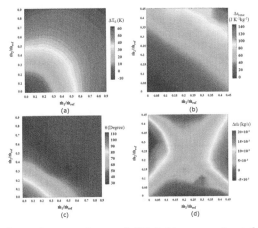

Figure 6. Profiles of source terms (a)Total temperature decrease ΔT_0, (b) specific entropy increase in the rotor Δs_{rotor}, (c)'virtual cascade' angle θ and (d) 'mass migration' value $\Delta\dot{m}$.

2.2   Rotor Unsteadiness and Unsteady Source Terms

Equations (4)-(7) show how to calculate quasi-steady source terms (J_{qs} in Eq. (8)) from performance maps. Another main feature of TURBODYNA is that rotor unsteadiness is considered. To capture rotor unsteadiness and describe the effect of flow field inertia, the lag equation can be used to calculate unsteady source terms:

$$\frac{\partial J}{\partial t} = \frac{1}{\zeta}(J_{qs} - J) \tag{8}$$

where ζ is the characteristic time of flow field in rotors. With the increase of pulsation frequency (increase of $\partial J/\partial t$), the difference between J_{qs} and J also increases and the lag effect becomes more significant. Oppositely, for a quasi-steady case ($\partial J/\partial t = 0$), J degenerates to J_{qs} and Euler equation degenerates to steady performances maps. In Eq. (8), lag effect is determined by ζ. In mean-line model, as the length of blade chamber line (l) as well as the rotor inlet, outlet velocities are known, ζ can be estimated by:

$$\zeta = 2\frac{l}{u_{InChamber} + u_{OutChamber}} \tag{9}$$

where $u_{InChamber}$ and $u_{OutChamber}$ are inlet and outlet velocity components which are tangential to chamber line. If steady CFD results are available, ζ can be calculated from CFD post processing. Streamlines in the rotor can be easily get and ζ can be calculated from:

$$\zeta = \int_{streamline} \frac{1}{u} dx \tag{10}$$

where u is the velocity magnitude.

2.3   Junction

Junction treatment is another unique feature in TURBODYNA. The junction appears when two limbs merge together at the rotor inlet. Figure7 gives an example of a simplified junction geometry: flow direction is from left to right, two symmetric limbs merge into a single one. As twin-entry turbine mainly works at unequal conditions, total pressure in separated limbs are normally different. Hence, a significant mixing process happens at the junction with mass, momentum and energy are conserved during the process. Traditional 1D modelling applies Benson constant pressure junction model. The model applies mass conservation accompanied with following assumptions (5, 6):

$$p_1 = p_2 = p_3, \quad \rho_1 = \rho_2 = \rho_3 \tag{11}$$

P_{01}=1.6 atm, T_{01}=340k Limb1	Limb3 P_3=1.0 atm
P_{02}=1.2 atm, T_{02}=340k Limb2	

Figure 7. Junction example.

To illustrate the problem of the model, artificially giving operation conditions: P_{o1} $=1.6\times10^5$pa, $P_{o2}=1.2\times10^5$pa, $T_{o1}=T_{o2}=340$K and $P_3=1.0\times10^5$pa, applying mass conservation equation and constant pressure condition (Eq. (11)), weakening constant density to $\rho_3=0.5\times(\rho_1+\rho_2)$, all variables can be solved and results are shown in Table1. It is clear $\rho_3=\rho_1=\rho_2$ is not satisfied even as a part of model assumption. In addition, although mass conservation is guaranteed, both momentum and energy are not conserved.

Table 1. Steady junction results based on Benson constant pressure model.

	P_0 (pa)	T_0 (K)	p (pa)	ρ (kg/s)	u (m/s)
Limb1	1.60×10^5	340.0	1.00×10^5	1.17	293.0
Limb2	1.20×10^5	340.0	1.00×10^5	1.08	186.2
Limb3	1.37×10^5	338.5	1.00×10^5	1.13	241.8

Different from traditional 1D modelling, TURBODYNA uses Riemann solver (Roe Algorithm) to calculate aerodynamic variables beside the junction. Riemann solver was developed to solve spatial discontinuity problems (like shock wave) and all conservation relationships are guaranteed. Table2 shows results of Riemann solver. Pressure and density do not keep constant through the junction. Mass, momentum and energy conservation are all satisfied. Entropy increase of the mixing process, which has been left in section2.1 (Eq. (6)), is:

$$dS_{mix} = S_{\lim b3} - \frac{1}{1+2} S_{\lim b1} + \frac{2}{1+2} S_{\lim b2} \quad ds_{mix} = S_{\lim b3} - \frac{1}{1+2} \tag{12}$$

Table 2. Steady junction results based on Riemann solver.

	P_0 (pa)	T_0 (K)	p (pa)	ρ (kg/s)	u (m/s)
Limb1	1.60×10^5	340.0	1.07×10^5	1.23	271.9
Limb2	1.20×10^5	340.0	0.91×10^5	1.00	229.6
Limb3	1.37×10^5	338.5	1.01×10^5	1.13	250.7

3   RESULTS

This section compares TURBODYNA with validated 3D CFD to illustrate the accuracy of TURBODYNA. CFD is chosen as the baseline rather than experiment because current testing does not measure total to total efficiency, which is crucial to generate source terms in TURBODYNA. The quasi-steady source terms are shown in Figure 6a- Figure 6d. Pressure ratios of two limbs vary from 1.5 to 3.5. 3D CFD setup and validation are first introduced followed by a steady case. Then, unsteady results of TURBODYNA and 3D CFD are discussed.

3.1   CFD setup and validation

The CFD domain is shown in Figure 3. Mesh generation and numerical calculations are carried out by using Ansys CFX 16.1. For turbulence model, URANS with SST is used. The details of the mesh are shown in Figure 7a and Figure 7b. Mesh information is given in Table3. y^+ for rotor is around 15 while for volute is around 50. Mesh near the volute outlet and rotor inlet are refined to provide a better calculation on mixing. For boundary conditions, total pressure and total temperature are specified at turbine inlet and static pressure is fixed as atmospheric pressure at turbine outlet. The rotation speed of the rotor is set to 42000 rpm and the time step is 5×10^6 s; meaning that for each time step, the rotor rotates 1.2°. The time step independence has been verified in previous studies (14) where a double entry turbine is investigated.

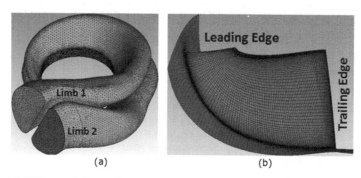

(a)   (b)

Figure 7. CFD mesh for twin-entry turbine (a) volute mesh, (b) rotor mesh.

Table 3. – CFD Mesh.

	Elements Number	y+
Rotor	5,000,000	15
Volute	500,000	47

CFD validation includes both steady and unsteady parts. Figure 8a and Figure8b show steady part validation. Normalized efficiency and MFP at full admission conditions are compared at three speed lines. CFD matches with testing well. At high velocity ratio (or low pressure ratio), CFD overestimates efficiency. The torque sensor is calibrated at U/C=0.65 and the torque at U/C=0.9 is only 15% of the value at U/C=0.65, it is difficult to measure the torque correctly at high velocity ratio. In addition, at high velocity ratio, secondary flows are very strong, this affects CFD accuracy as well. For unsteady validation, turbine inlet pulsating boundary conditions are shown in Figure 9 (a). Abscissa is time normalized by pulsating frequency (33Hz in this case). Ordinate is pressure ratio that varies from 1.3 to 3.8. Pulsating waves in two limbs are out of phase and the phase difference is 180°. Normalized mass flow rate comparison between testing and CFD is shown in Figure 9 (b). CFD generally has a good agreement with testing on mass flow rate. At the maxmia points (t_{nor}=0.3 and t_{nor}=0.7), CFD underestimate the value as testing is calibrated at normalized mass flow rate around 0.6. Minimum mass flow rate is close to zero. As hot-wire measures mass flow rate magnitude but not direction, from t_{nor}=0.2 to 0.4, reverse flow happens at limb 2 inlet. Correspondingly, CFD forces to change 'inlet' boundary to 'wall' during this period and numerical oscillation is obvious in this region.

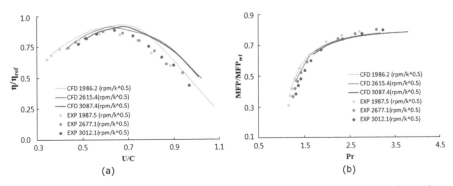

Figure 8. CFD steady validation at full admission conditions (a)total-static efficiency, (b) MFP.

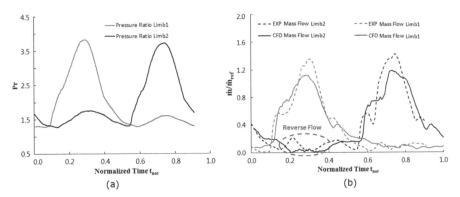

Figure 9. CFD unsteady validation (a)inlet condition (out of phase pulsating wave, 33Hz), (b) unsteady mass flow rate.

3.2   Steady case

The steady case illustrates effects of source terms in TURBODYNA and how does TUR-BODYNA simulate the twin-entry turbine. 1D computational domain is illustrated Figure 2. The length of volute and exit pipe are 0.28 m and 0.12 m respectively. The volume of the rotor follows the 3D geometry, which is about 1.5×10^4 m^3). The rotor length is around 0.06 m. The mesh is uniformly distributed and every 0.01m consists of 10 elements. Boundary conditions for the steady case are given in Table 4. From 3D steady CFD simulation, $\dot{m}_1$, $\dot{m}_2$, T_{O3} and P_{O3} are 0.424 kg/s, 0.109 kg/s, 285 k and 1.15×10^5 pa respectively.

Table 4. Boundary conditions for the steady case.

	P_0 (pa)	T_0 (k)	p (pa)
Limb1	3.5×10^5	340	
Limb2	2.0×10^5	340	
Exit Pipe			1.0×10^5

TURBODYNA steady results are shown in Figure10a - Figure10c. $\dot{m}_1$, $\dot{m}_2$, T_{03} and P_{03} match with CFD results correctly. As source terms distribution in the rotor is assumed to be uniform, total temperature linearly decreases from 340K to 285K. Flow in Limbs are assumed to be isentropic, hence there is no total pressure increase in limbs. Through the rotor, total pressure decreases to 1.15×10^5pa with entropy linearly increases in the rotor. Mass flow rates are also correctly calculated. As Δn Eq. (7) is a small value, Figure11 enlarges mass flow rates variation from L=0.278m to L=0.286m where the 'mass migration' is applied. It clearly shows 0.004kg/s mass flow rate has been linearly extracted from limb2 and injected into limb1. As this process is assumed to be isentropic and adiabatic, the 'mass migration' does not affect total temperature and total pressure in each limb.

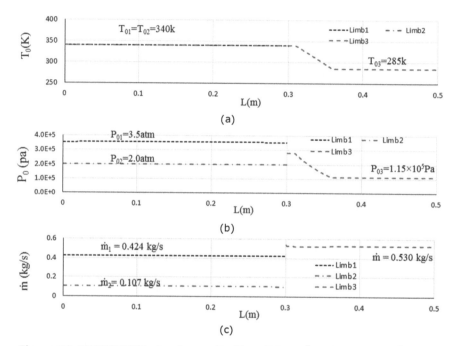

Figure 10. TURBODYNA steady results (P_{01}=3.5×10⁵pa, P_{02}=2.0×10⁵pa, p₃ =1.0×10⁵pa) (a)total temperature, (b) total pressure and (c)mass flow rate.

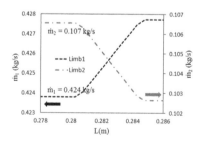

Figure 11. TURBODYNA steady results on mass migration.

106

3.3 Unsteady cases

Two numerical unsteady cases have been carefully designed to compare 1D results against 3D CFD simulations. Table5 shows limbs inlet boundary conditions for two cases and Figure 12 shows the shape of pulsating. The shape is similar to testing results in Figure 9: pulsating wave in each limb is identical but with 180° phase difference, two sine functions consists of one pulsating wave period with one magnitude five times to the other. Total temperature at inlet are calculated from isentropic assumption: where inlet entropy is fixed to 2220 J/(kg·K). At outlet, static pressure is specified and reflection boundary condition is adopted.

Table 5. Numerical cases for 1D modelling accuracy investigation.

	Pr	Frequency
Case 1	1.75 - 3.25	30 Hz
Case 2	1.75 - 3.25	60 Hz

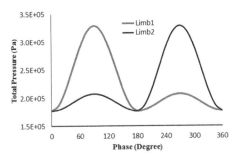

Figure 12. TURBODYNA steady results on mass migration.

Both CFD and TURBODYNA results have been post-processed via a time-shifting based on the time needed for pressure waves propagate through volute. Normalized mass flow rates comparison for Case1 are shown in Figure 13. Abscissa is the time normalized by pulsating period. Ordinate is normalized mass flow rate measured at limbs outlet. Mass flow rate in each limb is affected by pressure ratios in both limbs: increase of local pressure ratio increases mass flow rate in the same limb but decreases mass flow rate in the other limb. From t_{nor}=0.00 to t_{nor}=0.25 and from t_{nor}=0.50 to t_{nor}=0.75, pressure ratios increase in both limbs and mass flow rates in both limbs are determined by competition of their swallowing capacity. As limb1 has higher swallowing capacity, from t_{nor}=0.50 to t_{nor}=0.75, mass flow rate in limb1 increases. However, during same time, mass flow rate in limb2 decreases. Traditional 1D modelling cannot predict this swallowing capacity difference for a symmetric limb geometry. However, TURBODYNA has good agreement with CFD and accurately captures the feature. Figure14a and Figure14b compare total temperature decrease and entropy increase for Case1. The former determines turbine instantaneous power and the later determines turbine instantaneous efficiency. ΔT_0 variation is almost symmetric to t_{nor}=0.5. This implies, at unequal admission conditions, turbine specific power do not significantly distinguish which limb has higher pressure ratio. However, flow structures in the rotor is different as limb1 connects to the rotor shroud while limb2 connects to rotor hub. Thereby there is an obvious entropy increase difference in Figure 14b. Again, TURBODYNA agrees with CFD well and accurately captures the entropy

107

increase difference. To further explain the advantage of TURBODYNA, numerical results based on a similar twin-entry turbine by applying traditional 1D modelling are illustrated in Figure 15 (15). In reference (15), pulsating shape is similar to Figure 12 but with a much smaller pressure ratio range (P_r varies from 1.3 to 2.5). Pulsating frequency is also around 30Hz. It is clear that classic 1D modelling obviously underestimates both mass flow rate and torque. Particularly, during decrease part of pulsating wave, 1D results are only 50% of testing.

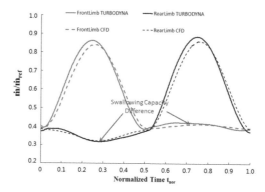

Figure 13. Mass flow rates comparison in Case1 (30Hz),dash line for CFD and solid line for 1D.

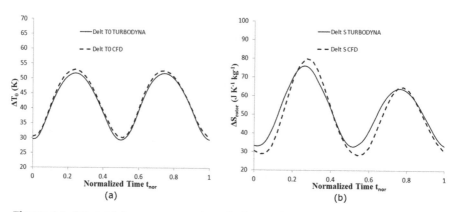

Figure 14. ΔT_0 and Δs_{rotor} comparison in Case1 (30Hz), a) ΔT_0 and b) Δs_{rotor}.

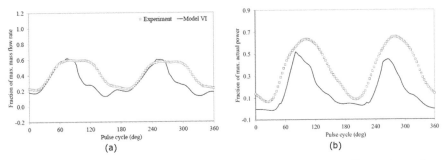

(a)

(b)

Figure 15. Traditional 1D modelling results (15), a) mass flow rate and b) torque.

In TURBODYNA, rotor geometry is taken into computation domain rather than a spatial discontinuity ('interface' as in Figure 1). As a result, the pressure wave transportation and reflection can be reproduced more accurately. In the rotor, there are three physical waves: two pressure waves are transported by sound speed plus/minus flow field velocity while entropy wave is transported only by flow field convection. As spatial discontinuity neglects difference of waves, it introduces numerical waves which affect the accuracy of pressure wave prediction. By contrast, TURBODYNA can capture physical waves accurately. A monitor is placed just at the outlet of rotor with the static pressure probed for case 1. Pressure waves coming from the rotor inlet and reflected from the exit pipe outlet interact at this point and complicate the prediction. The monitored static pressure variation at the point is shown in Figure 16. Even though the pressure variation is very weak, TURBODYNA captures the pressure waves accurately.

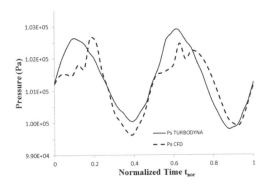

Figure 16. Rotor outlet static pressure comparison in Case1 (30Hz).

Effects of pulsating frequency on twin-entry turbine performance are illustrated by comparing Case1 with Case2. Figure 17 show normalized mass flow rates in both TURBODYNA and CFD. Mass flow rate is mainly determined by pressure ratio across the rotor and its response to pulsating wave is close to quasi-steady. Thereby in both 1D and CFD, there is no significant mass flow rate difference between 30Hz and 60Hz. In Case2, a slightly higher mass flow rates maxmia at $t_{nor} = 0.25$ and $t_{nor} = 0.75$ are observed. This is due to slightly higher unsteady pressure ratio across the rotor at higher pulsating frequency and the phenomena is accurately predicted in Figure 17. Total temperature variation across the rotor is affected by both unsteady flow structures and local pressure wave expansion or compression. Hence, during pulsating wave, ΔT_0 variation is normally more 'unsteady' compared to mass flow rate. In Figure 18, at 30Hz, ΔT_0 varies from 30.5K to 53.0K. At 60Hz, the variation range increases from 29.3K to 54.9K. This unsteady feature is also captured by TURBODYNA, where the range increases from 29.7K- 51.8K to 28.0K-53.0K.

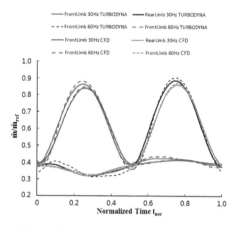

Figure 17. Effect of pulsating frequency on mass flow rate.

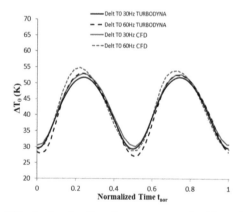

Figure 18. Effect of pulsating frequency on turbine shaft work.

4   CONCLUSIONS

Two limbs of twin-entry turbines deliver pulsating unsteady flows with 180° phase difference leading to significant mixing at the rotor inlet. Such effects cannot be reproduced in classic 1D modelling and hence the predictions by the latter are less satisfactory in twin-entry turbines. In fact, the difference between traditional 1D modelling with experimental results can be 50% (15). To overcome the difficulty, a new 1D modelling strategy (TURBODYNA) has been introduced in the paper. Differing from traditional 1D modelling, the volume of the rotor is considered and meshed. The effects of the steady performance maps are explained as different kinds of 'source terms'. 'Entropy increase' and 'shaft work' are applied to model entropy generation and enthalpy variation. 'Virtual cascade' are applied to generate swirling flows and control total mass flow rate. A new mechanism refers to 'mass migration' is applied to reallocate mass flow rate in each limb. To consider the time scale of flow field in the rotor, the dynamic equation (lag equation) is applied to describe the inertial effect due to flow field in rotor. At the junction, rather than applying constant pressure model, Riemann solver is directly used to solve mass, momentum and energy conservation equations during the mixing process. TURBODYNA has applied to a twin-entry turbine to assess turbine performance at different pulse unsteadiness. Comparisons between TURBODYNA and validated 3D CFD results illustrate that TURBODYNA takes advantage over traditional 1D modelling and TURBODYNA provides great agreement with validated CFD on twin-entry turbine unsteady performances. In addition, TURBODYNA also provides correct prediction on pulsating frequency effects when rotor unsteadiness becomes even more significant.

NOMENCLATURE

C_p	Heat Capacity	[J/(kg·K)]
P_{oi}	Total pressure in limb 'i'	[Pa]
T	Temperature	[K]
T_{oi}	Total temperature in limb 'i'	[K]
ΔT_o	Total temperature decrease	[K]
$\Delta \dot{m}$	Mass flow rate migration	[kg/s]
Δs	Specific entropy increase	[J/(kg·K)]
Ω	Shaft rotating speed	[rpm]
a	Volute outlet flow angle	[Degree]
$\dot{S}$	Source terms	[-]
$\dot{m}_i$	Mass flow rate in limb 'i'	[kg/s]
η_{tt}	Total to total efficiency	[-]
η_{ts}	Total to static efficiency	[-]
ρ	Density	[kg/m^3]
τ	Torque	[N·m]
θ	Cascade angle	[Degree]
ζ	Flow field characteristic time	[s]
p	Static pressure	[Pa]
p_r	Pressure ratio	[-]
s	Specific entropy	[J/(kg·K)]
u,v	Velocity	[m/s

111

REFERENCES

[1] IEA, "Emissions from Fuel Combustion: Highlights", Paris: International Energy Agency.

[2] Leduc, P., Dubar, B., Ranini, A. Monnier, G. "Downsizing of Gasoline Engine: An Efficient Way to Reduce CO_2 Emissions", *Oil & Gas Science and Technology*, 58 (1), 2006, pp. 115–127.

[3] Watson, N., Janota, M., "Turbocharging The Internal Combustion Engine", *Macmillan International Higher Education*, 1982.

[4] Chen, H., Hakeem, I., Martinez-Botas, R.F., "Modelling of A Turbocharger Turbine Under Pulsating Inlet Conditions", *Journal of Power and Energy*, 210(5), 1996, pp. 397–408.

[5] Winterbone, D.E., Pearson, R.J. "Theory of Engine Manifold Design: Wave Methods for IC Engines", Wiley-Blackwell, 2000.

[6] Costall, A.W., "A One-Dimensional Study of Unsteady Wave Propagation in Turbocharger Turbines", *Ph.D. thesis, Imperial College London*, 2008.

[7] Costall, A.W., Mcdavid, R.M., Martinez-Botas, R.F., Baines, N.C., "Pulse Performance Modeling of A Twin Entry Turbocharger Turbine Under Full and Unequal Admission", *Journal of Turbomachinery*, 133, 2011, pp. 021005.

[8] Chiong, M., Rajoo, S., Romagnoli, A. Costall, A., Martinez-Botas, R.F., "One Dimensional Pulse-Flow Modeling of A Twin-Scroll Turbine", *Energy*, 115, 2016, pp. 1291–1304.

[9] Yang, B., Martinez-Botas, R.F. "TURBODYNA: Centrifugal/Centripetal Turbomachinery Dynamic Simulator and Its Application on A Mixed Flow Turbine", *Journal of Engineering for Gas Turbines and Power*, 141(10), 2019, pp. 101012.

[10] Roe, P. "Approximate Riemann Solvers, Parameter and Difference Schemes", *Journal of Computational Physics*, 372, 1981, pp. 357–372.

[11] Yang, B., Newton, P., Martinez-Botas, R.F., "Unsteady Evolution of Secondary Flows in A Mixed Flow Turbine", *Global Propulsion and Power Forum, Zurich, Switzerland, 2017*, pp. GPPF–2017–188.

[12] Poinsot, T.J., Lelef, S.K., "Boundary Conditions for Direct Simulations of Compressible Viscous Flows", *Journal of Computational Physics*, 101(1), 1992.

[13] Thompson, K.W., "Time Dependent Boundary Conditions for Hyperbolic Systems", *Journal of Computational Physics*, 68(1), 1987, pp. 1–24.

[14] Newton, P., Martinez-Botas, R.F., Seiler, M. "A Three-Dimensional Computational Study of Pulsating Flow Inside A Double Entry Turbine", *Journal of Turbomachinery*, 137(3), 2015, pp. 031001.

[15] Chiong, M., "1d Gas Dynamic Modelling of Twin Entry and Variable Geometry Turbine for Automotive Turbochargers", *Ph.D. thesis, Universiti Teknology Malaysia*, 2015.

A system-level study of an organic Rankine cycle applied to waste heat recovery in light-duty hybrid powertrains

A. Pessanha[1], C.D. Copeland[2], Z. Chen[3]

[1,2]University of Bath, Bath, UK
[3]Jaguar Land Rover Powertrain Research, Coventry, UK

ABSTRACT

The global transportation sector produces approximately 20% of human-induced greenhouse gas and pollutant emissions. Road transport accounts for almost 75% of that amount. As a result, the automotive industry has invested heavily in powertrain electrification as a long-term strategy for reducing the environmental impact of urban mobility. However, in-vehicle energy storage and access to charging infrastructure are still large bottlenecks to widespread uptake of battery electric vehicles (BEV). Therefore, hybrid electric vehicle (HEV) architectures are a short-to-medium-term solution that will enable a gradual transition to a cleaner transportation system.

Current HEV platforms continue to rely on the internal combustion engine (ICE) for a significant proportion of tractive effort. Therefore, optimising the efficiency of the thermal power plant remains a key challenge. In fact, modern ICE's only convert up to 40% of the energy released in the combustion process into useful mechanical work. The remainder is lost in the form of high-enthalpy exhaust gases and engine cooling. Waste heat recovery (WHR) aims to reclaim a proportion of this energy to increase the overall efficiency of the vehicle and has been identified as a key enabler of real-world emissions reductions by the UK Advanced Propulsion Centre (APC).

The organic Rankine cycle (ORC) has been identified as a promising technology for WHR in HEV powertrains. Basic thermodynamic principles were used to identify three key parameters that determine ORC power output: condenser pressure, evaporation pressure and fluid superheating. Matlab was used to develop a semi-dynamic model that takes into account the thermal inertia of the entire system to predict ORC performance under transient operating conditions. This was coupled with a Simulink vehicle model capable of capturing the impact of WHR integration, i.e. increased cooling load, on fuel consumption to provide realistic estimations of ORC energy recovery over a drive cycle.

The study revealed a key trade-off between condenser heat rejection and ORC power output that depends on system architecture, operating point and working fluid selection. Hence, the study proposes a revised paradigm for ORC development that focuses on vehicle-level integration from the start of the design process by prioritising fuel economy over ORC power output.

1   INTRODUCTION

The transportation sector produces over 20% of global anthropogenic greenhouse gas emissions (1). Strict regulatory efforts combined with increased consumer pressure have already led to the development of novel powertrain technologies such as direct injection, engine downsizing, forced induction and variable valve timing that allowed automotive manufacturers to achieve significant improvements in light-duty vehicle fuel consumption and emissions performance over the past 15 years. However, the maximum thermal efficiency of modern internal combustion engines continues to plateau at approximately 40% and road transport still accounts for roughly 75% of the industry's carbon footprint (1). The remaining energy released during the combustion process is rejected to the environment in the form of high-enthalpy exhaust gases and coolant waste heat dissipated by the vehicle's cooling system.

Waste heat recovery (WHR) aims to reclaim a proportion of this lost energy to improve vehicle efficiency and reduce fuel consumption. At a time when the industry's move towards widespread adoption of battery electric vehicles (BEV) still faces challenges related to energy storage technology and infrastructural limitations, hybrid electric vehicle (HEV) powertrain architectures that incorporate an element of WHR are identified by the UK's Advanced Propulsion Centre as a key enabler of real-world emissions reductions (2). The WHR technological landscape features a number of technologies suited to the recovery and conversion of exhaust/coolant heat into mechanical or electrical energy. Conventional vehicles (CV) offer limited potential for the latter due to restrictions imposed by the on-board electrical system (3). Mild-hybrid (MHEV) and full-hybrid (FHEV) architectures feature greater energy storage capabilities that offer a wider scope for on-board power generation. The Organic Rankine cycle (ORC) has been identified as a promising strategy for energy recovery from low-to-medium temperature heat sources and is already a well-established WHR technology for industrial, marine and rail applications (4). However, implementing an ORC in a light-duty passenger vehicle poses particular challenges. Power generation systems are typically designed for constant operation at a single design point that provides the heat recovery system with constant inlet boundary conditions. In contrast, an exhaust-based ORC system in a passenger vehicle will operate under highly transient waste heat characteristics, i.e. mass flow and temperature, over a typical driving cycle. Responding to thermal fluctuations is one of the key challenges for heat-to-power systems that often leads to poor transient performance due to operation in off-design conditions (5).

Furthermore, WHR strategies are often designed as add-on systems so interactions between the recovery cycle and the base vehicle can be highly penalising – these include: added weight, increased exhaust gas backpressure and greater cooling demand. Vehicles equipped with active shutter grille (ASG) technology are particularly sensitive to the latter due to the impact of cooling drag on road-load and fuel consumption (3). In some cases, poor system integration can even lead to decreased charge air cooling efficiency (6). Ultimately, the combined effect of poor off-design performance and vehicle integration can be a significant reduction of the real fuel-saving potential of heat-to-power WHR.

Flik et al (6) analysed the WHR potential of a steam Rankine cycle (RC) in a heavy-duty Diesel truck equipped with exhaust gas recirculation (EGR) and water charge air cooling (WCAC). Their simulations computed fuel consumption for a number of different WHR system architectures at a steady-state engine operating point including the impact of increased exhaust backpressure, cooling fan power consumption and charge air temperature. A maximum fuel economy improvement of 6.9% was achieved when placing the RC condenser in the high-temperature cooling loop due to minimal impact on WCAC effectiveness and fan power consumption. Horst et al (3) analysed the energy recovery performance of an exhaust-driven steam RC with a conventional spark-ignition (SI) engine and found that fuel economy improvement over a dynamic motorway driving scenario was reduced from 3.4% to 1.3% when interactions between the WHR system and the base vehicle were taken into account. Their study identifies increased on-board power demand in future HEV powertrain architectures as

114

a key enabler for improved WHR performance due to greater electrical energy deployment flexibility and storage capacity.

Both publications demonstrate the need to perform vehicle-level analysis when evaluating the fuel-saving potential of a heat-to-power WHR system. However, to the author's best knowledge, the influence of these interactions is often underestimated or ignored in available literature. This paper aims to demonstrate the benefits of performing coupled vehicle/WHR analysis from the early stages of the ORC design process in order to maximise fuel-saving potential. Section 2 outlines the main working principle of the heat recovery system alongside the theoretical framework used to model its dynamic response to fluctuating heat source characteristics. Section 3 analyses four key design trade-offs that affect heat recovery performance, highlighting the need for system-level optimisation. Section 4 presents an optimisation procedure used to identify the optimum ORC design that minimises fuel consumption over a drive cycle. The results presented highlight the main shortcomings of an ORC passenger vehicle application and identify key areas of future improvement.

2 ORC WASTE HEAT RECOVERY

This section outlines the working principles of the ORC and the modelling approach employed to predict waste heat recovery performance over a drive cycle.

2.1 Working principle
The Organic Rankine is a closed-loop power generation cycle that converts heat into useful work by compression, evaporation and expansion of an organic fluid. Figure 1 shows the four stages of a basic ORC layout with a single working fluid, heat source and expander. More complex designs can employ internal recuperation or multi-stage heat addition to increase energy recovery efficiency. However, this comes at the expense of added system complexity, mass and cost – hence, the basic layout was selected for this study to minimise the impact of vehicle integration.

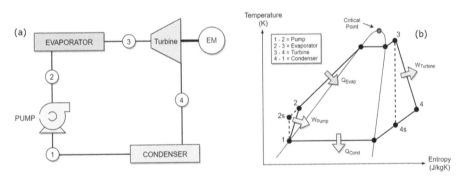

Figure 1. ORC system schematic and temperature-entropy diagram.

Working fluid and expander selection are key aspects of ORC design. Organic fluids are divided into three categories according to the gradient of their saturated liquid/vapour line: wet, dry or isentropic. These categories play a critical role in expander selection. Wet fluids expand into the two-phase region enclosed by the saturation curve so radial turbines become susceptible to blade damage due to liquid formation. On the other hand, dry/isentropic fluids always expand into the superheated vapour region but require greater heat rejection in the condenser. Radial turbines have been identified as the preferred expander solution for power generation (7) coupled with an electric machine. Therefore, this study disregarded wet working fluids and focused on a basic ORC layout with a radial turbine expander and dry/isentropic working fluid.

2.2 Modelling approach

An enthalpy-based approach was used to model the behaviour of the working fluid at each stage of the cycle. Fluid properties play a key role in ORC performance so the open-source CoolProp (10) fluid database was coupled with the simulation routine to obtain the specific enthalpy values outlined in equations 1-4 below. To begin, a pump raises the pressure of the working fluid. The work input required for process 1 → 2 is given by equation 1:

$$W_{\text{Pump}} = \dot{m}(h_2 - h_1) = \frac{\dot{m}(h_{2s} - h_1)}{\eta_{\text{Pump}}} \tag{1}$$

The liquid working fluid is evaporated in a heat exchanger to become a saturated or superheated vapour. The rate of heat input into the system in process 2 → 3 is given by equation 2:

$$Q_{\text{Evaporator}} = \dot{m}(h_3 - h_2) \tag{2}$$

The expansion stage returns the working fluid to condenser pressure – the work output extracted in process 3 → 4 is given by equation 3. The turbine is mechanically coupled to a generator that converts mechanical work into electric power, as per equation 4.

$$W_{\text{Turbine}} = \dot{m}(h_3 - h_4) = \dot{m}\eta_{\text{Turbine}}(h_3 - h_{4s}) \tag{3}$$

$$W_{\text{Generator}} = \eta_{\text{Generator}} W_{\text{Turbine}} \tag{4}$$

Lastly, the superheated vapour is cooled in a condenser that closes the cycle and returns the fluid to the original saturated liquid state. Heat rejection in process 4 → 1 is given by equation 5:

$$Q_{\text{Condenser}} = \dot{m}(h_4 - h_1) \tag{5}$$

The net work output of the system is given by the difference between pump work and expansion power, as per equation 6. It is also important to define two key metrics that characterise the energy recovery performance of the system: cycle thermal efficiency (equation 7) and benefit-to-cost ratio (BCR, equation 8).

$$W_{\text{Net}} = W_{\text{Turbine}} - W_{\text{Pump}} \tag{6}$$

$$\eta_{\text{Thermal}} = \frac{W_{\text{Net}}}{Q_{\text{Evaporator}}} \qquad BCR = \frac{W_{\text{Net}}}{Q_{\text{Condenser}}} \tag{7}(8)$$

Thermal efficiency is conventionally used to assess the thermodynamic performance of any heat-to-power system. BCR aims to characterise the recovery cycle from a vehicle point of view by comparing useful power output to the additional cooling demand imposed on the vehicle and its impact on drag – section 2.3.3 will provide further detail on the impact of ORC heat rejection on vehicle road load.

Equations 1 to 5 provide a quasi-static framework for evaluating the performance of the pump, evaporator, turbine and condenser. However, this paper aims to evaluate

ORC performance over a drive cycle so modelling thermal lag dynamics in the heat addition stage of the cycle is critical to predicting energy recovery performance with respect to transient waste heat characteristics (5). This study employs a hybrid simulation methodology proposed by Horst et al (3) that combines a dynamic calculation of evaporator bulk temperature with quasi-static models of the pump, turbine and condenser.

$$Q_{\text{Exhaust} \to \text{Evaporator}} - Q_{\text{Evaporator} \to \text{Fluid}} = \frac{dT_{\text{Evaporator}}}{dt} \times (mc)_{\text{ORC}} \tag{9}$$

$$(mc_p)_{\text{ORC}} = (mc_p)_{\text{Evaporator}} + (mc_p)_{\text{Condenser}} + (mc_p)_{\text{Pump}} + (mc_p)_{\text{Turbine}} + \sum (mc_p)_{\text{Fluid,avg}} \tag{10}$$

As per equation 9, the evaporator was represented as an aggregate heat capacity that incorporates the thermal inertia of the entire system in order to predict component warm-up over a drive cycle. The thermal inertia of the working fluid varies greatly with respect to temperature and state so the model considers average fluid state at each stage of the cycle. The rate of change of evaporator bulk temperature is proportional to the net power transfer across the evaporator, i.e. heat transfer from the heat source to the metal compared to that extracted by the fluid. For a given evaporator heat transfer coefficient and area, the energy transferred to the working fluid is proportional to the difference between evaporator surface temperature and the average temperature of the working fluid over the evaporation process:

$$Q_{\text{Evaporator} \to \text{Fluid}} = UA_{\text{Evaporator}} \left(T_{\text{Evaporator}} - T_{\text{Fluid,Avg}} \right) \tag{11}$$

Model control strategy is set such that the ORC is only engaged once evaporator temperature is above the average fluid temperature for the selected vapour outlet conditions. This relationship can be combined with equation 2 to give a dynamic prediction of the working fluid mass flow over time:

$$Q_{\text{Evaporator} \to \text{Fluid}} = Q_{\text{Evaporator}} \to \dot{m} = \frac{Q_{\text{Evaporator} \to \text{Fluid}}}{h_3 - h_2} \tag{12}$$

The mass flow obtained using equation 12 is combined with equations 1, 2, 3 and 5 to give dynamic predictions of pump work, turbine power and condenser load at each simulation time step.

2.3   Vehicle integration
The ORC simulation tool developed in Matlab was coupled with a Simulink full-vehicle model that computes fuel consumption over a dynamic drive cycle. Importantly, the coupled models capture the negative interactions between the heat recovery system and the base vehicle to give a realistic estimation of energy-saving potential.

There are three integration effects that need to be taken into account: increased vehicle mass, additional exhaust system backpressure and extra cooling demand. All of these factors cause an increase in the baseline energy demand of the vehicle for a given driving profile and increase the total ORC energy output required to generate net fuel savings.

2.3.1 *Additional vehicle mass and thermal inertia*
Vehicle mass is directly linked to road load and the torque output required from the powertrain. In turn, increased torque demand leads to greater energy consumption so extra vehicle mass leads to an inevitable increase in fuelling. Table 1 shows the component mass values used in the simulation. The electric machine and additional piping/coolant charge were not taken into account so a total of 30kg was used in the simulations presented in later sections of this paper.

Table 1. ORC component thermal inertia.

Component	Mass (kg)	Heat Capacity (J/kgK)
Evaporator	10	510
Condenser	3.8	900
Pump	1	900
Turbine	1	900
Working Fluid	7	As per equation 9

The mass and specific heat capacity values shown above were also used to calculate the overall thermal inertia of the system in the dynamic model of evaporator bulk temperature (equation 8).

2.3.2 *Exhaust system backpressure*
The integration of an evaporator in the exhaust system generates an additional pressure drop that increases engine backpressure and fuel consumption.

This phenomenon was captured in the vehicle model using a correlation proposed by Risse (8) and Mazar (9) that estimates a 2% increase in fuel consumption per 100mbar of additional backpressure for a turbo-charged four-cylinder spark ignition engine – this value was used as a constant for this paper.

2.3.3 *Condenser integration and cooling demand*
Figure 2 shows three condenser integration options that were considered for this study. Each scenario supplies the condenser with coolant at different temperatures. Scenarios (a) and (b) place the condenser at different points of the vehicle's high-temperature cooling loop while scenario (c) exploits the low-temperature cooling circuit used for the charge air cooler.

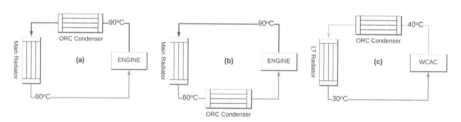

Figure 2. Condenser integration scenarios.

There are complex interactions between ORC cooling demand and road load. As shown in Figure 2, this study focuses on system architectures that add the condenser to the vehicle's existing cooling circuits. As a result, this arrangement increases the heat rejection performance required from the vehicle frontal cooling package when the ORC is engaged.

For vehicles equipped with ASG technology, additional cooling demand leads to an increase in aerodynamic drag because shutter opening is increased to allow greater airflow through the radiators. The additional airflow through the engine compartment increases the vehicle's drag coefficient (11) and aerodynamic drag force, as per equation 13. ORC heat rejection can also lead to a requirement for faster fan speed and/or coolant flow that increase parasitic loads on the powertrain – however, these effects were not taken into account for the results presented in this study. Since drag coefficient is directly linked to ASG opening position, ORC heat rejection has a direct impact on aerodynamic drag and road load.

$$F_{\text{Aerodynamic}} = \frac{1}{2} \rho_{Air} A_{Frontal} C_D v^2 \tag{13}$$

$$A_{\text{Frontal}} = \frac{Q_{Condenser}}{SD_{Radiator}} \quad \text{Where} \quad SD_{Radiator} = \text{Specific Dissipation (W/m}^2) \tag{14}$$

Modelling the dynamics described requires detailed information about the vehicle's cooling system and active shutter grille control strategy. Therefore, a simpler approach was developed to capture the impact of cooling drag on road load that modifies the frontal area of the vehicle with as a function of heat rejection in the condenser. The increase in frontal area is calculated based on the additional radiator area that would be required to dissipate the condenser load of the ORC system at each time step of the simulation – the concept is illustrated in Figure 3a.

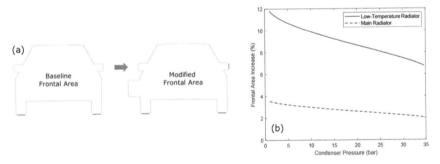

Figure 3. Cooling demand modelling approach *(fluid = R245ca).*

To ensure realistic simulation results, the vehicle model calculates the reserve heat dissipation capacity of the relevant cooling circuit throughout the simulation and disengages the ORC when cooling demand is too large.

Cooling circuit choice becomes an important consideration because equation 12 uses radiator specific dissipation to estimate the impact on frontal area. Figure 3b shows that rejecting a fixed condenser load via the main engine cooling loop leads to a lower increase in frontal area compared to rejecting the same load via the low-temperature circuit due to higher specific dissipation values.

119

3   KEY DESIGN PARAMETERS

This section outlines the impact of three ORC design parameters on energy recovery performance: working fluid selection (i.e. fluid properties), evaporator pressure and condenser pressure. The sub-sections that follow will demonstrate the key trade-offs associated to each variable and their subsequent impact on ORC energy recovery performance over an entire WLTP drive cycle.

3.1   Fluid properties

Table 2 identifies eleven dry/isentropic working fluid candidates considered for this study – evaporation temperature (T_{Evap}) and latent heat capacity (L) were evaluated at atmospheric pressure.

Table 2. Working fluid thermodynamic properties.

Fluid	T_{Evap} (K)	T_{Max} (K)	L (kJ/kg)	P_{crit} (bar)	Type
R245ca	298.4	450	204.2	39.41	Dry
R245fa	288.2	440	196.8	36.51	Dry
R236ea	279.3	412	165.3	34.20	Dry
R113	320.7	525	144.3	33.92	Dry
R141b	305.2	500	222.7	42.12	Dry
R114	276.7	507	135.9	32.57	Dry
R1233ZD	291.4	550	194.6	36.23	Dry
R11	296.9	625	181.4	43.94	Isentropic
R142b	264.0	470	223.3	40.55	Isentropic
R124	261.2	470	165.8	36.24	Isentropic
R1234ZE	254.2	420	195.6	36.37	Isentropic

For a sub-critical cycle, critical pressure is a determining factor for energy recovery performance. To avoid operation in the supercritical region, fluid pressure cannot exceed the critical value – this imposes a limit on evaporator pressure. Therefore, for a fixed condenser pressure, maximum pressure ratio across the cycle is directly proportional to critical pressure.

However, Table 2 also shows that evaporation temperature changes for different fluids so condenser pressure for a fixed coolant temperature is dependent on fluid choice. Figure 4 compares ORC performance over a WLTP drive cycle for two fluids to illustrate the effect of fluid properties on energy recovery:

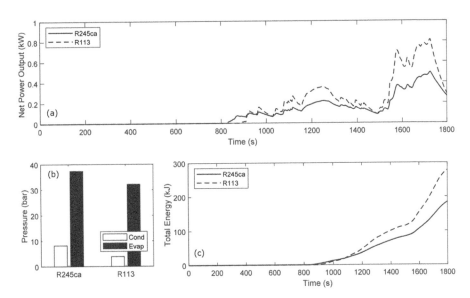

Figure 4. Effect of fluid choice on drive cycle performance *(superheat = 0K;* $P_{Evap} = 0.95P_{Crit}$; $\eta_{Pump} = 0.55$; $\eta_{Turbine} = 0.5$; $UA_{Evap} = 370W/K$; $T_{Coolant} = 60°C$).

Figure 4a shows the average net work output using R113 as the working fluid is significantly higher than R245ca – as a result, the total energy recovered shown in Figure 4(c) is almost 50% greater. Even though R245ca has a higher critical pressure that allows a higher maximum evaporator pressure, R113 requires a significantly lower condenser pressure for the same coolant temperature. Hence, the overall expansion ratio across the turbine is greater for R113 so work output is higher.

Fluid latent heat capacity plays a key role in the heat addition and heat rejection stages of the ORC. A relatively high latent heat capacity increases the specific enthalpy gain in the evaporator, as per equation 2. This leads to a reduction in the mass flow rate required to achieve a given vapour superheating that causes a reduction in mass flow through the system and turbine work, as per equations 3 and 12. Likewise, high latent heat also causes an increase in condenser load that penalises the vehicle's cooling system and aerodynamic drag – this concept is explored in further detail in sections 3.2 and 3.3.

3.2   Evaporator pressure
The pressure set in the evaporator has a significant impact on the heat addition and expansion stages of the ORC. The model was run at constant evaporator temperature for a sweep of pressure ratio values to isolate the influence of evaporator pressure on cycle performance:

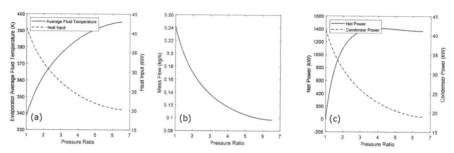

Figure 5. Effect of evaporator pressure on cycle performance (*fluid = R245ca; superheat = 0K; η_{Pump} = 0.55; $\eta_{Turbine}$ = 0.5; UA_{Evap} = 370W/K; $T_{Coolant}$ = 60°C*).

As shown in Figure 5a, for fixed condenser pressure, high pressure ratio leads to an increase in average fluid temperature in the evaporator. This has two significant impacts: (i) reduced heat transfer to the fluid, as per equation 9, which causes (ii) a reduction in the mass flow required for the chosen operating point, as per equation 10. The impact on net power and condenser load is shown in Figure 5c – despite the reduction in mass flow, net power increases up to a maximum due to a greater turbine expansion ratio. On the other hand, condenser load monotonically decreases with increasing pressure ratio so both thermal efficiency and BCR are improved.

To demonstrate the impact on energy recovery over an entire drive cycle, Figure 6 compares the WLTP performance of an ORC based on R245ca for two simulation scenarios: (i) evaporator pressure set to maximum permitted pressure and (ii) evaporator pressure set 50% of maximum pressure. Scenario (i) produces a relatively higher average fluid temperature in the expander so ORC "light-off" takes longer than the lower pressure alternative – fluid flow begins later in the cycle (Figure 6b) so the ORC produces power for a shorter period of time (Figure 6c). Despite benefitting from double the expansion ratio, Figure 6c shows that net power output for scenario (i) is very similar to (ii) so the total energy recovered is effectively the same for both cases, as shown in Figure 6d. This is because scenario (ii) operates the ORC for a longer period of time and produces higher mass flow rates that compensate for the lower enthalpy drop across the turbine. Mass flow tends to be higher in scenario (ii) for a number of reasons. Even though Figure 6a shows that evaporator bulk temperature is lower for scenario (ii) because the ORC begins to extract heat earlier in the cycle, lower average fluid temperature in the evaporator means that the temperature difference in equation 11 is actually higher – therefore, mass flow rate is higher for the specified turbine entry conditions and power output increases as well.

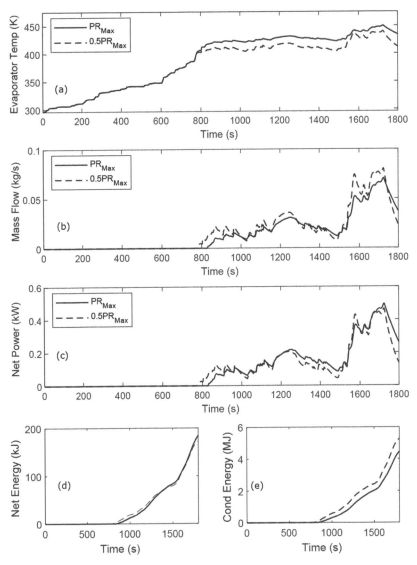

Figure 6. Effect of evaporator pressure on drive cycle performance (*fluid = R245ca; superheat = 0K; η_{Pump} = 0.55; $\eta_{Turbine}$ = 0.5; UA_{Evap} = 370W/K; $T_{Coolant}$ = 60°C*).

Despite similar performance from the point of view of total energy, Figure 6e shows that the total energy rejected to vehicle's cooling system in the heat rejection stage of the cycle is considerably higher for the lower-pressure scenario due to lower cycle efficiency. Therefore, operating at higher pressure is a better option from a vehicle-level standpoint because BCR is higher and the additional cooling burden placed on the vehicle is relatively lower.

3.3 Condenser pressure

Condenser pressure has similar impacts on overall cycle performance as those described in section 3.2 – coolant temperature is a determining factor so integration with the vehicle's cooling architecture plays a key role in energy recovery performance. Figure 2 showed three potential ORC integration scenarios that supply the condenser with coolant at (a) 90°C, (b) 60°C or (c) 40°C. Figure 7 compares scenarios (a) and (c) to illustrate the impact of condenser integration on ORC performance:

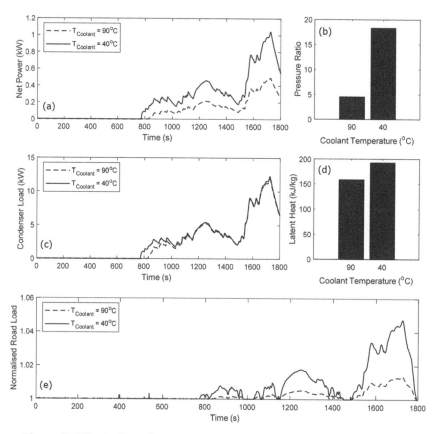

Figure 7. Effect of condenser pressure on WLTP performance *(fluid = R245ca; superheat = 0K; $P_{Evap} = 0.95P_{Crit}$; $\eta_{Pump} = 0.55$; $\eta_{Turbine} = 0.5$; $UA_{Evap} = 370W/K$).*

As expected, net power output (Figure 7a) is significantly higher for the low-temperature variant because lower condenser pressure increases the expansion ratio available across the turbine (Figure 7b).

However, the fluid's latent heat capacity is also higher (Figure 7d) at lower pressure so specific condenser load increases. Even though this translated to a minimal increase in condenser load (Figure 7c) for the chosen operating point, impact on road load is significantly higher (Figure 7e) because scenario (c) rejects heat via the low-temperature radiator.

3.4   Operating point selection

Sections 3.2 and 3.3 focused on the impacts of evaporation and condensation pressure for fixed values of vapour superheating. However, vapour superheating is an equally important variable with a large impact on ORC performance. Mapping the response of the system as a function of both pressure ratio and superheating provides valuable insights on ORC performance. Table 3 lists the simulation parameters used to produce the ORC performance maps presented in Figure 8:

Table 3. ORC simulation parameters.

Parameter	Value
Working Fluid	R245ca
Turbine Efficiency	0.5
Pump Efficiency	0.55
$UA_{Evaporator}$ (W/K)	370
Evaporator Bulk Temperature (°C)	165
Coolant Temperature (°C)	60
Condenser Pinch Point (K)	5

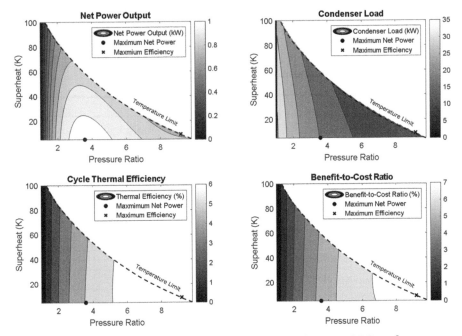

Figure 8. Effect of pressure ratio and vapour superheat on ORC performance.

125

Comparing the performance of the cycle at the maximum net power and maximum thermal efficiency operating points highlights a key trade-off. For the example shown above in Figure 8, maximum net power output is 1.01kW with a corresponding condenser load of 19.8kW. At the maximum thermal efficiency operating point, both net power and condenser load reduce to 0.78kW and 10.5kW respectively due to a decrease in mass flow and evaporator heat input. However, it is interesting to note that the 23% reduction in power is accompanied by a much larger 47% reduction in condenser heat rejection that improves that improves BCR. Again, the option that provides greater fuel reduction over an entire dynamic duty cycle is not immediately obvious. Furthermore, system performance is also sensitive to operating temperature, so the response surfaces shown above in Figure 8 only apply to a bulk temperature of 165°C (as per Table 3). Hence, optimising the performance of the cycle at a single operating temperature will inevitably lead to sub-optimal performance whenever operating conditions are off-design. In light of this, it becomes evident that full-vehicle optimisation is necessary to identify the ORC operating point that maximises fuel reduction over an entire drive cycle.

4   FULL-VEHICLE DRIVE CYCLE OPTIMISATION

Earlier sections of this paper aimed to demonstrate that the design and integration of an ORC system in a light-duty vehicle is a complex multi-variable problem that is highly sensitive to four key design variables: working fluid selection, condenser coolant temperature, pressure ratio and vapour superheating. A genetic algorithm (GA) optimisation tool was employed in conjunction with the coupled vehicle/ORC model to find the optimum combination of working fluid choice, condenser location, pressure ratio and vapour superheating that minimises fuel consumption over a full WLTP drive cycle – Table 4 lists the fixed parameters used:

Table 4. Optimisation simulation parameters.

	Parameter	Value
Vehicle	Mass (kg)	Representative of a mid-size sedan
	Frontal Area (m^2)	
	C_D	
	Baseline WLTP Fuel Mass (kg)	
ORC	Mass (kg)	30
	Turbine Efficiency	0.5
	Pump Efficiency	0.55
	Generator Efficiency	0.95
	UA$_{Evaporator}$ (W/K)	370
	Maximum Fraction of Critical Pressure	0.95
	Condenser Pinch Point (K)	5

The algorithm was bound by a set of pressure constraints that ensure supercritical conditions are avoided at all times. Table 5 specifies the maximum possible pressure ratio possible for each working fluid at each condenser coolant temperature to ensure the optimisation procedure does not exceed 95% of critical pressure.

Table 5. Optimisation pressure constraints.

Fluid	Maximum Pressure Ratio		
	$T_{Cond} = 40°C$	$T_{Cond} = 60°C$	$T_{Cond} = 90°C$
R245ca	17.7	9.82	4.50
R245fa	11.5	6.53	3.07
R236ea	8.30	4.70	2.30
R113	31.8	17.8	8.24
R141b	24.8	13.7	6.55
R114	7.91	4.68	2.42
R1233ZD	13.7	7.63	3.70
R11	19.7	11.6	5.63
R142b	6.41	3.89	2.05
R124	5.05	3.07	1.60
R1234ZE	3.92	2.41	1.26

Optimisation results are shown in Table 6. Despite recovering a total of 375kJ over the entire cycle, total fuel mass increased by 0.07% compared to the baseline so incorporating the optimised ORC heat recovery system did not produce any net positive impact on fuel consumption.

Table 6. GA optimisation final results.

ORC Design Variable	Value
Working Fluid	R11
Superheat (K)	24.7
Pressure Ratio	8.8
$T_{Coolant}$ (°C)	60
Total Energy Recovered (kJ)	*375*
Change in Fuel Mass (%)	*+0.07*

Figure 9 highlights the influence of evaporator bulk temperature on ORC performance. Once "light-off" is completed after approximately 750 seconds, the change in bulk temperature becomes proportional to the balance between the heat input from the exhaust and that extracted by the ORC working fluid. The dynamic temperature profile shown in Figure 9(A) means that working fluid mass flow rate requires continuous adjustment to provide the specified evaporator exit vapour superheating, as shown in Figure 9(D). This has a direct impact on the net power and condenser load profiles shown in (b) and (c) respectively.

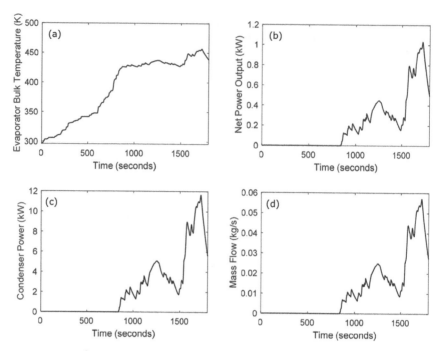

Figure 9. Optimised ORC drive cycle performance.

Although power output exceeds 1kW at the end of the drive cycle when evaporator bulk temperature reaches its highest, low average power output in the first 1500 seconds of the drive cycle prevents the system from generating any real fuel savings over the WLTP as a whole. The highly transient nature of the mass flow through the system also leads to low average pump and turbine efficiency. Low turbine efficiency is especially penalising due to its direct impact on turbine power and condenser heat rejection. In practical terms, higher turbine efficiency allows a greater enthalpy exchange across the expansion stage that gives an increase in power output and reduced heat rejection to ambient. The resultant improvement in BCR is conducive to a greater potential for fuel-saving due to a reduced aerodynamic cooling drag penalty. Hence, maximising net power output and reducing condenser load are equally important avenues for achieving real energy savings at a vehicle level. Condenser load can be reduced by employing more complex ORC architectures that use internal recuperation to pre-heat the working fluid before the evaporator using residual vapour superheat after the expansion stage (4). Another potential strategy could focus on minimising working fluid flow rate fluctu-ation over the drive cycle to narrow the operating window of the turbomachinery and increase average isentropic efficiency over time.

5   CONCLUSION

This study presented a simulation methodology that models the energy recovery per-formance of an ORC system over a dynamic driving duty cycle. Coupling the model to a genetic optimisation algorithm identified the combination of system architecture (condenser location and working fluid) and operating point (pressure ratio and vapour superheating) that minimised fuel consumption over a WLTP drive cycle. Despite recovering a total of 375kJ, the waste heat recovery system led to an overall increase in fuel consumption of 0.07% compared to the baseline vehicle due to increased vehicle mass, exhaust backpressure and cooling demand. The results highlight that performing simulation capable of capturing the vehicle-level penalties of installing an ORC system is critical at all stages of the waste heat recovery design process. Finally, simulation results were used to identify potential strategies for improved ORC performance that will be explored in future work.

LIST OF ACRONYMS & SYMBOLS

Acronym/Symbol	Description
APC	Advanced Propulsion Centre UK
ASG	Active Shutter Grille
BCR	Benefit-to-Cost Ratio
BEV	Battery Electric Vehicle
CV	Conventional Vehicle
EGR	Exhaust Gas Recirculation
FHEV	Full Hybrid Electric Vehicle
GA	Genetic Algorithm
HEV	Hybrid Electric Vehicle
ICE	Internal Combustion Engine
MHEV	Mild Hybrid Electric Vehicle
ORC	Organic Rankine Cycle
PR	Pressure Ratio
RC	Rankine Cycle
SI	Spark-Ignition Engine
WCAC	Water Charge Air Cooler
WHR	Waste Heat Recovery
W	Mechanical Power (W)
Q	Thermal Power (W)
m	Mass (kg)
$\dot{m}$	Mass Flow Rate (kgs^{-1})
T	Temperature (K)
P	Pressure (Pa)

(Continued)

(Continued)

Acronym/Symbol	Description
L	Working Fluid Latent Heat Capacity (Jkg^{-1})
η	Isentropic Efficiency
h	Specific Enthalpy (Jkg^{-1})
c_p	Specific Heat Capacity ($Jkg^{-1}K^{-1}$)
U	Evaporator Heat Transfer Coefficient ($Wm^{-2}K^{-1}$)
$A_{Evaporator}/A_{Frontal}$	Evaporator Heat Transfer Area/Vehicle Frontal Area (m^2)
ρ_{Air}	Air Density (kgm^{-3})
v	Vehicle Speed (ms^{-1})
C_D	Vehicle Aerodynamic Drag Coefficient
$SD_{Radiator}$	Radiator Specific Dissipation (Wm^{-2})

REFERENCES

[1] The International Council on Clean Transportation, "2017 Global Update – Light Duty Vehicle Greenhouse Gas and Fuel Economy Standards", Washington, 2017.

[2] Advanced Propulsion Centre UK, "Towards 2040: A Guide to Automotive Propulsion Technologies", 2018.

[3] T.A. Horst, W. Tegethoff, P. Eilts, J. Koehler, "Prediction of dynamic Rankine Cycle waste heat recovery performance and fuel saving potential in passenger car applications considering interactions with vehicles' energy management", *Energy Conversion and Management*, vol. 78, pp. 438–451, 2014.

[4] C. Sprouse, C. Depcik, "Review of organic Rankine cycles for internal combustion engine exhaust waste heat recovery", *Applied Thermal Engineering*, vol. 51, pp. 711–722, 2013.

[5] M. Jimenez-Arreola, R. Pili, F. Dal Magro, C. Wieland, S. Rajoo, A. Romagnoli, "Thermal Power Fluctuations in Waste Heat to Power Systems: An Overview of the Challenges and Current Solutions", *Applied Thermal Engineering*, vol. 134, pp. 576–584, 2018.

[6] M. Flik, S. Edwards, E. Pantow, "Emissions Reduction in commercial vehicles via thermomanagement", in *Proceedings of the 30th Wiener Motorensymposium*, Vienna, Austria, 2009.

[7] F. Zhou, S. Joshi, R. Rhote-Vaney, E. Dede, "A Review and Future Application of Rankine Cycle to Passenger Vehicles for Waste Heat Recovery", *Renewable and Sustainable Energy Reviews*, vol. 75, pp. 1008–1021, 2017.

[8] S. Risse, "Motornahe thermoelektrische Rekuperation der Abgasenergie an einem turboaufgeladenen direkteinspritzenden Ottomotor", *Dissertation*, University of Dresden, Germany, 2012.

[9] B. Mazar, "Gesamtsystemoptimierung eines thermoelektrischen Generators fur eine Fahrzeuggruppe", *Dissertation*, University of Dresden, Germany, 2010;

[10] Bell, I., CoolProp Team. (2010). CoolProp. Available: http://www.coolprop.org/. Last accessed 31st Oct 2019.

[11] Bouilly, J., Lafossas, F., Mohammadi, A., and Van Wissen, R., "Evaluation of Fuel Economy Potential of an Active Grille Shutter by the Means of Model Based Development Including Vehicle Heat Management," *SAE Int. J. Engines* 8(5): 2015, doi: 10.4271/2015-24-2536.

Designing a bespoke high efficiency turbine stage for a key engine condition through pulse utilisation

C. Hasler

Cummins Turbo Technologies – Performance Team, UK

ABSTRACT

CO_2 legislation is becoming more stringent year on year, this is especially clear with the new EU 2025-2030 targets. As such commercial vehicle OEMs must focus on reducing engine fuel consumption. A substantial proportion of this reduction will need to come from the powertrain of which the air handling system is a key component. The increase in on-engine performance of the turbocharger can greatly impact the system's brake thermal efficiency and thus fuel consumption and CO_2 production.

This paper covers the methodology and design of a bespoke, high-efficiency turbine stage for a key engine running condition through the application of pulse utilisation. The design process uses a single design point, alongside four points of interest, generated from the high-frequency engine data of a key engine fuel burn point. The optimisation of both the turbine housing and wheel are interlinked forming a synergy in the flow of energy through the stage. This methodology allows Cummins Turbo Technologies to maximise cycle average turbine stage efficiency at the exact point which the engine requires.

1   INTRODUCTION

In 2025-2030 significant changes in the EU CO2 legislation are due to begin, with a 15% reduction rising to a 30% reduction in vehicle fleet CO2 emissions [1]. These will require that commercial vehicle OEMs make improvements to both the vehicle (aerodynamics, rolling resistance etc...) and the powertrain [2]. As air handling suppliers, turbocharger companies will be required to provide state of the art performance to aid OEM's powertrain efficiency improvements. A project was started to further improve the process of both matching and tailoring the turbine stage to a customer's engine.

This paper will discuss the improvements made to CTT's turbine stage design process along with an overview of the reasoning behind it. As a precursor, it is important to know and understand the pulse utilisation terminology which will be used throughout this paper. Much of the pulse terminology is similar to that described in [3].

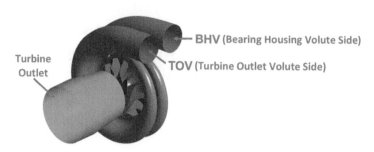

Figure 1. Description of volute terminology.

Figure 1 illustrates the bearing housing volute side (BHV) and turbine outlet volute side (TOV) on a turbine stage model. Figure 2 shows the inlet conditions for an equal admission (EA) point. The same inlet pressure is observed in both BHV and TOV. This condition will occur very briefly during the pulse when each bank of the manifold has the same pressure, as highlighted in Figure 5.

Figure 3 shows the inlet conditions for an un-equal admission (U-EA) point. At this point, there are different inlet pressures for both the BHV and TOV. Un-equal admission occurs through most of the engine pulse although at differing levels of scroll pressure ratio (SPR).

$$SPR = \frac{TOV\ Total\ Pressure}{BHV\ Total\ Pressure}$$

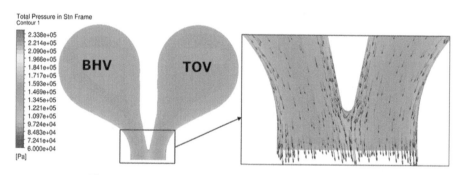

Figure 2. Example equal admission inlet condition.

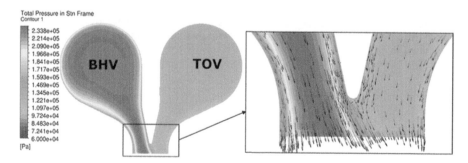

Figure 3. Example un-equal admission inlet condition.

Figure 4 shows the inlet conditions for a single admission (SA) point. In single admission, there are significantly different inlet pressures for both the BHV and TOV. Single admission occurs when the SPR is so large that the flow from the pressurised volute completely blocks or even enters the non-flowing volute.

The first graph within Figure 5 shows a typical example of a pressure pulse for a heavy-duty 6-cylinder diesel engine. The data presented in the additional graphs within Figures 5 and 6 are calculated from this pressure pulse.

Traditionally turbine stages were designed at a steady-state inlet condition which was an averaged pressure from the target engine pulse, as shown as the dashed grey line in graph one of Figure 5. This method, used through industry, was a clear simplification of a complex problem but had generally shown to be a good approximation [4]. The project discussed in this paper found that the previously used approaches were now generally outdated and by using the latest analysis methods, enabled by substantial improvements in computational power in recent years, a much better product could be designed.

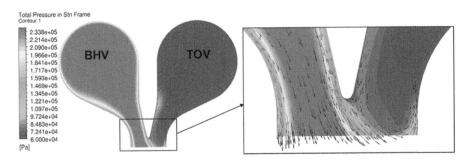

Figure 4. Example single admission inlet condition.

As shown by Figure 5 the turbine stage is only in an equal admission condition for a fraction of its operation, the available energy at this point is very low. Clearly, there is a disparity between the inlet conditions at the averaged pressure condition previously used and the point with the maximum available energy.

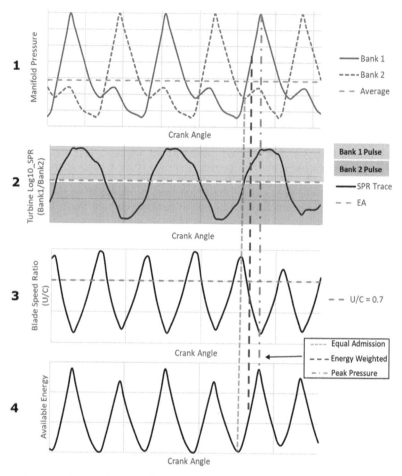

Figure 5. Example high-frequency data for a hypothetical engine.

Figure 6 shows two methods of representing the proportion of the available energy in the pulse at each turbine blade speed ratio (U/C). The definition of U/C can be found in many turbomachinery books [5]. The optimum efficiency of an ideal radial flow turbine such as this should occur at a U/C≈0.7 [4, 5]. The left-hand plot shows the accumulated available energy through the pulse highlighting where 50% of the energy is and the U/C=0.7 point as shown in [6]. The right-hand plot shows the instantaneous available energy against U/C as shown in [7]. Both of these plots show the effect of changing the wheel diameter whilst maintaining wheel speed in rpm and all inlet parameters such as pulse shape and temperature. Increasing the diameter of the turbine moves the U/C at which the majority of available energy is found.

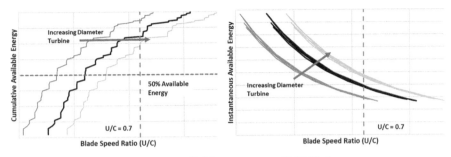

Figure 6. Available energy vs U/C Plots.

As seen in graph three in Figure 5 the U/C varies in a sinusoidal fashion where the highest U/C is found at the equal admisison point and the lowest at the peak pressure point. Thus, increasing the turbine diameter moves the U/C=0.7 point to a more un-equal admission inlet condition. This then means the pulse will move over the peak efficiency point, giving a higher efficiency where there is more available energy. To gain the full benefit the stage then needs to be optimised at this new un-equal admission design point.

It should be noted that generally there is a limit to both the inducer and exducer diameter of a turbine wheel subjected by mechanical limits or inertia. Centrifugal stress increases with the square of the radius of the turbine wheel for a fixed wheel speed. Both [8] and [9] refer to a trade-off between turbine stage efficiency and inertia. It is stated that a 40% change in inertia is equivalent to 7-8% change in turbine efficiency. Diameter increases then should be carefully considered to ensure the efficiency benefits outweigh the stress or inertia increases.

CTT now employs a method to calculate a single point which is weighted to represent a design point that will capture, on average, the most available energy over the pulse. This inlet condition is then denoted as the energy weighted point and is a key element of the pulse utilisation design methodology. Alongside this design point, four other points of interest are used to ensure a good understanding of the performance trade-offs. These five points can better describe the shape of the pulse and thus the on-engine behaviour of the turbine stage. High-frequency turbine inlet pressure is required to calculate these points of interest. They are described in Table 1 and shown in Figure 7. It should be noted that these points are all at the same turbo speed. The X-axis of Figure 7 is in Log_{10} SPR, this is for ease of visualisation, where zero denotes equal admission. The method used in this paper follows the method shown by [10] but most others have plotted a similar figure before as seen in [3, 6].

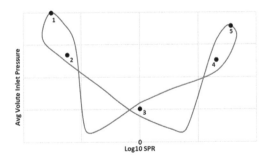

Figure 7. 5 key points of interest plotted on an interaction plot (demonstration – not project key points).

Table 1. Description of 5 key points of interest.

Point No	Description
1	Peak Pressure – BHV
2	Energy Weighted - BHV
3	Equal Admission
4	Energy Weighted - TOV
5	Peak Pressure – TOV

As can be seen, there is an energy weighted point for both BHV and TOV. It has been found through this project that the fundamental efficiency trends are the same for both points. As such it is recommended that one is selected as the design point and the other is left as a point of interest. This reduces the complex work of balancing multiple design points. Although, if this is found not to be the case, both points should be used to find the best compromise.

As a baseline for this project, a current product turbine stage was found to be a good match to the target pulse's averaged equal admission pressure point. The baseline, although not designed for this specific application, was designed using the traditional method as previously described. This paper covers the computational fluid dynamics (CFD) method and selected steps that were taken to design a bespoke stage using the pulse utilisation methodology.

2   METHOD

Historical turbine stage design would use an average pressure, equal admission design condition. As seen in Figure 2, equal admission gives a uniform flow field into the turbine wheel. As such a designer could easily simulate this condition in a wheel or segment only analysis without the requirement for a 3D housing model by applying the whirl angle and upstream pressure loss representative of the volute geometry. The whirl angle in equal admission for a radial turbine is a simple velocity triangle as shown in [4]. This simplification allows a designer to work only on the wheel and not consider the housing in the design. Removing the housing reduces the mesh count in the analysis and can save time when doing multiple optimisation loops.

Figure 8 shows an example of the CFD domain that is simulated using ANSYS CFX in conjunction with BladeGen. This single segment model allows manipulations of the blade geometry to be made and analysed quickly without the time-consuming process of building a full wheel model and user meshing. The required whirl angle can easily be added to the wheel domain inlet as denoted by point A.

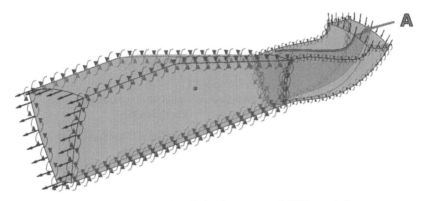

Figure 8. Example of single segment CFD model.

Observing the un-equal admission point in Figure 3 there is a non-uniform flow field in terms of both pressure and whirl angle. Changes to the wheel geometry can affect scroll separation and therefore the flow leaving the housing. Thus, the housing is required to generate the appropriate wheel inlet conditions. Figure 9 shows the full stage single segment model. The addition of a housing does, of course, mean that this quick, simple analysis will be more computationally expensive. But without a housing, the required condition could not be accurately represented. The use of a single blade segment set up for the wheel, on the other hand, is extremely useful. This method allows modifications to the blade geometry at a fundamental level allowing for quick iterative optimisation.

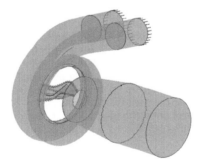

Figure 9. Example of full stage single segment CFD Model.

Analysis Type - Steady State

Fluid Models	**Inlet Boundary Conditions**

Heat Transfer - Total Energy inc. Viscous Work Team

Flow Regime - Subsonic

Turbulence Model - Shear Stress Transport

Inlet Mass & Momentum - Total Pressure (Stable)

Wall Function - Automatic

Inlet Heat Transfer - Total Temperature

Turbulence Intensity - Medium

Stage Interfaces

Intel - Genral connection

Hsg-Whl - Stage (Mixing-Plane) with Specific Pitch Angles

Whl-Exit - Stage (Mixing-Plane) with Specific Pitch Angles

The CFD setup for the full stage single segment model is as follows:

The domains were meshed using a combination of tetrahedral and prismatic elements sized to give a resolution required to achieve the analysis objectives with the available computing resources. A global element size of 2.5mm was used but face element sizes ranging from 0.1mm to 2mm were applied to areas requiring higher resolution. This meant a housing size of ~7million elements, a wheel periodic size of ~0.75million elements and a diffuser periodic of ~0.15million elements.

As shown in Figure 9 the wheel domain uses a single segment model, as such, rotational periodic interfaces are used to calculate the flow through the full blade passage. The single segment also requires a stage mixing-plane interface between the housing and the wheel. A mixing-plane circumferentially averages the flow around the interface. Thus, effects such as influences from the tongue are lost. This is clearly a downside although is a requirement for this analysis else the circumferential location of the segment would significantly change the result. With that said, the mixing-plane does maintain the axial flow differences and so represents the un-equal admission condition well.

3   AREA REACTION

In literature, Degree of Reaction (DoR) is generally defined either of two ways. Definition 1 is taken from [4,10,11]:

$$Degree\ of\ Reaction\ (DoR) = \frac{Static\ Enthalpy\ Change\ in\ the\ Rotor}{Total\ Enthalpy\ Change\ in\ the\ Stage}$$

In the referenced literature it states that if the stage inlet velocity and stage outlet velocity are of equal magnitude then the total enthalpy term can be substituted for static enthalpy. This is due to the dynamic component of the total enthalpy being equal at both inlet and outlet.

Definition 2 taken from [12,13]:

$$Degree\ of\ Reaction\ (DoR) = \frac{Static\ Enthalpy\ Change\ in\ the\ Rotor}{Static\ Enthalpy\ Change\ in\ the\ Stage}$$

Thus the definition which should be used is dependant on the stage inlet and outlet velocities. These velocities are controlled by several parameters such as expansion ratio (ER), housing inlet geometry, wheel geometry, housing outlet geometry as well as others. Table 2 shows the dynamic component of the total enthalpy for three of the points of interest at the stage inlet, wheel outlet and stage outlet. For this case, it can be seen that the dynamic enthalpy components are similar at the stage inlet and outlet. In this case, the stage has a fairly large diffuser for an automotive application (area ratio of 2.4 at a half angle of 10°). For a stage with a smaller diffuser then there could be a larger discrepancy. As such the terminology used in this paper will use definition 1 for DoR.

Table 2. Dynamic component of total enthalpy at various locations.

Inlet condition	Dynamic Enthalpy % @ Stage Inlet	Dynamic Enthalpy % @ Wheel Outlet	Dynamic Enthalpy % @ Stage Outlet
Equal	1.27	5.11	2.09
Energy Weighted	1.74	6.33	1.65
Peak Pressure	1.97	8.09	2.60

On test, it is fairly difficult to quantify the static enthalpy changes across the turbine wheel. This would require knowledge of the static pressures and temperatures around the circumference of the turbine wheel inlet as well as in the turbulent swirling exit flow. In order to simplify this, the ratio of the housing critical area divided by the wheel throat area can be used as a surrogate for DoR, called area reaction. As shown in Figure 10 there is a linear relationship between both methods of calculating DoR (generated from CFD) and the equal admission area reaction (geometric areas). Thus, using this term in place of either DoR definition is adequate. It can also be seen that for this case there is little difference between either definition of DoR.

In theory, the area reaction should be based on the aerodynamic areas instead of the geometric areas. In equal admission these are very similar, so for ease, the geometric areas can be used instead. The ideal turbine design has a DoR (and by extension, area reaction) which achieves an optimum combination of turbine wheel inlet incidence and outlet swirl, minimising stage loss [4].

Figure 10. Comparison of DoR to area reaction at EA condition.

A benefit of using area reaction is that it can be used as a driving design parameter allowing a designer to both size and then optimise the stage. As mentioned in section 2, the previous wheel or blade only analysis does not consider the housing but uses a set whirl angle. This method, although quick, does not include most influences from the housing design. The addition of the housing allows the correct sizing of the stage in conjunction with the reaction optimisation. Designing in unison like this improves the synergy of the flow through the entire stage and allows the designer to understand the trade-offs in loss through each section.

This project designed a bespoke stage at an un-equal admission, energy weighted point. Un-equal admission does not use both volutes evenly (see Figure 3), as such, the aerodynamic area within the housing available to the flow will be reduced. This will have the impact of creating a difference between the geometric and aerodynamic areas and as such the area reaction. This will be even more extreme if the stage goes into single admission (Figure 4) where one volute may be entirely blocked, thus halving the housing aerodynamic area.

Unfortunately, this poses an issue for the use of area reaction as it will not be easy to calculate the aerodynamic areas under un-equal admission. However as shown by Figure 11 the DoR vs area reaction correlation still stands even at un-equal admission, although at a slightly different relationship. This relationship change is due to the disparity between geometric and aerodynamic areas. If instead the ratio of aerodynamic areas was plotted this would have the effect of shifting the points left (on Figure 11). If plotted it is likely that these would all align on the same line as EA. Although the area reaction method is a simplification, it does not impact the design of a stage using these terms as the stage is being optimised at a set inlet boundary condition. Even though the design point is in un-equal admission it will drive the design to a point which gives the best DoR at the right flow regardless of the actual geometric area reaction's relationship.

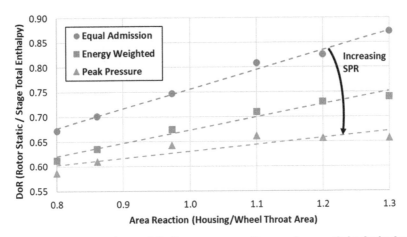

Figure 11. A comparison of DoR vs area reaction under partial admission conditions.

The flow of a fixed geometry turbine stage is controlled by several parameters [4], two of which are in area reaction: the critical area of the housing and the throat area of the wheel. By calculating the effective area of the stage, the housing and wheel combinations that provide the correct flow and reaction required can be generated. There may be some level of iteration required to provide the flow to within a set percentage (±2% may be a good target). The simplest way to maintain an estimate stage flow is to use the equivalent area calculation. It should be noted this method generally works well near to the design point but can reduce in accuracy at off-design conditions.

$$Equivalent\ Area\ f(ER) = \frac{1}{\sqrt{\frac{1}{Housing\ Critical\ Area^2} + \frac{1}{Wheel\ Throat\ Area^2}}}$$

As area reaction is increased the housing area must increase and the wheel area reduces. Area reaction was varied from 0.8-1.3 when running this analysis. This is a good spread and a peak area reaction was found for the design point as will be seen in Figure 14. Although in hindsight, a lower area reaction may be desirable to fully understand the efficiency trade-offs over all five key points. Five housing areas were set, the wheel areas varied to generate the required area reactions.

There are several ways to vary the throat area of a turbine wheel. Firstly, changing the exducer diameter of the turbine wheel. This affects the height dimension of the throat area whilst maintaining other parameters, thus the same wheel casting can be used. A second and in this case preferable method is to change the blade angle of the blade. This has the effect of changing the width dimension of the throat area. See Figure 12 to visualise turbine wheel throat area. A key benefit of changing the blade angle is for prototype DOEs such as this, where each wheel has the same profile and thus can fit in the same housing.

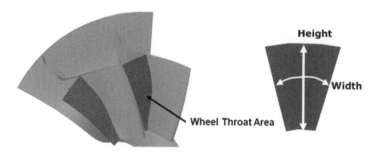

Figure 12. Turbine wheel throat area.

The five wheels of varying throat area created for this area reaction DOE have the same inducer diameter (94mm), exducer diameter (77mm), length (39mm), inducer tip width (11mm) and the number of blades (11). In Figure 13 there are multiple reactions for each of the shown housing areas. These denote the iterations required to find a reaction close to the target flow. The housings were set because using the single blade method, it was found to be quicker to generate and mesh a change in wheel area rather than a change in housing area.

4   ANALYSIS RESULTS

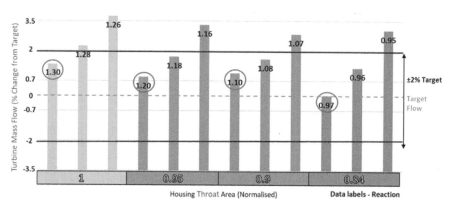

Figure 13. Selection of critical areas/reactions to maintain set stage mass flow.

The five key points shown in Table 1 were run for the range of reactions. This data starts to show some clear trends. Figure 14 shows the BHV points on the left and TOV points on the right of the 0 $\log_{10}$ SPR line which represents equal admission. It should be noted that as shown in Figure 5, the equal admission point runs at around a U/C of 0.8, the energy weighted points around 0.7 and the peak pressure points at around 0.6. A U/C of approximately 0.7 was selected for the design point as conventionally it is the point of highest efficiency for a radial inflow turbine with no inlet blade angle [4]. If 0.7 U/C was targeted for equal admission it would drive the un-equal admission points to very low U/Cs through the selection of a smaller turbine wheel diameter. This then somewhat explains the deltas in efficiency seen between each inlet condition.

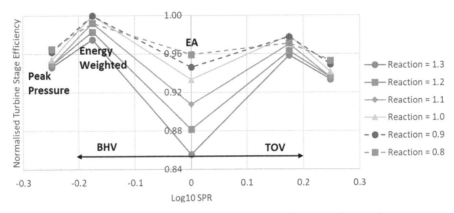

Figure 14. Effect of reaction on efficiency for the 5 key points.

From Figure 14 as reaction reduces, efficiency keeps increasing for all points except the energy weighted point, where a peak area reaction is found. This is much clearer to visualise in Figure 15, where only the BHV points and equal admission are plotted. An area reaction between 0.9-1 gives the highest efficiency at the energy weighted point. The peak pressure inlet condition shows a very small increase in efficiency with reducing area reaction. The equal admission condition shows a much larger swing in the efficiency of around 11% between an area reaction change of 0.8-1.3.

The engine will respond to the flow and efficiency of the turbine through the pulse. The five points of interest are a guide as to how the engine may respond as they represent the pulse better than an average equal admission point. If an increase in all five points is observed, it can be expected that there will be an increase in on-engine efficiency. Considering the data presented in Figure 15 it would be expected that either the reaction of 0.8 or 0.9 stage would perform best on the engine. Unfortunately, it is impractical to create a single value from these points as each point is only a snapshot of the pulse. To understand the pulse cycle average turbine stage efficiency, data must either be extracted from an engine simulation model or from engine testing. A cycle average turbine stage efficiency equation can then be used as shown in [14], defined as the turbine work through the cycle divided by the available energy through the cycle.

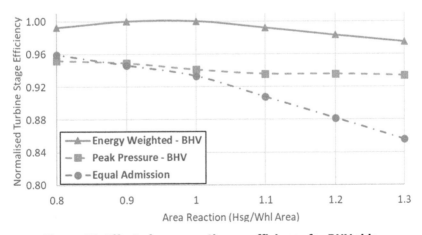

Figure 15. Effect of area reaction on efficiency for BHV side.

143

Hardware was created for 3 of these reactions: 0.9, 1.0 and 1.1 and tested on the target engine. The apparent on-engine turbine stage efficiency trend was as anticipated when considering the efficiency trends of the five points of interest. The best reaction was shown to be 0.9 which was around 1% better than the reaction of 1.0 and 4% better than the reaction of 1.1. This clearly shows that tailoring area reaction to the key points of interest is of great importance to the performance on the engine.

5 METHOD COMPARISON

To validate the CFD results the hardware created for engine testing was first tested on a turbine dynamometer as described in [15]. Also, see the section in [16] on hydraulic dynamometers for a further review of this kind of test apparatus. Figure 16 shows a comparison between the CFD results generated using the above method and data generated from the turbine dynamometer. The top two plots are data generated under equal admission (EA). The bottom two plots are generated under single admission (SA). The single admission mapping is generated using a plate to block one inlet leg.

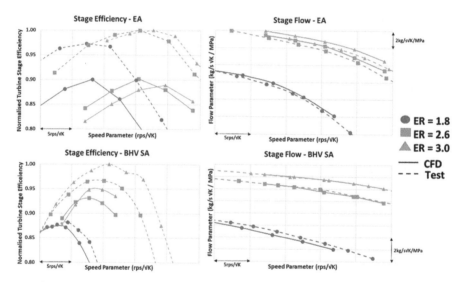

Figure 16. Comparison between CFD and Dynamometer performance.

As can be seen from Figure 16 there is an offset between the CFD and test turbine stage efficiency. This is due to factors such as differences from the test cell geometry, measurement methods and modelling of the physics. Despite the offset, the trend between expansion ratios in consistent. There is very good alignment in turbine flow parameter. The stated observations are applicable for both the equal and single admission cases. This data shows that the CFD predicted both efficiency trend and flow well. As such it can be concluded that the studies conducted using this CFD methodology are valid.

The new method of using an un-equal admission, energy weighted design point has driven the stage to a different design. Using the area reaction study, it was possible to select a stage which could be expected to give the best overall cycle average turbine stage efficiency. Figure 17 shows the comparison against two other design methods using test data taken at the BHV points of interest. Each point is running at the required speed of the target engine pulse and thus the turbine U/Cs are as mentioned in section 4.

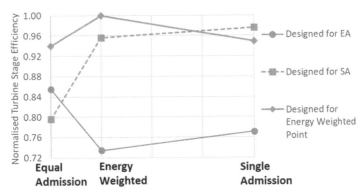

Figure 17. Turbine stage efficiency trends for different design methods (CTT Dynamometer).

Note that the single admission point is not the same as the peak pressure point shown in previous figures. This point is often run as an extreme case of SPR and as such gives a larger picture of the potential on-engine running conditions. A second point to note is that the dynamometer used allows the turbine stage to be run to conditions not pos-sible on a gas stand. This gives the ability to produce much wider turbine maps and again a better picture of potential performance. These are important when considering the large variance of both U/C and SPR observed in real-life engine conditions.

The "designed for equal admission" data is from the current product baseline that was mentioned in the introduction of this paper. As discussed already, this stage has not been optimised for these operating conditions. A significantly lower turbine stage effi-ciency is seen, even at the equal admission point of interested. Observing the trend of the equal admission design, there is a clear bias towards this equal admission point as it is nearest to its original design point. Due to the pulse of the target engine, the equal admission is around a U/C=0.8, thus this is not the peak efficiency of this stage. It is probable that a stage designed for an equal admission average of this pulse (but at a U/C=0.7) should be more efficient than the energy weighted design at equal but would suffer a similar reduction at the other two points shown.

The "designed for single admission" data is from a stage designed before the conception of this project. The initial thinking was that the most available energy was found at high SPR (see Figure 5) such as single admission, thus maximising the efficiency in this area would mean more energy could be extracted. Further learning showed that in fact the turbine rarely passed into single admission in this application and therefore this design was not designed at the correct condition. With that said the performance at energy weighted was significantly better than that of the equal admission design but it suffered around a 16% reduction in turbine stage efficiency in equal admission.

These three turbine stages were also tested on the target engine. Compared to the energy weighted design, the single admission design showed around a 6% reduction in cycle average efficiency. This suggests that there is a weighting towards the lower SPR efficiencies such as the energy weighted point and equal admission. Figure 17 shows that the single admission design does not perform as well under these conditions. The equal admission design as expected does not perform as well as either of the others on

145

the engine, with around a 15% reduction in cycle average efficiency. This efficiency of the baseline may be improved by redesigning it to the specific pulse condition instead of using a current product match. It is expected that the change would not be sufficient to improve the stage enough to produce a higher cycle average efficiency than the energy weighted design. Unfortunately, the effect on the specific fuel consumption (SFC) of the target engine cannot be shared. With that said, as expected the energy weighted design greatly improved the SFC over the other two designs.

6   CONCLUSIONS

It was the aim of this paper to give an overview of CTT's current pulse utilisation design process for developing a bespoke stage for a given engine operating condition. Throughout the introduction, the background for this method was covered. It was shown that an equal admission inlet condition at an averaged pulse pressure was not a representative condition of the point with the most available energy. Instead, an un-equal admission design point is created which characterises a point weighted to give the most available energy. The baseline used to measure the benefits of this project was the closest match within the current product catalogue at the time.

The CFD method used through the project and outlined within this paper required the addition of a turbine housing into the single segment analysis. This ensured that the correct un-equal admission conditions were presented to the turbine wheel, an issue for the single segment or wheel only analysis. A further benefit is a synergy in the design between the housing and the wheel. Using an area reaction term as a design lever within this process allows a designer to optimise both the housing and wheel sizing in conjunction. An optimum reaction was found in CFD and was shown to give the best cycle average efficiency on the engine, validating the CFD analysis.

This new design was compared to designs generated using two other methods, a single admission design and the current product baseline. It was shown that pulse utilisation method gave the best on-engine cycle average turbine stage efficiency. This new method created a product which delivered the required performance to the customer at their target operating condition.

ABBREVIATIONS

CTT	Cummins Turbo Technologies
OEM	Original Equipment Manufacturer
BHV	Bearing Housing Volute
TOV	Turbine Outlet Volute
EA	Equal Admission
U-EA	Un-Equal Admission
SA	Single Admission
SPR	Scroll Pressure Ratio
EGR	Exhaust Gas Recirculation
CFD	Computational Fluid Dynamics
DOR	Degree of Reaction
ER	Expansion Ratio
DOE	Design of Experiment
SFC	Specific Fuel Consumption

ACKNOWLEDGEMENTS

I would like to acknowledge the significant amounts of work that both Mr Vishal Seeburrun and Dr Kai Zhang have done in furthering CTT's understanding into pulse utilisation and the creation of our proprietary methods for calculating energy weighted design points. I would also like to thank Mr Robert Dewhirst, Mr Stephen Hughes, Mr Jamie Archer and Mr Atif Mahmood for their expert knowledge and support throughout this project.

REFERENCES

[1] "Regulation (EU) 2019/1242 of the European Parliament and of the Council of 20 June 2019 setting CO2 emission performance standards for new heavy-duty vehicles and amending Regulations (EC) No 595/2009 and (EU) 2018/956 of the European Parliament and of the Council and Council Directive 96/53/EC", Published 25th July 2019, weblink: https://eur-lex.europa.eu/eli/reg/2019/1242/oj

[2] Delgado O., Rodriguez F., Muncrief R., 2017, "Fuel efficiency technology in European heavy-duty vehicles: baseline and potential for the 2020–2030 timeframe", The International Council on Clean Transportation, white paper

[3] Walkingshaw, J., Iosifidis, G., Scheuermann, T., Filsinger, D., Ikeya, N., 2015. "A Comparison of a Mono-, Twin-, and Double-Scroll Turbine for Automotive Applications", ASME Turbo Expo 2015, Volume 8. GT2015–43240

[4] Watson N., Janota M.S., 1982, "Turbocharging the internal combustion engine", ISBN: 978-0-333-24290-2.

[5] Whitfield A., Baines N.C., 1990, "Design of radial turbomachines", ISBN: 0–582–49501–6

[6] Palenschat T., Marzolf R., Muller M., Martinez-Botas R.F., 2017, "Design process of a radial turbine for a heavy duty engine application taking into consideration load spectra and unsteady engine boundary conditions", Aufladetechnische Konferenz 2017, Dresden.

[7] Zhang J., Zangeneh M., 2015. "Increasing Pulse Energy Recovery of Radial Turbocharger Turbines by 3D Inverse Design Method". ASME Turbo Expo 2015, Volume 2C. GT2015–43579.

[8] Filsinger D., Leonard T.M., Spence S., Early J., 2014, "A numerical study of inlet geometry for a low inertia mixed flow turbocharger turbine", ASME Turbo Expo 2014, Volume 2D. GT2014–25850.

[9] Roclawski, H., Bohle, M., Gugau M., 2012. "Multidisciplinary design opitimization of a mixed flow turbine wheel", ASME Turbo Expo 2012. GT2012–68233.

[10] Dale A., 1990, "Radial, vaneless, turbocharger turbine performance", PhD thesis, Imperial College, London, UK.

[11] Japikse D., Baines N.C., 1994, "Introduction to turbomachinery", ISBN:0-933283-10-5.

[12] Dixon S.L., 2005, "Fluid Mechanics and Thermodynamics of Turbomachinery", ISBN: 978–07506–7870–4

[13] Rogers G.F.C, Cohen H., Saravanamutoo H.I.H, 2001, "Gas Turbine Theory", ISBN:978–0–13–015847–5.

[14] Lee S.P., Rezk A., Jupp M.L., Nickson A.K., "The Influence of Pulse Shape on the Performance of a Mixed-Flow Turbine for Turbocharger Applications", International Journal of Mechanical Engineering and Robotics Research Vol. 7, No. 2, March 2018

[15] Nikpour B., 1990, "Measurement of the performance of a radial inflow turbine", PhD thesis, UMIST, Belfast, UK.

[16] Szymko S., 2006, "The development of an eddy current dynamometer for evaluation of steady and pulsating turbocharger turbine performance", PhD thesis, Imperial College, London, UK.

Development and validation of a high-pressure compressor stage

R.D. Lotz

BorgWarner Emissions, Thermal and Turbo Systems USA

ABSTRACT

This paper presents the development and validation of a high-pressure compressor stage at BorgWarner Turbo Systems. The new stage was targeted for use on Genset applications, providing a pressure ratio similar to a two-stage turbo from a single stage. The overall design process of setting development targets, executing the design, and verifying performance will be described. Numerical optimization combined with computational fluid dynamics was used extensively, final design selection for testing also employed more detailed performance map CFD predictions. The initial design candidates were tested at BorgWarner's Technical Center in Arden, North Carolina. Performance of the stages exceeded initial expectations such that significant redesign and upgraded materials were required to safely test the entire performance envelope of the stage. Full performance maps proved to not only decisively surpass the baseline stage, but also exceeded performance targets significantly. An additional assessment using conjugate heat transfer analyses was conducted to better understand heat transfer mechanisms and its consequences on component temperatures and material selection. Overall, the new stage shows significant improvement in pressure ratio and efficiency over current medium/high pressure compressor stages.

1   INTRODUCTION

The configuration of a turbo charger compressor for a commercial vehicle (CV) Diesel engine typically comprises of a radial impeller with a vaneless diffuser and a single exit volute. To control surge, almost all stages are equipped with a casing treatment, typically a bleed slot downstream of the impeller leading edge, and a recirculation volume that discharges the bleed air back into the inlet upstream of the impeller inducer. Single stage turbos are favoured for their simplicity and lower cost, two stages are usually only employed if the required pressure ratio or mass flow range cannot be accommodated by a single stage. (2,3)

The initial match for a new engine generally utilizes an existing compressor stage that was scaled or trimmed to approximately meet the specific operating range of the application. New developments are typically only commenced if the match requires performance or efficiency beyond what the existing stage can deliver. They seldomly start as clean sheet designs but are derived from an existing design. Performance optimization is either accomplished by making manual modifications combined with detailed CFD simulations and testing, or through numerical optimization. (5)

The present design effort focusses on a turbo for a Genset, and as such deviates from typical requirements for commercial vehicle turbos, especially in requiring a very high pressure ratio combined with only moderate map width. In effect, instead of being able to interpolate between existing designs, this effort made it necessary to extrapolate into unexplored design space. Furthermore, while the employed CFD methodology yields acceptable accuracy at the typical CV turbo operating conditions, its accuracy decays significantly at high speed and high pressure ratio conditions. (5)

If the objective of high pressure ratio from a single stage can be achieved, operating conditions, especially temperatures and rotation speed, will exceed those of the second stage of a two-stage turbo without interstage cooling. While high temperature materials like Titanium are capable of surviving these conditions, not all applications can tolerate the additional cost. Consequently, it was important to determine how much performance would need to be restricted in order to use less expensive Aluminium for the impeller. Conjugate heat transfer analyses on turbo chargers have been performed in the past but require calibration against physical testing. (24,25) The methodology used for this work centers on comparative parameter analysis to gain understanding in the relative heat flows and importance of modelling parameters.

2   METHODOLOGY

BorgWarner has invested considerable effort in the last years to perfect a design approach based on numerical optimization. (2,4,5,6) This approach has been so successful that the current compressor portfolio for commercial applications comprises entirely of numerically optimized designs. The methodology is based on a consistent approach of clearly defining performance goals, executing a well-established design process, and concluding the development with an unbiased evaluation based on the goals set out in the beginning.

2.1   Geometry generation

The geometry was generated using a commercial design software (5), parametrized to allow batch modification of Bezier points prescribing blade angles, and hub and shroud passage contours. Designs are generally constrained to be flank millable, such that blade angles only need to be specified at hub and shroud. Blade thickness is primarily determined by modal requirements, it is set at the outset of the optimization and only modified at the end of the optimization to meet structural requirements. About 50 design variables are required to model the impeller.

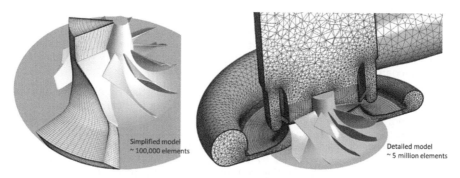

Figure 1. Representative computational mesh for simplified (left) and detailed (right) CFD models.

2.2   Numerical performance prediction

Two different approaches to CFD analysis were used in the present work. Numerical optimization with evolutionary algorithms requires many thousands of design evaluations to develop a Pareto front of near optimal designs. (6) This requires a CFD analysis that is fully automated and very fast to be able to run the required analyses in a reasonable time frame. This methodology is referred to as "Simplified CFD" in the following text.

Given the limitations of the simplified approach, a slower but more accurate methodology, to be referred to as "Detailed CFD", was also employed. Detailed CFD analyses were primarily used to make performance evaluations of promising designs retrieved from the optimization and to help select the final set of designs to be tested on the gas stand. Figure 1 contrasts the geometry and level of complexity of the two approaches.

2.2.1 *Simplified CFD*

This approach is largely similar to the one described by Lotz. (4) The model comprises a single passage through the impeller and vaneless diffuser. All other components of the stage, i.e. ported shroud and outlet volute, are neglected. The mesh is of all hexahedral topology with mesh spacing adequate for high Reynolds number near wall modelling, resulting in an overall cell count of around 100,000 for the passage. Steady Reynolds-averaged Navier-Stokes equations are solved iteratively, and turbulence closure is provided by the Spalart-Almaras (7) model. Our previous paper contended that results from this approach, while not predictive for absolute stage performance, were adequate to represent relative performance differences between designs, and consequently were sufficiently accurate for numerical optimization. (4) For the present design effort, the important operating points are generally located in the upper right side of the performance map. This corresponds to an area of higher uncertainty for the predictions, so it was advisable to revisit this contention. Figure 2 compares gas stand measurements with CFD predictions on a high-speed stage that was used as one of the starting points of this development. Corresponding speed lines are shown with consistent line styles.

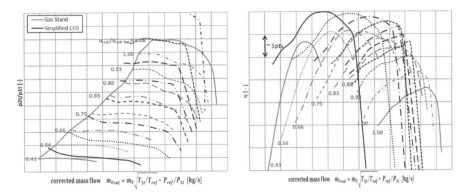

Figure 2. Comparison of simplified CFD results to gas stand test.

The simulated map is generated with the same settings used to run the operating points of an optimization. Mid map, predictions for both pressure and efficiency are generally too high, but overall shape of the speed lines and efficiency brows are captured. Near surge there is a non-physical increase in pressure ratio, but its onset does not correlate well with the actual measured surge line, not surprising given the lack of casing treatment in the CFD model. Convergence near surge also shows some unsteadiness, which becomes more pronounced the closer to surge the operating condition is.

151

At high speed, the discrepancies become more severe. The highest speed line was so unsteady that no useful information could be extracted from any of the operating points, predictions from the second highest speed line did converge but pressure and efficiency were overpredicted by more than 15%.

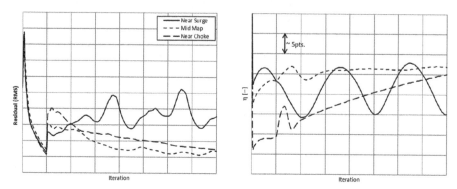

Figure 3. Convergence behaviour of simplified CFD analyses for operating points on a mid-high speed line.

Figure 3 shows the convergence behaviour on a speed line between the very high speeds and the mid map conditions typically used for more conventional design efforts. Instability near surge is more pronounced, but from mid map to choke the calculations converge more reliably. The entire speed line is shifted towards lower mass flow rates while the transition to choke is less gradual. The lower pressure ratio target points were still close to the flow conditions where relatively stable CFD predictions were possible. Outcomes from the optimizations however need to be treated with caution and checked with a more reliable prediction method, and ultimately, against physical measurements.

2.2.2 *Detailed CFD*
This approach is typically used for stage performance predictions, involving calculations of an entire performance map. Execution speed is still important to ensure acceptable turnaround times of no more than a few days for each design. A typical performance map contains about 60 operating points on 6-7 speed lines.

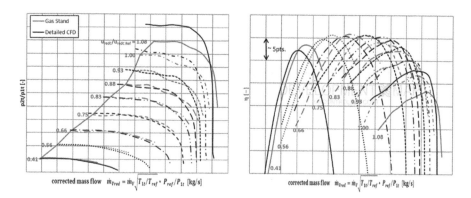

Figure 4. Comparison of detailed CFD results to gas stand test.

As shown in Figure 1, the model comprises the impeller, inlet including casing treatment, and outlet volute geometry. The mesh is of all hexahedral topology for the rotating components and unstructured tetrahedral with prism layers for all other regions. The geometry of the impeller is simplified slightly by removal of fillets and the retaining nut at the nose of the impeller to facilitate hexahedral meshing. Rotating and stationary domains are connected using "frozen rotor" interfaces. Cell counts are generally around 5 million for the entire domain, of which approximately 2 million are used for the rotating impeller. Near wall mesh spacing is adequate for high Reynolds number near wall modelling. Steady Reynolds-averaged Navier-Stokes equations are solved iteratively (18), and turbulence closure is provided by the k-ω (16) model with Shear Stress Transport (SST) extension. (17) Predictions from this setup have been benchmarked against gas stand tests numerous times and have proven to provide repeatable results.

Figure 4 presents comparisons of performance map predictions on the same stage as the previous benchmark for the simplified CFD methodology. Predictions are substantially closer to reality, as would be expected from the more detailed model. Speed line shape is generally represented well at low to medium pressure ratio, significant deviations only appear on the top speed line. Efficiency is less accurate, peak efficiency is predicted at higher mass flow than on the test. Surge prediction is not possible with either methodology, calculations will still converge at mass flow rates that trigger surge on the test stand.

For each promising design generated from of an optimization, a full CFD map using the "Detailed CFD" methodology was created. Selection of test geometries for physical testing on the gas stand was based on performance predictions from the CFD maps.

2.3 Optimization setup
The approach used for this development is substantially similar to the one described in Reference 4. The numerical optimization was driven by a commercial software package (8). A hybrid scheduler combining evolutionary and gradient based algorithms (9,10,11) was employed, with typically several thousand design evaluations to achieve an optimized geometry. Optimization only considers aerodynamic performance; structural requirements are addressed in a separate design loop.

2.4 Mechanical qualification
Mechanical qualification of new stages is accomplished once the aerodynamic optimization is completed, without significantly altering aerodynamic surfaces. For the present work, thermal and speed requirements were the determining factors in material selection for the impeller and housing.

2.5 CHT modelling
Conjugate heat transfer (CHT) modelling was employed to assess heat load and relative heat flow within the stage. Figure 5 shows a cross-section of the model with the individual components that were modelled.

Each component was meshed individually and then connected with general grid interfaces (GGI) (22) to the neighbouring components in contact. In order to achieve a reliable match between fluid and solid domains, the hexahedral fluid mesh from the performance prediction CFD model was replaced with an all tetrahedral mesh with additional detail like fillets, retaining nut and the cavity between the compressor wheel back wall and the bearing housing. The resulting model contains about 7.9 million fluid and 5.4 million solid elements and is shown in Figure 6.

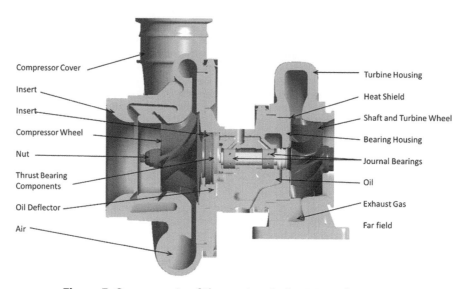

Compressor Cover

Insert

Insert

Compressor Wheel

Nut

Thrust Bearing
Components

Oil Deflector

Air

Turbine Housing

Heat Shield

Shaft and Turbine Wheel

Bearing Housing

Journal Bearings

Oil

Exhaust Gas

Far field

Figure 5. Components of the conjugate heat transfer model.

The following items present major challenges to accurately model heat transfer mechanisms in a complete turbo charger stage:

1) Heat transfer to the oil in the bearing housing. The oil flow is highly complex, it has to travers several very narrow passages, and it mixes with air on the drain side of the bearing housing. Attempting multi-phase modelling to capture the air/oil mixture was beyond the scope of this effort. Additionally, heat is generated in the bearings, which changes with speed and thrust load between compressor and turbine, as well as heat transfer from the turbine side hot exhaust gas, and heat transfer from or to the compressed air on the compressor side.

2) Contact resistance between metal components. Published values (23) display a wide range of plausible values, which are highly dependent on surface finish, contact pressure, surface corrosion, and many more.

3) Heat transfer to the surroundings, be it convective or radiative. On a gas stand, far field conditions can be controlled reasonably well, but on an engine there is much larger variation.

4) Inaccuracy of CFD results for both compressor and turbine side flow. As shown in the CFD methodology section, CFD overpredicts pressure, and with that temperature, significantly at high speed. If the calculated heat transfer from the CFD model is not corrected to realistic values, components temperatures in the CHT model will also be too high.

For the present effort, far field, oil and the turbine side exhaust flow were modelled using convection boundaries. Turbine flow heat transfer and bulk temperatures were extracted from a separate CFD analysis. Corresponding values for oil and far field were determined based on reasonable guesses. Heat flux between metal components was modelled using contact resistances. Accurate predictions of component temperature will typically require calibration against physical measurements. This data was not available at the present time, so for the purpose of this study, plausible ranges of parameters were assembled for a maximum and minimum heat transfer scenario to provide upper and lower bounds of component temperatures for the impeller and housing.

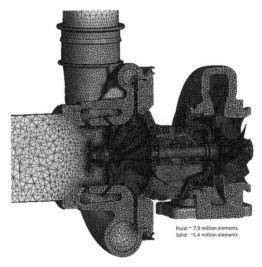

Fluid: ~ 7.9 million elements
Solid: ~5.4 million elements

Figure 6. Computational mesh of the conjugate heat transfer model.

Air flow through the compressor, as well as temperatures in all components shown in Figure 5 were calculated using a fully coupled conjugate heat transfer analysis. Analysis settings on the fluid side were equivalent to the detailed CFD methodology described above, while the solids were treated as immersed in the same commercial flow solver. (22)

2.6 Gas stand testing
Given the limitations of numerical performance predictions, physical testing remains necessary to determine compressor performance. All geometries were tested on hot gas stands conforming to SAE J1826-199502. (12) Performance maps presented herein are physical tests of the respective geometries, tested at the BorgWarner Technical Center in Arden, North Carolina. (13)

3 PERFORMANCE DEVELOPMENT

3.1 Development targets
Design targets are set by the systems performance team, generally trying to identify key operating points representing the specific requirements for a specific application. In this case, the design is intended for genset application. The pressure ratio required of the design is high, and an additional stretch goal of reaching a peak pressure ratio of 6.0 without using a vaned diffuser was even more challenging. Map width requirement is moderate, mainly driven by the need of supporting multiple AC frequencies with the same hardware. A sketch of the baseline stage map topology is shown in Figure 7, including approximate locations of the key mode points as well as a dashed outline of an acceptable map topology. Since a Genset needs to operate at constant RPM to provide a consistent AC frequency, the lug line of possible operating points at varying load is essentially a straight line, represented in the sketch as two dash-dot lines. The baseline compressor does not meet the two higher pressure mode points. The targets generally emphasized improving stage performance at the higher pressure ratios.

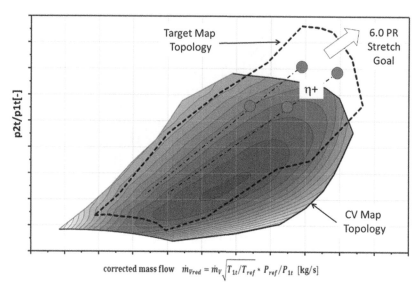

Figure 7. Sketch of the map of the baseline overlaid with target map topology, including design objectives and constraints.

3.2 Design and results

The impeller initially selected for this project was a legacy design. It was developed at a time before numerical optimization, but was the only option in the current portfolio of CV stages with anywhere near the pressure ratio potential for this application. It fell short of meeting the two high pressure targets presented above and was no longer competitive in the efficiency it could achieve. Instead of trying to evolve the baseline design, a high speed aftermarket stage was selected as the starting point, being aware that substantial changes to both the aerodynamic as well as structural characteristics would be necessary.

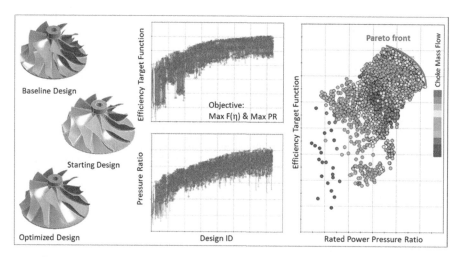

Figure 8. Comparison of the Phase 1 baseline and optimized impeller geometries.

The geometry was parametrized as described in the methodology section, yielding approximately 50 design variables. Geometric constraints were applied at the inducer shroud and exducer hub in order to ensure the impeller could be machined and assembled into a one-piece cover. Impeller designs with only full blades as well as designs with splitter blades were included in the optimization. CFD evaluations were performed representing each of the key modes, as well as a single low speed operating condition to anchor the spine of the map to align with the Genset's lug line. This resulted in a total of 5 CFD evaluations for each design. Initially, the optimization was run in "goal seek" mode, trying to increase both pressure ratio and mass flow rate at the same time. Once a substantial increase in pressure ratio had been achieved, the optimization objectives could be shifted to a non-complementary objectives setup, allowing the optimization algorithm to develop a Pareto front from the trade-off between the two objectives. The objectives were pressure ratio and a weighted average efficiency at the four main mode points. The results were constrained by a minimum mass flow rate at high speed, and pressure ratio at the low speed operating point.

The numerical optimization was initiated from a set of designs created by random perturbation of the starting design with a Latin hypercube algorithm. The optimization was subsequently run, with continuous adjustments to constraints and objectives, until the rate of change in the objectives had levelled off, as shown in the two middle graphs in Figure 8. Approximately 5000 design evaluation are necessary to reach that point. Several designs from the Pareto front where extracted, and a full CFD map using the detailed CFD methodology was created. An example Pareto front is shown on the right side of Figure 8. On the left is a comparison of the baseline legacy impeller, the starting geometry and one of the optimized designs. This process was repeated for several configurations with differing number of blades, splitter or full blades, and inducer trims. Finally, several designs covering differing trade-offs between pressure ratio, efficiency and surge stability were selected for gas stand testing. Given the uncertainty in CFD predictions, it was not clear what level of performance was going to be achieved by the new stages.

No constraints for structural integrity were included in the optimization. Thermal consideration drove material selection for this stage, therefore typical mechanical constraints became a secondary concern. Several prototypes were built and benchmarked against the baseline impeller. Initial impellers were manufactured from Aluminium, but testing showed such high pressure, and associated high temperatures, that Titanium impellers were needed to safely explore the full capability of the stages.

Table 1. Comparison of design parameters.

Parameter	Baseline	Starting Design	New Stage
Blading	7 Full/7 Splitter	10 Full	10 Full
D_2	Same	Same	Same
D_1/D_2	0.75	0.75	0.77
EIAR	0.54	0.63	0.47
Blade Wrap	57.8°	55.1°	54.9°
Backsweep	26.7°	28.9°	15.2°
L/D_2	0.40	0.33	0.42

Table 1 provides a comparison of important design parameters of the baseline stage, as well as the design used to seed the optimization and the final, winning configuration. Performance maps of the winning design are shown in Figure 9. Results are normalized with consistent reference values. Results show the optimized design having much higher specific pressure, while retaining almost the same map width as the baseline. In fact, at top speed the stage surpasses the stretch goal of 6.0 PR, without requiring a vane diffuser. While it only improves peak efficiency by about 2 percentage points, efficiency remains higher at higher pressure ratios, resulting in substantially better performance at the key operating points.

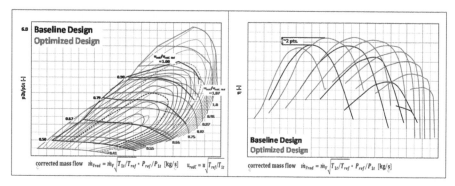

Figure 9. Comparison of baseline and optimized impeller performance.

3.3   Performance evaluation

The final step in stage development is scoring the performance test outcomes relative to the development targets set at the beginning of the development. A consistent use of metrics for targets and relative importance is expressed in a weighted average efficiency (WAE), which is generally a better representation of stage performance for a specific application than peak efficiency.

Performance at the key mode points is summarized in Table 1. Given that the baseline stage is not able to reach the two high power mode points, the performance assessment should be primarily based on the development targets, which were worked out by the performance team as being both realistic and competitive. The new stage surpasses the efficiency targets at every point, with a weighted average 1.8 points higher than requested.

Table 2. Performance Outcomes, difference of gas stand measured efficiency vs. efficiency targets set at the outset of the development.

Design point	Weight (%)	Baseline	New Stage
Key Mode 1	30	-	+2.0
Key Mode 2	30	-	+3.0
Key Mode 3	20	-2.2	+0.2
Key Mode 4	20	-3.4	+1.5
WAE		-	+1.8

3.4 Thermal management

While CFD predictions showed substantially improved performance of the new stages, the lack of fidelity of the predictions at high pressure and speed made the implications of the high performance potential apparent only once results from physical testing became available. Figure 10 shows a performance map overplayed with temperature contours as well as a photo of the compressor cover. At peak pressure, discharge temperature surpass 300ºC, hot enough to cause discoloration to the cast iron compressor cover.

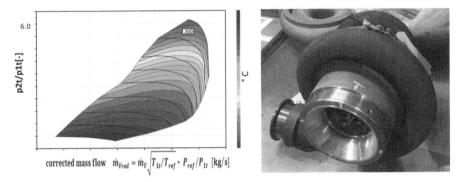

Figure 10. Compressor map with discharge temperature contours, image of the compressor cover after the performance test.

The CHT model is described in the methodology section. Since detailed temperature measurements on the gas stand were not available yet, for this contribution we looked at minimum and maximum plausible values of heat transfer to oil and surroundings, as well as the various contact resistances between individual solid components. The compressor side fluid flow was run at conditions that provide a realistic heat load based on the performance gas stand measurements. The compressor wheel was modelled both as Titanium (as used in the final performance demonstrator) as well as Aluminium.

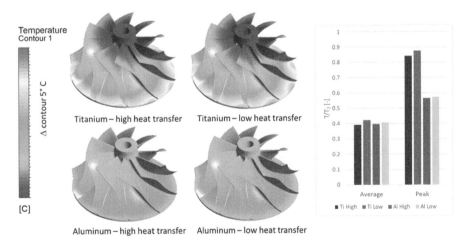

Figure 11. Compressor impeller temperature predictions from the CHT analysis using maximum and minimum heat transfer assumptions.

Figure 11 shows temperature contours of the compressor wheel, using either Aluminium or Titanium material properties, and using maximum or minimum plausible heat transfer coefficients and contact resistances. The heat transfer path on a compressor wheel is relatively simple, primarily comprising convective heat transfer from the compressed air and heat conduction into the shaft. From the shaft, the heat can either transfer to the oil, or is conducted through the bearings or seals into the bearing housing. Due to lower conductivity, temperature gradients in the Titanium wheels are larger, peak temperature reaching about 85% of T_2 compared to only about 58% with Aluminium. Based on this analysis, the influence of the higher thermal conductivity is much larger than the plausible range of heat transfer through the shaft in determining wheel temperatures. As expected, the highest wheel temperatures occur near the exducer.

Figure 12 Shows the outcome for the compressor housing. Due to the high temperatures at the stagnation points, a cast iron cover had to be used, so only variation of heat transfer assumptions are considered. The heat transfer path for the housing includes convective heat transfer from the compressed air on the inside of the housing, heat conduction through the contact with the bearing housing and the V-band clamping the two components together, and convection and radiation to the surroundings. While peak temperature is predominantly influenced by the stagnation temperature near the tongue of the volute, the range of average temperatures for the scroll has a plausible range of about 90% to 96% of T_2, depending on assumptions for heat transfer to the surroundings and conduction into the bearing housing. Quantitatively, the calculated heat contours correlate with the heat induced discoloration of the tested compressor housing.

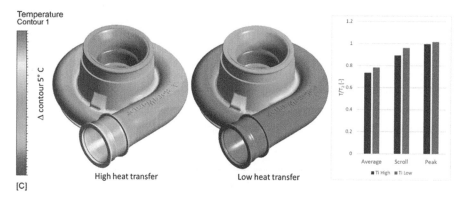

Figure 12. Compressor cover temperature predictions from the CHT model using maximum and minimum heat transfer assumptions.

4   DISCUSSION AND CONCLUSIONS

The optimization methodology employed for the present wheel development was very reliant on numerical simulations, specifically capturing differences in performance between designs, and making designs changes based on these differences to improve performance. Despite the documented inaccuracies of the CFD predictions, the development resulted in higher performance than was expected when setting performance targets. Furthermore, a peak pressure ratio over 6.0 was achieved without the use of a vane diffuser and its commensurate consequences for cost, complexity, reduced map width, and high cycle fatigue implications. The resulting stage represents not only a substantial improvement over existing stages in BorgWarner's portfolio, but offers opportunities of employing turbo chargers derived from commercial vehicle applications to new applications in marine, Gensets, and similar spaces.

The high specific pressure ratio of the new stage also imposes higher heat loading on the turbo. Plausible temperatures were derived to aid suitable material selection for the components and additional validation will be required to verify the assumptions made for this analysis.

ACKNOWLEDGMENTS

The author would like to thank BorgWarner Emissions, Thermal and Turbo Systems for permission to publish the results presented in this paper.

NOMENCLATURE

D_1	- Diameter at inducer shroud	D_2	- Diameter at exducer hub
EIAR	- Exit – Inlet area ratio	$F(\eta)$	- Efficiency target function
η	- Efficiency	L	- Impeller aero length
$\dot{m}$	- mass flow rate	P	- Pressure
PR	- Pressure ratio (compressor)	T	- Temperature
U	- Circumferential speed		
Subscripts			
1	- Compressor in	2	- Compressor out
C	- Corrected	red	- Reduced
ref	- Reference	s	- Static
t	- Total	t-t	- Total to total
t-s	- Total to static		

REFERENCES

[1] Development of a High-efficiency commercial diesel turbocharger suited to post Euro VI emissions and fuel economy legislation, Watson et. al., IMechE, 2014
[2] Further development of a high-efficiency compressor stage suited to post Euro VI emissions and fuel economy legislation, Lotz et. Al., IMechE, 2018
[3] Baines, N., 2005, "Fundamentals of Turbocharging", Concepts ETI, Inc., ISBN 0-933283-14-8.
[4] Mueller, L., Alsalihi, Z., and Verstraete, T., 2013, "Multidisciplinary Optimization of a Turbocharger Radial Turbine", Journal of Turbomachinery, 135, 021022-1-9.
[5] Lotz, R. D., 2017, "Aerodynamic Optimization Process for Turbocharger Compressor Impellers", GT2017-64365, Proceedings of ASME Turbo Expo 2017, June 26-30, 2017, Charlotte, North Carolina, USA

[6] Lotz, R. D., "Optimization of a Turbo Charger Compressor using AxCent and modeFrontier" ESTECO User Meeting presentation, Detroit, Michigan, 2015.

[7] AxCent User Manual, Concepts NREC.

[8] Merkle, C.L., et al., 1992, "The Relationship Between Pressure-based and Density-based Algorithms", AIAA 92-0425. 30th Aerospace Sciences Meeting & Exhibit, January 6-9, 1992, Reno, NV.

[9] Spalart, P. R. and Allmaras, S. R., 1002, "A One-Equation Turbulence Model for Aerodynamic Flows", AIAA Paper 92-0439.

[10] modeFrontier User Manual, Esteco

[11] Turco, A., 2011, "HYBRID: Description", Esteco Technical Report 2011-003

[12] Turco, A., 2011, "HYBRID: Benchmark Tests", Esteco Technical Report 2011-004

[13] Deb, K. et. Al., 2002, "A Fast and Elitist Multiobjective Genetic Algorithm: NSGA-II", IEEE Transactions of Evolutionary Computation, Vol. 6, No. 2.

[14] SAE Engine Power Test Code Committee. Turbocharger Gas Stand Test Code. March 1995, 01. SAE J 1826.

[15] Considerations for gas stand measurement of turbocharger performance, Schwartz, J.B, and Andrews, D.N, IMechE, 2014

[16] Wilcox, D.C., "Multiscale model for turbulent flows", In AIAA 24th Aerospace Sciences Meeting. American Institute of Aeronautics and Astronautics, 1986.

[17] Menter, F.R., "Two-equation eddy-viscosity turbulence models for engineering applications", AIAA-Journal., 32(8), pp. 1598–1605, 1994.

[18] ANSYS CFX-Solver Theory Guide, Turbulence and Wall Function Theory

[19] Wallin, S., and Johansson A., "Modelling streamline curvature effects in explicit algebraic Reynolds stress turbulence models", International journal of Heat and Fluid Flow, 23(5), pp. 721–730, 2002.

[20] Smirnov, P.E., and Menter, F.R. "Sensitization of the SST turbulence model to rotation and curvature by applying the Spalart-Shur correction term", ASME Paper GT 2008-50480, Berlin, Germany, 2008.

[21] xANSYS CFX-Solver Theory Guide, Basic Solver Capability, Immersed Solids

[22] ANSYS CFX-Solver Theory Guide, GGI and MRF Theory

[23] Mills, A. F., Heat Transfer, CRC Press, 1992, ISBN 0-256-07642-1

[24] Heuer, T, et al., "Thermomechanical Analysis of a Turbocharger Based on Conjugate Heat Transfer", GT2005-68059, Proceedings of ASME Turbo Expo 2005, June 6-9, 2005, Reno, Nevada, USA pp. 829–836

[25] Baines, N., et al., "The Analysis of Heat Transfer in Automotive Turbochargers", J. Eng. Gas Turbines Power. Apr 2010, 132(4): 042301

The smallest CV VNT™ developed for Euro VI+ & Japan PPNLT light duty commercial applications with extreme braking

S. Ikeda[1], O. Senekl[2], V. Kalyanaraman[3]

[1]Garrett – Advancing Motion, Saitama, Japan
[2]Garrett – Advancing Motion, Brno, Czech Republic
[3]Garrett – Advancing Motion, Torrance, USA

ABSTRACT

Challenging emissions and fuel consumption legislations across the globe drive continuous evolution of VNT™ turbocharger technologies. In the light duty (100 kW – 140 kW engines) commercial vehicles segment, this challenge is further complicated by high cost-sensitivity and unique packaging constraints. Moreover, due to engine down-sizing, light duty commercial vehicle powertrains need to expand their range to cover a portion of medium duty vehicle applications as well.

Garrett – Advancing Motion has developed the GT17V Gen 3 VNT™, which is Garrett's smallest CV VNT™ turbocharger to meet customers' performance, emission and reliability targets. This paper presents the key technologies and development methodology of GT17V turbocharger.

1   INTRODUCTION AND BACKGROUND

The global trend of engine downsizing for light duty commercial vehicles segment requires new product range which is smaller than conventional CV VNT™ turbochargers. Especially for 4~5L diesel engine, GT20V~GT22V turbochargers were general frame size as recently as a few years ago which were regulated by Euro VI and Japan post new long term (PNLT), Figure 1. However, the engine downsizing trend has set new target to turbocharger to have higher efficiency at low flow region maintaining equivalent or more engine power density and achieving new emission regulations such as Euro VI+ and Japan post post new long term (PPNLT). At the same time, CO_2 emission has to meet the 10.35km/l target (Japan 2015 fuel emission standard, GVW 3.5~7.5t, maximum loading weight 1.5~2.0t). Significant upgrades of powertrain and chassis are necessary to meet these challenging targets. The turbocharging system plays a major role in powertrain optimization for both emission control, fuel economy and allowing higher EGR rate. Furthermore, since the vehicle system requires exhaust brake operation, turbocharger needs to be durable under high temperature, vibration and exhaust pulsation conditions.

In response to these challenging targets, Garrett has developed the GT17V Gen 3 VNT™ turbocharger dedicated to Euro VI+ and Japan PPNLT emission level in 3~4L engine frame size. This turbocharger has new aerodynamics with +5% turbine efficiency and +3% compressor efficiency against conventional design, new VNT™ cartridge architecture with Fork & Block design for extreme dynamic loading on kinematic components under 5.5bar braking condition, introduction of upgraded one-piece rotor bearing system with integrated thrust and journal lubrications for low friction and high thrust loading capability to achieve the durability target.

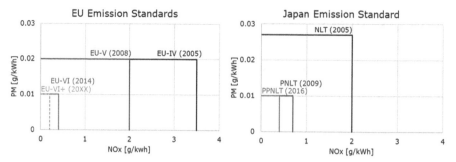

Figure 1. 15 years of regulated CV emissions reductions.

2   GENERAL PRODUCT FEATURES

GT17V Gen3 product benefits from the high efficiency of passenger vehicle (PV) segment's turbochargers and the robustness of the kinematic mechanism used in commercial vehicle (CV) segment's turbochargers. The high efficiency performance is reached by compressor and turbine aerodynamics with low friction rotor bearing system. The robustness of the kinematic mechanism includes the Fork and Block architecture at the main arm for VNT operation and capable for the extreme exhaust brake operation. Also, the oil sealing performance has been developed considering the higher negative pressure condition in compressor stage during exhaust brake operation and in turbine stage during the high idling operation which were the lessons and learnt from past applications' experience.

3   ACHIEVING PERFORMANCE TARGET

For the trend of engine downsizing, achieving emission regulation and fuel emission standard, the performance requirement to turbocharger became more challenging. Higher pressure ratio, higher efficiency and wider operating range in compressor stage to supply the sufficient boost to engine. And extremely higher aerodynamic and mechanical efficiency in turbine stage to reduce engine pumping loss and better fuel consumption.

The typical engine lug lines at full load operation on compressor map is shown in Figure 2. The black plot assumes 4~5L engines power density, Euro VI and Japan PNLT. Point A indicates the rated point which is put on high efficiency region of compressor and also ensure the margin from choke line to avoid over speeding. Point B indicates torque point and point C is low-end torque which requires the margin from surge line to avoid unstable compressor operation. On the other hand, the blue plot assumes 3~4L engine power density, Euro VI+ and Japan PPNLT. By increasing engine power density +20~30%, rated point A shifts to A' which is high pressure ratio side maintaining equivalent air flow. With respect to the torque point and low-end torque, engine requires +30~40% higher pressure ratio as point B' and C' requiring even lower air flow. In the meantime, the fuel emission standard requires real driving operating mode which means low~mid engine speed at low~mid load condition is critical area needs to be improved. Thus, the high compressor efficiency requires at also low flow and low pressure ratio region as point D'. Based on these requirements, the compressor needs to be upgraded to cover multiple operating points.

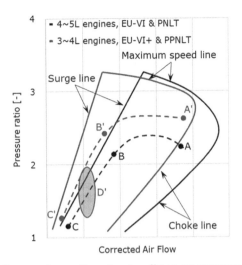

4

- 4~5L engines, EU-VI & PNLT
- 3~4L engines, EU-VI+ & PPNLT

Maximum speed line

Surge line

Pressure ratio [-]

3

A'

B'

2

B

D'

C'

1

C

Choke line

Corrected Air Flow

Figure 2. Comparison of engine lug lines on compressor map.

3.1   Compressor stage

The design of compressor involves trade-off parameters among not only perform-ance target but also durability (Low Cycle Fatigue and High Cycle Fatigue), pack-aging (wheel diameter and length) and manufacturability. Key tools used to develop new compressor are Computational Fluid Dynamics (CFD), for perform-ance optimization of aerodynamic design, and Finite Element Analysis (FEA) for mechanical design. For the LCF evaluation, it is essential to have a good under-standing of customer duty cycle.

The design path selected for this GT17V Gen3 was to design a new wheel so-called C288 (Figure 3) and compressor outer diameter 49mm. The blade shape of this wheel has been optimized to improve effi-ciency at low~mid flow region and low~mid pressure ratio which is not only scaling effect from baseline C241 design 56mm of GT20V Gen2 product. The per-formance of C288 49mm is depicted in Figure 4 comparing to C241 56mm which shows that target engine lug line is fully covered within the compressor map width and the efficiency was improved by 5 pts at key operating points.

Figure 3. C288 compressor wheel.

165

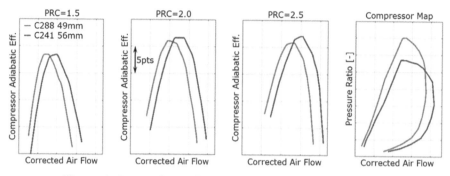

Figure 4. Comparison of compressor wheel performance. GT17V Gen3 vs. GT20V Gen2.

3.2 Turbine stage

The design of new turbine wheel is coupled to variable nozzle geometry involves similar trade-off to those already described in the compressor stage section. The right compromise has to be taken between the aerodynamic performance and mechanical constraint, in particular that of High Cycle Fatigue which restricts max speed, temperature and expansion ratio.

For GT17V Gen3, the aerodynamic design configuration has come from passenger vehicle segment product so-called T282 43mm diameter turbine wheel (Figure 5) and FR02 cambered nozzle vane design. The material of turbine wheel is Inconel 718C which is capable up to 780degC of continuous turbine inlet temperature. The performance of turbine stage measured on gas stand is shown in Figure 6 comparing to baseline T232 47mm with conventional cambered nozzle vane of GT20V Gen2 product. It shows that massive benefit in low~mid flow efficiency, in line with target setting.

Figure 5. T282 turbine wheel.

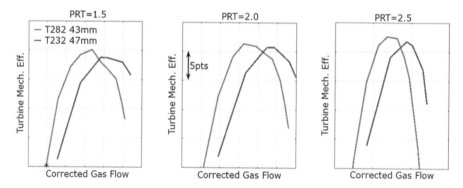

Figure 6. Comparison of turbine performance. GT17V Gen3 vs. GT20V Gen2.

3.3   Rotor bearing system

Also the bearing system is key enabler to contribute improving mechanical efficiency that essentially lower friction bearing gives as higher turbine mechanical efficiency. For GT17V Gen3, the low friction semi-floating journal bearing so-called Z bearing was introduced which has come from passenger vehicle segment product [1]. As shown in Figure 7, the power loss of Z bearing is 25% lower than the conventional bearing so-called S bearing with separate thrust bearing of GT20V Gen2 product. This lower power loss was achieved by optimization of radial clearance between turbine shaft and bearing inner diameter which is corresponding to the optimization of critical oil film thickness, optimization of thrust clearance between thrust pad of bearing and spacer which is counter component, and reducing oil flow amount by 60% against S bearing [2][3]. Since the introduction of Z bearing to CV VNT™ turbo was first time in GT17V Gen3, the trade-off among performance, reliability and durability had to be well balanced in order to survive the target vehicle life requirement from customer which is described in following chapters.

Bearing Power Loss, Z BRG vs. S BRG

Turbo speed [rpm]

Figure 7. Comparison of bearing power loss.

4   ACHIEVING RELIABILITY TARGET

The recent light duty commercial diesel engines require relatively severe operating conditions compared to medium duty commercial diesel engine and medium duty turbochargers as higher speed to meet power density and higher braking pressure for better vehicle braking capacity. In the meantime, the general vehicle life target is 300,000~500,000km which is much longer life requirement than passenger vehicle segment. Considering these usage condition, key reliabilities that LCF, HCF, rotor shaft motion, oil sealing and thrust capacity need to be secured.

4.1   Low Cycle Fatigue

The mechanical design speed limit of GT17V Gen3 turbocharger's C288 49mm compressor wheel is increased by 14% and the T282 43mm turbine wheel is 5% against GT20V Gen2 with respect to the allowable stress as Garrett's design standard. However, the actual speed limit is restricted by the duty cycle, target life and failure rate. During the development stage of both compressor and turbine wheels, the LCF life prediction was conducted based on the representative duty cycles, and confirmed that longest target life of light duty application 500,000km could be achieved. The methodology of LCF life prediction is using the coupon test data, the stress data of hub and bore based on FEA and the equivalent duty cycle based on rain flow counting of measured duty cycle which means the predicted fatigue life is realistic based on practical S-N and cycle data. As a part of wheel validation, the actual fatigue durability test is conducted on the gas stand to quantify the design and the validity of fatigue life.

4.2   High Cycle Fatigue

The HCF analysis, aimed at avoiding fatigue failure of turbine wheel induced by the excitation of pressure disturbances passing the nozzle vanes which is a particularly critical element of the design of a new turbine stage. The methodology followed by Garrett involves a number of calculation stages, both aerodynamics and mechanics. The standard FEA is applied in order to get modal resonances of the wheel, including the blade and the back-disk, while the CFD is used on solving unsteady flow equations in the turbine stage including vane and wheel passages. CFD result is used on forcing wheel resonance by pressure. The forces response results of the turbine wheel are then plotted in Goodman diagram with respect to the alternating strain and the centrifugal strain are compared, in such a way that the strains must stand within material fatigue limit. Furthermore, in order to understand the critical operating conditions in terms of nozzle vane position, turbine expansion ratio and exhaust gas temperature, Garrett runs light prove test (LPT) which practically measures the forced response and run HCF killer test to validate the HCF robustness [4][5].

The Campbell diagram of T282 turbine wheel is shown in Figure 8. In order to have robust HCF resistance, the blade thickness and hub line were optimized for uniform distribution of vibratory stress as much as possible and also considered the number of nozzle vanes as 12 vanes to reduce the excitation loading against 11 blades turbine wheel. Finally, the LPT resulted the HCF strain at 1st mode 12th order is low enough against material limit and the rest of higher resonance modes were not visible within an allowable design speed.

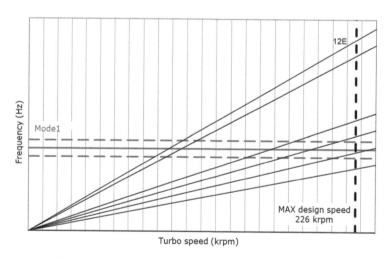

Figure 8. Campbell diagram of T282 turbine wheel.

168

4.3   Oil sealing

The oil sealing requirement is becoming severe in modern commercial vehicle application. In terms of the compressor side, the engine sucks compressor pressure at motoring condition and the exhaust braking condition increases the pressure inside center housing. In such condition, the delta pressure across the seal ring becomes significantly negative value which means compressor side pressure is much lower than inside center housing. On the other hand, the severe condition of turbine side is the engine idle condition which is low turbo speed and low exhaust pressure. As a worst case, the turbine pressure becomes lower than inside center housing by combination of engine valve timing, combustion and nozzle vane position depends on the ECU tuning for emission and fuel consumption target. Furthermore, the trend of engine oil viscosity is becoming lower that using 10W-20 and 5W-30 even in commercial vehicle engines compared to conventional 10W-30. This means the oil flow amount inside center housing becomes higher at high oil temperature condition and the risk of oil leakage becomes higher as well. Essentially, the oil leakage happens the worse condition of the oil proximity at sealing area and negative delta pressure combination.

The design target to improve oil sealing capacity is reducing the risk of oil proximity at sealing by enlarging the distance between the end of journal bearing and seal ring at turbine side. As a result, the width of oil cavity slit is 2 times larger and the length of turbine shaft is +1.6mm longer than base product for passenger vehicle segment. On the other hand, applied the oil deflector between the end of journal bearing and seal ring to restrict the oil splash on seal ring. And both compressor and turbine have twin seal rings for better blow-by capability. The turbocharger section view, oil sealing architecture is shown in Figure 9. The oil seal rig test resulted significant improvement compared to GT20V~GT22V turbochargers which are predecessor of GT17V Gen3 of light and medium duty applications. The capability of oil pressure at same seal delta condition can be 2 times improved which covers aforementioned usage condition on vehicle.

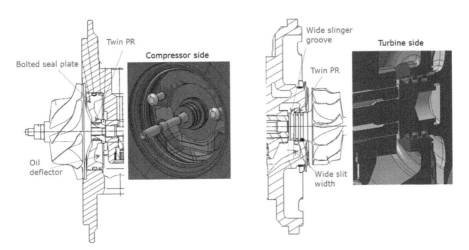

Figure 9. Section views of oil sealing architecture.

4.4   Rotor shaft motion

Since the GT17V Gen3 has a longer turbine shaft for oil seal improvement, the center of gravity of total rotor system has to be changed. In order to quantify the new turbine shaft design in terms of rotor shaft motion characteristic, the shaft motion simulation was conducted back to back with the original shaft design. The shaft motion simulation outputs the orbital fluctuation based on the boundary conditions based on rotor components geometry, mass, lubrication oil viscosity, temperature and pressure. As shown in Figure 10, the shaft motion behavior of longer turbine shaft was equivalent with original shaft design and it was quantified under the HTHS 2.4mPa.s condition.

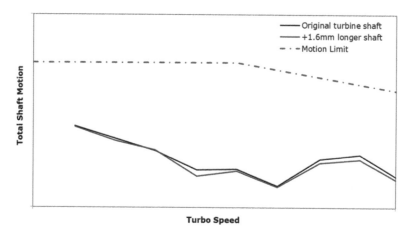

Figure 10. Shaft motion simulation.

4.5   Bearing thrust capacity

As introduced in previous term, since the Z bearing has come from passenger vehicle turbocharger, the thrust capacity was most important matter under the high thrust load condition by exhaust braking operation of commercial vehicle application. Essentially, the journal lubrication is enabled by oil film which is formed by hydrodynamic pressure between bearing thrust pad and counter components as turbine shaft. This means that certain rotor speed and oil pressure are required to function the journal lubrication. However, the commercial vehicle application has specific severe condition for thrust lubrication that the low rotor speed, low oil pressure and high thrust load

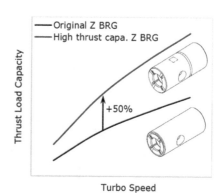

Figure 11. Thrust capacity comparison.

which is realized by the exhaust braking operating condition at low~mid engine speed. This condition has exceeded allowable thrust capacity of original Z bearing design and led the wear of thrust pad.

In order to increase the thrust capacity, the design target was enlarging the thrust pad area so that the thrust load per unit area can be reduced. The outer diameter of thrust pad was increased till the outer diameter of journal bearing itself maintaining the inner diameter of thrust pad which has increased the thrust pad area by 30% and the thrust capacity has improved by 50% as shown in Figure 11. Finally, the bearing was qualified under up to 5.5bar exhaust braking condition.

5   ACHIEVING DURABILITY TARGET

The wear of kinematic components especially the nozzle vane arm and pocket of unison ring which is generally caused by sliding movement under high exhaust pressure and aero loading condition, and fluttering movement under high vibration condition. To response the durability target 500,000km with 5.5bar braking condition, the durability of GT17V Gen3 was further improved over the previous generation GT20V Gen2 turbocharger. The contact stresses on the kinematic interface of the nozzle vane mechanism were reduced, resulting in lower wear rates compared to preceding generation product. This was vital in ensuring that any drift in vane position and hence the turbine flow was minimal enough at the end of life to meet emissions and performance targets. Some of key enablers for limiting wear are the Fork & Block (F&B) architecture, optimized kinematics through the analysis and thermal decoupling.

5.1   Fork & block architecture

The critical component for kinematic wear is the main arm and main arm pocket of unison ring because the main arm is the bridging the dynamic forces of both total aero loading from exhaust gas and the actuator movement for nozzle vane operation. In order to reduce the contact stress on main arm, GT17V Gen3 has introduced F&B architecture instead of conventional standard radius head design. The fork is a part of the center housing assembly and rigidly welded to internal crank shaft which is a drive shaft converts actuator movement to the nozzle vane operation. On the other hand, the block belongs to the unison ring assembly on cartridge assembly side. The block is engaged by the pin which is welded on unison ring and some clearance between block hole inner diameter and pin outer diameter gives freedom of block rotation according to the nozzle vane operation as shown in Figure 12.

Figure 12. Fork & Block architecture.

By the F&B design introduction, the sliding area increased +33% and reduced the wear amount -20% observed from exhaust brake durability test on engine compared to predecessor GT20V Gen2 turbocharger.

5.2   Kinematic design optimization

In addition to the F&B design introduction, the optimization work has been done for the roller count and its location in order to have most stable unison ring retention. The wear of individual vane arm and unison ring pocket becomes severe as the vibration level increases. The engine downsizing trend requires relatively higher engine speed operation hence the higher vibration frequency is given to turbocharger which leads the acceleration of kinematic components' wear.

A multi-body dynamics simulation was conducted to understand the behavior of unison ring retention and load propagation at individual rollers under given vibration condition. Figure 13 shows the simulation result that the orbital fluctuation of unison ring center position among different design combination of roller counts and its pitch circle diameter. From this simulation, the design of roller counts increased from 2 to 4 and the pitch circle diameter increased to φ73.6mm.

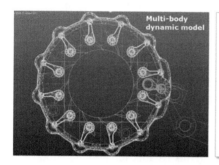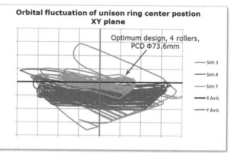

Figure 13. Simulation result of multi-body dynamic simulation.

5.3   Thermally decoupled cartridge assembly

The VNT cartridge assembly is centered with respect to turbine wheel by an elastic ring on its inner diameter (not shown). As shown in Figure 14, it's axially abutted against a flange, but can freely expand radially (green radial arrow) under thermal shock conditions. The sealing between turbine housing outlet and the part forming the wheel contour is taken by twin piston rings, which allow a proper sealing with no direct mechanical interaction between VNT cartridge assembly and turbine housing during the relative axial expansion event of the VNT cartridge assembly (green axial arrow) or deformation of the turbine housing. Finally, the VNT cartridge assembly is engaged in rotation by an axial pin (not shown). The key feature of this installa-

Figure 14. Function installation of VNT cartridge assembly decoupling.

tion is to allow a complete thermo-mechanical decoupling of VNT cartridge against turbine hosing, avoiding reliability issues such as the vane sticking if the deformation or oxidation of the housing occur.

6   SUMMARY AND OUTLOOK

The design concept and the methodology used to develop the new GT17V Gen3 has been presented as totally new products which achieved a balance between the high aerodynamic performance, lower mechanical efficiency loss and the robust durability for commercial vehicle life requirement in a smallest packaging among Garrett's CV VNT turbochargers. The result is that the turbocharger contributed reducing fuel consumption by 2%, increasing engine power density 30% and reducing VNT wear by 60% with better controllability.

In Garrett's experience, Asian region especially Japan is leading light ~ medium duty commercial vehicle development and population in the market because of the aggressive emission regulations and the unique logistic industry in a small island country compared to other global regions. Looking at next evolution of emission, Japan requires further reduction of fuel consumption by 13.4% (truck) in 2025 against 2015 regulation and also Euro VI+ would be spread globally. Which means further enhancement on turbocharging system, both on performance and durability, Garrett to continue to deliver the differentiated solutions to response to the demands of commercial diesel engine industry.

REFERENCES

[1] P. Barthelet, et al., "The new family of VNT^{TM} turbocharger developed by Honeywell Turbo Technologies for Eu6 and beyond", *ATK Conference*, Dresden, 2013.

[2] Thierry Lamquin, Kostandin Gjika, Honeywell, "Power losses identification on turbocharger hydrodynamic bearing systems: test and prediction", *ASME Turbo Expo*, Orlando, 2009.

[3] B. Remy[a,b], Thierry Lamquin[a], B. Bou-Said[b,] [a]Honeywell, [b]CNRS/INSA, "The modified Phan-Thien and Tanner model applied to turbochargers thrust bearing", *IMECHE*, London, 2014.

[4] A. Kulkarni, et al. "Turbine wheel high cycle fatigue reliability prediction", *Institution of Mechanical Engineers 9th Intl. Conference on Turbochargers and Turbocharging*, London, UK, 2011.

[5] A. Kulkarni, G. LaRue, "Vibratory response characterization of a radial turbine wheel for automotive turbocharger application", *GT2008-51335, Proceedings of ASME Turbo Expo 2008*, Berlin, Germany, 2008.

14th International Conference on Turbochargers and Turbocharging
Institution of Mechanical Engineers, ISBN: 978-0-367-67645-2

Integrated design optimisation and engine matching of a turbocharger radial turbine

P. Luczynski[1], K. Hohenberg[2], C. Freytag[1], R. Martinez-Botas[2], M. Wirsum[1]

[1]Institute of Power Plant Technology, Steam and Gas Turbines, RWTH Aachen, Germany
[2]Department of Mechanical Engineering, Imperial College London, UK

ABSTRACT

This paper presents a novel methodology for engine tailored optimisation of turbocharger turbine design. Both the turbine rotor and volute geometries for a turbocharger radial turbine were parameterised in order to enable CFD calculations for variations of predefined design parameters. The results of this analysis where used to develop and validate two approaches for computationally efficient and reliable prediction of radial turbine performance maps, quantified by total-to-static turbine efficiency and mass flow parameter. The first method utilises a meanline model which was calibrated to experimentally validated CFD data using a genetic algorithm. The second method makes use of an artificial neural network which was trained using the same CFD approach, to predict turbine performance as a continuous function of design and operating parameters. The modelling accuracy of both approaches was evaluated and compared. Finally, the meanline model was integrated into the calibrated 1D engine model of a turbocharged 1.6 litre gasoline engine. The meanline model was used to generate maps for a latin hypercube sample of four meanline design parameters. Five steady-state operating points and one transient operating point were simulated for each point in the sample, allowing the selection of optimised designs on the basis of fuel consumption and transient performance as objectives.

Due to the use of a design of experiment approach, the impact of turbine design parameters on the engine performance could also be evaluated separately. Finite Element Analysis of the turbine wheel was conducted simultaneously for the assessment of stress in individual turbine geometries. Three optimised turbine designs were selected to cater to different engine operating scenarios: ecological, sustainable and sport driving. The presented investigation clearly displays the methodology and benefits of engine integrated turbocharger design optimisation.

1   INTRODUCTION

For automotive manufacturers to meet environmental objectives and remain competitive in the internal combustion engine market, further development of turbocharging technologies is required. For similar engine performance, the technology of turbocharging enables a reduction of engine displacement and lower fuel consumption compared to the naturally aspirated engines. This significantly contributes to reduction of emissions and the development of sustainable, low carbon transport. While reducing emissions is of increasing relevance, it is important to provide drivability for a satisfying end-user experience, in particular a dynamic engine response at low rotational speeds. Two key aspects for this development are the aerodynamic optimisation of primary turbocharger components and improved matching between turbocharger and engine.

Within the literature, much research focuses on optimising the design point efficiency of the turbine and compressor, while reducing moments of inertia for improved transient performance. In a common approach for turbine optimisation, the turbine geometry is parameterised and optimised with regards to adiabatic efficiency and rotor inertia, while constraining the mass flow parameter and considering structural integrity through mechanical stress evaluation [1, 2, 3].

However, the above investigations do not assess performance of the whole, pulsating engine system in various operating modes including off-design operation and neglect the "matching" aspect of turbine design. For consideration of the trade-offs which occur when matching the turbine, modelling the turbocharger within the wider vehicle air system becomes inevitable. Instead of conventional processes based on the separate modelling and development of turbine designs, and subsequent matching with an engine by selection of a design, an integrated turbine optimisation at system-level can provide an advantage for engine performance. Several authors including Winkler and Ångström [4], Chen et al. [5], Zhuge et al. [6] and Pesiridis et al. [7] combined low order modelling of turbochargers with a 1D combustion engine model to improve overall system performance by modification of the compressor/turbine geometry. Gugau and Roclawski [8] presented a methodology to overcome the inaccuracies in matching a turbine to a full load pulsating engine flow characteristic. In the proposed approach, an improved steady-state matching quality was achieved using a commercial CFD software to generate extended turbine maps and to assess the turbine power output under unsteady flow admission. Nevertheless, this method was solely used for performance comparison of four different turbine geometries and hence stops short of truly optimising the design. Finally, Halamek et al. [9] developed a methodology which implemented a turbine meanline model to predict turbine maps as a function of key design parameters, and integrate these into an engine model to find an optimum match with regards to both transient performance and steady state fuel consumption.

The main objective of the present work was to develop a model based methodology for engine-tailored turbine design optimisation, in pursuit of improved fuel economy, while maintaining a reasonable dynamic response. This was performed in the 1D engine modelling software GT-Power, which was used to optimise the turbine design for a 1.6L turbocharged gasoline engine. Turbine geometry optimisation within GT-Power required the efficient generation of turbine performance maps for a given volute and rotor design. Both one- and three-dimensional calculation approaches were examined for this purpose.

The one-dimensional method utilises a meanline model of the single entry turbine, whereas the three-dimensional approach is based on the results of CFD simulations used to train an artificial neural network (ANN). To accommodate the low order approach, four geometric input parameters to the meanline model were used for optimisation. A 3D parametric model was developed to generate turbine geometries as a function of these parameters, which enabled a large number of designs to be calculated using CFD. The results of numerical simulations were used for training of the ANN and calibration/validation of the meanline model. Subsequently, the meanline model was integrated into the engine model to capture the impact of individual turbine geometric parameters on the engine performance. Latin hypercube sampling (LHS) was used to construct a design of experiment (DOE) for the geometric parameters. Maps for each point in the sample were generated and modelled within steady-state and transient engine models, allowing a multi-objective optimisation of the turbine design. In this study, the objective functions under optimisation were fuel economy and transient response. To achieve a high-level of process automatisation an in-house Python-based code, TORTOISE, was developed. Ultimately, three turbine designs were identified for the highest system performance for the following predefined operating scenarios: eco driving, sustainable driving and sport driving.

2   INTEGRATED TURBINE DESIGN OPTIMISATION

Within a typical industrial engine development process, turbocharger suppliers generate experimental steady-state performance maps on hot gas stands, for various compressor and turbine designs within their product portfolio. These maps are subsequently used to match a turbocharger to the engine full load curve for discrete values of rotational speed. In each of these operating points the turbine driven compressor has to ensure the necessary boost pressure to match a predefined value of engine torque, while the turbine utilises the energy of the exhaust gas. The typical main goal of the turbocharger matching process is to develop an engine system design with low fuel consumption, high low-end torque output and satisfying dynamic response. However, the chosen turbocharger comes from the supplier product catalogue and hence, is not necessarily optimally tailored to the individual engine requirements. The geometries of compressor and turbine are optimised for high performance at the component design point represented by efficiency and moments of inertia, but engine performance is still constrained by the quality of the matching.

In this paper a method for performance optimisation with consideration of the trade-off between engine fuel economy and transient performance is proposed, as graphically presented in Figure 1. The first step in the process is the fast generation of turbine characteristic maps of total-to-static efficiency $\Delta\eta_{t-s}$ and mass flow parameter MFP. In pursuit of this objective, 1D (meanline calibrated with CFD or experimental data) and 3D (training of ANN using CFD) methods were developed and evaluated with regard to their accuracy and their computational effort, as shown in the sections below.

The meanline code described in Section 2.3, was determined to be sufficiently accurate for the generation of turbine maps for different designs. A key advantage of the meanline model is its ability to extrapolate to the low and high power regions of the map, so that the maps can be directly integrated into GT-Power without the need for further processing. The low power region, in particular, is difficult to capture with CFD. Wide turbine maps are required since the pulsating inflow conditions of the turbine are explicitly modelled in GT-Power, which results in large excursions from the turbine design point operation.

For engine optimisation, four geometric parameters, r_2, A/R, r_{3t}, $d\theta/dz$ were sampled with a DOE method using LHS, and turbine maps were generated for each design. The characteristic maps of every turbocharger design were implemented into the GT-Power engine model, along with moment of inertia, which was calculated from the 3D rotor geometries. Since the optimisation of compressor geometry is not a part of this work, the characteristic maps for the baseline compressor geometry were used. A simultaneous optimisation of both compressor and turbine geometries would lead to further benefits with regard to engine performance.

Consequently, the engine brake-specific fuel consumption (BSFC) and the dynamic response evaluated as a load step at fixed engine speeds were determined for each turbine design. Considering these two objectives, optimisation leads to a pareto front as can clearly be seen in Figure 1. This shows that the two objectives are in conflict and require compromise. The turbine geometries on the pareto front form the basis for a final design selection depending on the predefined engine requirements. These requirements will generally consist of an upper limit for transient response time, and optimisation of fuel consumption. The chosen new turbine wheel designs were evaluated in FEM structural analysis at highest thermal and mechanical load in order to directly exclude geometries, which do not satisfy the maximum stress limits.

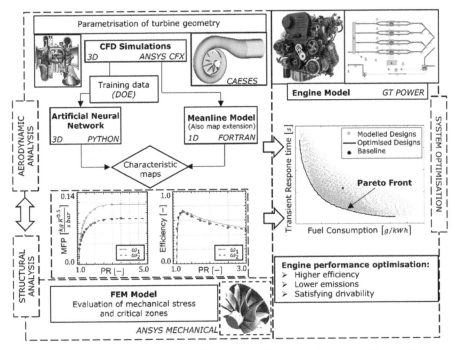

Figure 1. Process flow chart of integrated turbine design optimisation.

2.1 Parameterisation of turbine geometry

To enable the prediction of turbine performance for a given set of design parameters using CFD, a 3D parametric model of a corresponding radial turbine was developed. The stock turbocharger turbine of the investigated engine was used as a foundation for this, as it already provided a design which was matched to the compressor and the engine as a whole, and because the engine model was calibrated on the basis of engine testing which used this turbocharger. To enable the development of a 3D parametric model which recreates the stock turbine design for the baseline set of design parameters, detailed information of the interior geometry was required. Therefore, the turbine was scanned using CT at the Natural History Museum in London. The resulting single-entry nozzleless turbine with a radius r_2 of 19 mm and nine blades thus served as a baseline design for a subsequent analysis. The azimuthal cross section of the volute was parameterised, and varied around the azimuth to give a linear reduction of A_Ψ/r_Ψ ratio with azimuth angle Ψ [10]. This allows the creation of a series of cross sections which are swept around the azimuth angle. The radial turbine rotor was parameterised with a radial fibre design to minimize mechanical stress in material due to centrifugal forces. Therefore, the projection of hub and shroud curves onto the meridional plane (view in r-z coordinates, Figure 2), in addition to the camberline at the reference radius, fully define a zero-thickness blade which was thickened using a fixed axial blade section. The camberline was extracted from the stock turbo model at the tip of the blade. Its main input parameter is the camberline gradient at the exducer given by $d\theta/dz$ (cf. Figure 2). This parameter defines the exducer blade angle at a given radius by the function $\tan(\beta_3) = rd\theta/dz$. The two main input parameters parameters for the hub and shroud curves are the exducer shroud and hub radius r_{3t} and r_{3h}. The model was set up to generate a geometry as a function of the meanline design parameters. All input parameters are relative to the rotor tip radius r_2 which scales the entire geometry. More information on the turbine parametrisation method are provided by Hohenberg et al. [11].

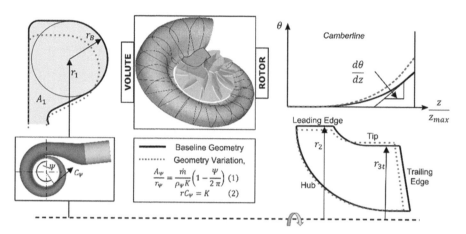

Figure 2. Parametrisation of volute and rotor geometries [11].

2.2   3D numerical model

The 3D CFD simulations play a key role both in the aerodynamic analysis and in the development of low order calculation approaches (cf. Figure 1). Since the whole characteristic maps for various turbine designs had to be calculated, the numerical setup was optimised with regard to simulation accuracy and required computational effort. Consequently, for turbine design and off-design operating point numerical settings were investigated. The comparison involved various resolutions of the boundary layer, characterized by the value of dimensionless wall distance y^+, unsteady/steady formulation of Reynolds-averaged Navier-Stokes equations and full circle (360°)/single passage (40°) rotor models. Furthermore, in single passage models with periodic boundary conditions, the peripheral averaging (mixing plane) volute-rotor interface was used in contrast to steady-state full-stage simulations, which utilised a Frozen Rotor-Stator interface. In the most accurate time-dependent URANS calculations the Transient Rotor-Stator Interface was applied. In all simulations low Reynolds kω-SST turbulence model was applied. The flow boundary conditions of these adiabatic calculations were defined by total pressure p_{01} at the volute inlet, static pressure p_3 at the rotor outlet and a value of rotational speed.

In pursuit of numerical set-up comparison, the total-to-static adiabatic efficiency η_{t-s} and mass flow parameter MFP were chosen as the evaluation parameters. The highly-resolved full-stage URANS simulations, which have been calibrated using the experimental data presented by Hohenberg et al. [11], serve as a reference. In line with expectations, the required computational time for URANS simulations was very high and it amounts to 1200 core hours on Intel Skylake Platinum 8160 CPUs at Cluster Aix-la-Chapelle (CLAIX-2018). Fundamentally, more pronounced absolute differences between individual models were observed in the value of efficiency then in the value of mass flow parameter. Hence, the deviation $\Delta\eta_{t-s}$ between URANS and RANS for 360° geometry with y^+ smaller than unity amounts to 0.96%/0.41% at design/off-design operating point (OP), with the reduction in calculation time equal to 54%, respectively. For the same mesh resolution the accuracy $\Delta\eta_{t-s}$ achieved by a single passage model is 0.57%/0.71% for design/off-design OP with the required computational effort lowered to 11%. These results prove that in off-design operating conditions full-stage simulations with Frozen Rotor interface lead to higher accuracy then single passage simulations with mixing plane, whereas in design OP the opposite is observed. However, a good compromise between accuracy - $\Delta\eta_{t-s}$ equal to 0.66%/1.41% at design/off-design OP - and calculation time reduced to 4% of the URANS reference value was found for the steady-state single passage model with y^+ smaller than five. Similar conclusions may be drawn for the differences in mass flow parameter, evaluated as a relative error $\Delta MFP/MFP_{ref}$ Consequently, the steady-state single passage model with y^+ smaller than five was chosen as a default approach for 3D CFD investigation, which allows for simulation of a single OP under 1h using single 48 CPU node on CLAIX-2018 Cluster.

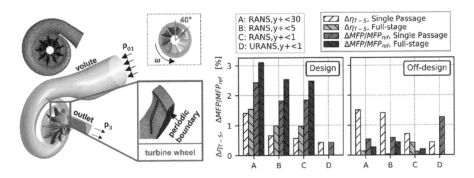

Figure 3. 3D adiabatic numerical model of turbocharger (left) and comparison of various numerical setups (right).

2.3 Meanline model

In its essence, the meanline modelling approach utilises 1D equations and velocity triangles to describe the flow in a turbine. Functions to account for sources of loss, based on physical processes in the turbine, are implemented to allow a physically-based prediction of turbine performance. This low-order modelling approach was developed for axial turbines and subsequently adapted to radial turbines by NASA in [12] and [13]. In the current work, an in-house meanline code, presented by Hohenberg et al. [11] was used for generation of turbine maps as a function of the turbine design. The main development targets for this approach was to ensure high degree of robustness, low convergence time and ease of integration with GT-Power.

The model was developed for a nozzleless radial turbine, such as the one used for the investigated engine. The turbine domain is split into the volute and the rotor, giving three stations: scroll inlet, volute-rotor interface, and rotor exit. At each station, the mass, momentum and energy conservation equations are resolved. The volute is modelled as a free vortex flow, with a pressure loss coefficient and a swirl loss as seen in the schematic in Figure 4. Four main sources of loss are considered in the rotor: passage, incidence, tip clearance and disk friction loss.

- Passage loss is incurred due to skin friction at the walls of the blade passage.
- Incidence loss is caused by imperfect flow incidence with the rotor leading edge.
- Tip clearance loss is the loss due to secondary flows which occur because of the clearance between the blade tip and the shroud.
- Disk friction is the frictional loss due to rotation at the back face of the rotor, which is usually relatively small.

As the losses are only approximations of the flow physics, and therefore semi-empirical, calibration coefficients are required for correct prediction of turbine performance for a given design.

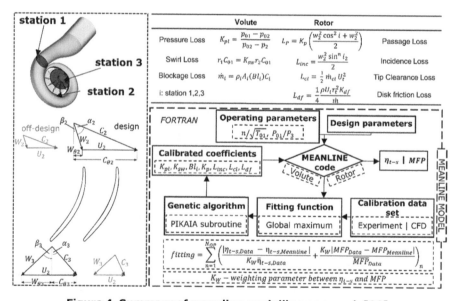

Figure 4. Summary of meanline modelling approach [11].

The calibration coefficients of the loss models in the code generally require tuning to a dataset if one is available. In the present work, this was done for using a set of CFD calculated points, spanning the required design and operational space. Calibration to these points was achieved using a genetic algorithm procedure (PIKAIA subroutine)[1] to determine the global minimum of a fitting function, consisting of the weighted sum of the absolute difference between data points and corresponding meanline model prediction for both values of η_{t-s} and MFP (Figure 4). Calibration across not only operating condition but also turbine designs, allows for improved performance of the model in the required design range.

2.4   Artificial neural network

In general terms, the ANNs are nonlinear models which can be used to map functions with multiple inputs and outputs based on a set of training data. Several successful applications of ANN in turbomachinery design have been reported [1]-[3]. In a common approach, an ANN is trained using a wide database containing experimental or CFD results to design more efficient geometries of turbomachines. However, the accuracy of the ANN method in this context strongly depends on the number and distribution of training samples in a design space. In one of the most sophisticated approaches presented by Mueller et al. the performance of a radial turbine was predicted by a self-learning system, which comprises a detailed 3D CFD solver, ANN and genetic algorithm. In a fast optimisation loop, the performance of a parameterised turbine design was predicted by the ANN, which was continously trained using the numerical data available in the database. Once the fast optimisation loop was accomplished, the optimised geometry was calculated using 3D CFD solver and the results were added to the database. Consequently, the amount of data available for training of the ANN was extended for every optimisation loop, which positively influences the accuracy of the prediction method. However, this method allows for an optimisation for a low number of operating conditions. Furthermore, it does not provide the possibility to apply statistical sampling methods to estimate the characteristics of the whole design space and to adjust the architecture of a ANN to the applied sampling.

For these reasons, in the presented paper a Latin Hypercube sampling (LHS) is used to construct a design of experiment for CFD simulations. As a result, the multidimensional design space given by three geometric parameters – $\frac{A}{r}$, $\frac{d\theta}{dz}$, r_{3t} – and two operating parameters – pressure ratio PR or velocity ratio $\frac{U}{c_{is}}$ and reduced rotational speed n_{red} – is represented by the sampling provided by a LHS. The parameter r_2 was considered by a simple scaling function and not included in the training set. Following the CFD simulations using the numerical model described in Section 2.2, the architecture of ANNs is adjusted to achieve a fitting error as low as possible for a training data set (LHS) and high prediction accuracy for a validation data set, as presented in Figure 5. Moreover, the LHS with 900 training samples was constructed in a way, which allows for its division into smaller LHSs with 225, 500 and 725 elements.

In contrast, only one hidden layer with 512 neurons is required for MFP calculation. According to common practise, a Rectified Linear Unit function was used as an activation function in both neural networks.

---

[1]*PIKAIA* is a general purpose function optimization FORTRAN-77 subroutine based on a genetic algorithm. It is available as a public domain software [15].

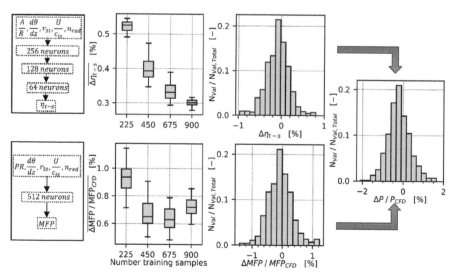

Figure 5. Evaluation of ANN accuracy for prediction of turbine performance.

The prediction error is evaluated using an additional set of 550 randomly generated samples in the design space. Considering determination of total-to-static efficiency, the average absolute error $\overline{\Delta\eta_{t-s}}$ amounts to 0.3% for the best LHS with 900 training samples, whereas in case of 225 training samples this error increases to over 0.52%. However, the corresponding results for MFP prediction show lower accuracy of ANN trained by LHS with 900 samples then by 450 or 675 samples. This could be explained with an overfitting of the first LHS and for this reason the most accurate LHS with 675 samples is used in following analysis. Finally, three histograms presented in Figure 5 depict the η_{t-s} and MFP error distribution for final ANNs and quantify the resulting error in determination of turbine power. This is less than 1% for 87% of validation samples.

2.5 Comparison of fast modelling approaches

The evaluation of prediction accuracy achieved by both meanline model and ANN approach was conducted using numerical results calculated for two speedlines of four different turbine designs. The chosen geometries strongly differed from the baseline turbine with respect to the design parameters, which were varied by up to 17%. The summary of the investigation is presented in Figure 6.

Both meanline model and ANN approach showed a good agreement with the CFD validation data set. In the case of meanline modelling, slightly higher rRMSE values for determination of η_{t-s} and MFP were obtained – 4.4% and 2.16%, respectively – than for corresponding ANN calculation - 1.87% and 1.84%. However, the ANN turbine map prediction is limited only to the range of the design space used for the training process. This can be clearly observed in characteristic map of geometry B (cf. Figure 5), in which the prediction error of ANN noticeably increases for the operating point with lowest pressure ratio and highest rotational speed, as it is defined outside the ANN design space. Consequently, for the description of turbine performance at low and high power regions of the map, an extrapolation procedure is needed. This could be realised by integration of meanline code into ANN prediction process. However, under consideration of good accuracy of 1D approach, in this work the turbine maps provided solely by the meanline model are directly integrated into engine model.

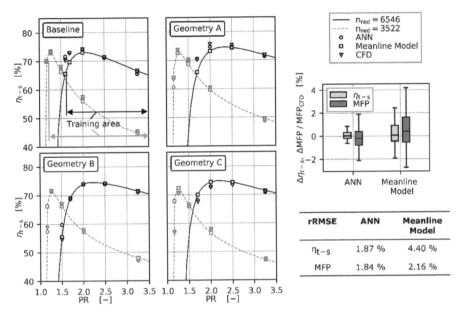

Figure 6. Comparison of meanline model and ANN turbine maps with CFD data.

3   SETUP OF ENGINE MODEL

Optimisation of the turbine design was performed for the turbocharged 1.6L Proton CAMPRO gasoline engine with variable valve timing. Engine dynamometer test data with the stock turbocharger was available from the Centre for Low Carbon Transport (LoCARtic) at Universiti Teknologi Malaysia. A 1D model of the engine was developed in GT-POWER and calibrated using the available data, for a range of steady and transient operating conditions. The engine was instrumented with pressure and temperature sensors at various key points, as well as air mass flow, fuel mass flow and turbocharger rotational speed sensors. A schematic of the engine is shown in Figure 7 with a close-up view of the turbocharger instrumentation.

To enable the proposed optimisation procedure, the turbine modelling methods were integrated with the GT-Power model through a standard lookup table procedure, requiring input of maps for the full operating range of the turbine. It should be noted that this approach does not use GT-Power's own extrapolation routine, but rather supplies the model with already extrapolated maps. These replaced the hot map of the stock turbocharger turbine which was used for calibration. For transient performance prediction, the rotor inertia was determined from the parametric design of the rotor and included in the model.

While the engine model was calibrated using available hot maps for the stock turbocharger, the turbine modelling methods outlined above calculate the aerodynamic performance, neglecting bearing friction and heat transfer effects. In addition, although the 3D geometry of the baseline turbine was based on the stock turbine design, it was recreated to allow parametrisation, leading to discrepancies in aerodynamic performance. The combined impact of these effects was found make little difference at the investigated steady-state operating points, causing a BSFC difference of 0.12% at 5000 RPM full load, and less than 0.05% for all other steady-state points.

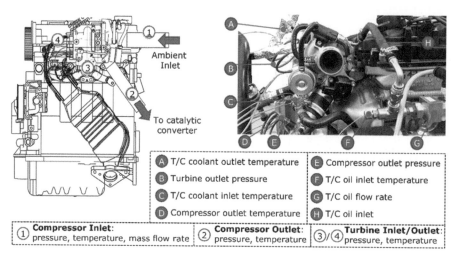

Figure 7. Scheme of gasoline engine test rig and numerical model in GT-Power.

The transient performance on the other hand was more significantly affected, with a difference of 0.6s in transient time to torque. All subsequent comparisons and optimisations are therefore made with respect to the meanline generated adiabatic map of the parametrised baseline geometry which was calibrated on the basis of CFD. To ensure the comparability of engine BSFC with different turbine designs, the steady-state model was set up with a torque targeting wastegate controller, so that the operating point could be defined by engine torque and speed.

Optimisation was performed using a latin hypercube sampling DOE of 1000 points, encompassing values of four turbine design parameters: r_2, A/r, r_{3t}, $d\theta/dz$. The parameters of the baseline design, and the parameter ranges used for the DOE sampling are shown in Table 1.

Table 1. Summary of parameter ranges used for engine optimisation.

Parameter	Baseline Design	Lower Limit	Upper Limit
r_2 [mm]	19.25	18	20
A/R [mm]	12	9	13
r_{3t} [mm]	16.15	15.15	18.15
$d\theta/dz$ [deg/mm]	7	5	9

The designs where implemented in the engine model at 5 different steady-state engine operating points (1500, 2000 & 5000 RPM at full load and 2000 and 3000 at 50% throttle), as well as a transient model at 1500 RPM.

Fuel consumption is generally measured over a drive cycle, however explicit modelling of this is computationally expensive. Simplification is possible by selecting operating points which are relevant to drive cycle operation, as was done by Rode et al. [14]. Consequently, evaluation of BSFC was done by implementing the turbine designs within the steady state engine model at 5 relevant engine operating points (1500, 2000 & 5000 RPM at full load and 2000 and 3000 at 50% throttle). Transient time to 100% torque was modelled at 1500 RPM.

4   RESULTS

The latin hypercube sampling of turbine designs was implemented to show the overall effect of the design parameters on the transient and steady-state performance of the engine. Figure 8 shows the results for each turbine design in the LHS, plotted on axes of relative ΔBSFC (at 2000 RPM full load) and transient response. A clear Pareto front of optimum designs emerges, showing the nature of the trade-off between the two objectives. To give more insight, three geometries were selected from the pareto front for the three criteria outlined in the introduction: eco driving, sustainable driving and sport driving (Geometries A, B and C respectively). Design selection was done for the 2000 RPM full load point, but analysis of the other steady-state points showed that the designs remain on, or very near to the pareto front.

The position of geometries A and C was set at -0.5% and 0.5% ΔBSFC on the pareto front. Design B was selected to represent an improvement to the baseline with respect to both ΔBSFC and transient response. The parameters of the selected geometries are shown in Table 2. As can be seen in Figure 8, Geometry B falls close to the baseline geometry which indicates that the baseline geometry is in fact well matched to the engine.

A comparison of percentage change in BSFC for each design and each operating point is shown in Figure 9, next to a chart comparing the transient time to torque results. It is evident that the operating point has an impact on the relative magnitude of BSFC variation, with the 5000 RPM full load point exhibiting changes of +/- 1.5% while the part load points show changes of less than +/-0.5%. In terms of transient response, Figure 8b shows the torque profile during the transient run, while Figure 9b shows the resulting time to 100% torque. Comparison of the torque profile displays a marked difference between the geometries after the initial torque jump due to throttle opening.

Table 2. Design parameters and inertia of selected turbine geometries.

Parameter	Design A	Design B	Design C
r_2 [mm]	20.2	19.0	18.8
A/R [mm]	11.9	11.4	9.9
r_{3t} [mm]	17.6	17.6	16.6
$d\theta/dz$ [deg/mm]	7.3	6.3	5.8
Inertia [kg mm^2]	9.0	7.2	6.8

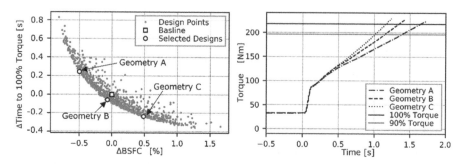

Figure 8. (a) All design points shown on axes of time to 100% torque and relative ΔBSFC with respect to the baseline. The baseline design and selected geometries are highlighted. (b) Transient torque profile of each selected geometry.

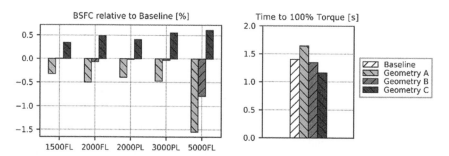

Figure 9. (a) Comparison of relative BSFC change for each selected design at each steady-state operating point. (b) Comparison of time to 100% torque for the baseline design and each selected design.

Figure 10 shows a comparison of speed lines of the three designs. It is evident, that when comparing Designs A, B and C, the mass flow parameter has the deciding impact in terms of aerodynamic performance. A lower MFP will result in a faster build-up of back pressure, meaning that the power available to the turbine increases more quickly, resulting in an improved transient response. At the same time, the higher back pressure of a turbine with lower MFP, will result in higher pumping losses, and thus have a negative impact on fuel economy. Another major factor impacting transient performance is the inertia of the rotor, which is mostly a function of rotor radius. Consequently lowering the rotor radius improves the transient performance by reducing both the mass flow parameter and the rotor inertia. This explains the differing slopes of the torque profiles seen in Figure 8b.

The efficiency map shows that the geometries selected from the pareto front have a higher efficiency than the baseline design, explaining why the baseline design does not lie on the pareto front. However, it is worth noting that the large difference in map efficiency between the baseline geometry and Geometry A (up to 5%), results in a relatively small change in overall engine performance.

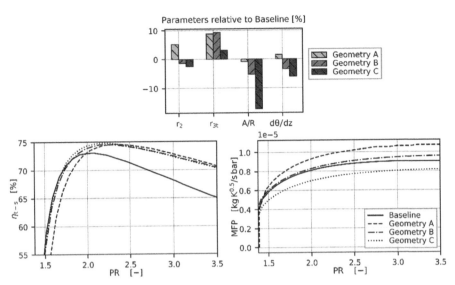

Figure 10. Maps of total to static efficiency and MFP for each design, at $n_{red} = 6546$ RPM/K^0.5.

Analysis of the sampling result also allowed the evaluation of individual geometric parameters independently. Figure 11 shows the impact of sweeping each design parameter, while keeping the others at baseline value.

It can be seen that the result for the rotor tip radius, determining size of the entire turbine, runs mostly parallel to the Pareto front, although it diverges towards the extremes. The A/r ratio line also runs parallel to the pareto front in the region of high transient response, but diverges as the fuel economy is improved. The impact of the exducer tip radius on the distance to the pareto front shows that it is an important parameter with regards to optimisation, which should generally be maximised. This is because increasing the exducer tip radius results in a higher efficiency at high power.

Overall it is evident, that modification of all the parameters together as opposed to sizing just one, can lead to a turbocharger turbine design for improved engine performance.

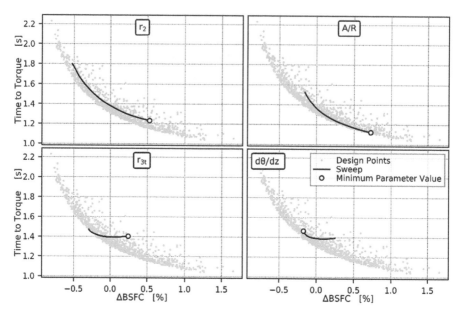

Figure 11. Impact of sweeping design parameters independently, with all other parameters held constant.

5   CONCLUSION

A methodology was developed for optimising turbocharger design for engine perform-ance, considering both engine transient response and fuel economy as competing objectives. The methodology utilises low order modelling methods for generation of turbine performance maps which are integrated into a 1D engine model in GT-Power, allowing optimisation of engine performance by the variation of turbine design param-eters. Two methods where analysed for generation of maps: a meanline model and an artificial neural network. A parametric turbine model was developed for the generation of turbine geometries as a function of meanline geometric parameters. This allowed a large sample of designs and operating points to be selected and run in CFD. In pur-suit of process automatisation and a reduction of user effort, the optimisation proced-ure was integrated into fully automated in-house code TORTOISE. The results of this were used on the one hand for calibrating the meanline model, and on the other for training of the artificial neural network. Both methods proved capable of predicting turbine performance as a function of design, however the meanline model code was more easily integrated with GT-Power, and was thus selected as the chosen method for the subsequent optimisation.

Maps for a latin hypercube sample of the design parameters r_2, A/r, r_{3t}, $d\theta/dz$ were generated and run within steady-state and transient models of a 1.6L turbocharged gasoline engine. The result confirmed that a compromise is needed between transient response and fuel economy. Three optimum designs, one with an improved transient response, one with an improved fuel economy, and one with an improvement in both objectives were selected and analysed. A total difference of 1% in fuel consumption between the two compromising designs, resulted in a 0.48s change in transient time to torque. Additionally, the method was used to show the performance impact of each design parameter independently.

NOMENCLATURE

A	Area [m^2]	n_{red}	Reduced Speed [m/s$\sqrt{K}$]
A/r	Ratio of cross sectional area to radius at scroll inlet [m]	p	Pressure [Pa]
MFP	Mass flow parameter [kg$\sqrt{K}$/s bar]	p_{01}	Total pressure at turbine inlet [Pa]
$BSFC$	Brake-specific fuel consumption [g/kWh]	p_3	Static pressure at outlet [Pa]
C	Absolute velocity [m/s]	r_b	Azimuthal cross section radius [m]
$d\theta/dz$	Camberline gradient at the exducer [deg/m]	r_2	Rotor inlet tip radius [m]
K	Constant	r_{3t}	Exducer tip radius [m]
$\dot{m}$	Mass flow rate [kg/s]	α, β	Inflow/outflow angle [rad]
U	Rotational Speed [m/s]	ρ	Density [kg/m^3]
U/C_{is}	Velocity ratio	η_{t-s}	Total-to-static isentropic efficiency [%]
PR	Pressure Ratio	Ψ	Azimuth angle [rad]
r	Radius [m]	θ	Angular coordinate [rad]
y^+	Dimensionless wall distance		
t	Time [s]		
x, y, z	Cartesian coordinates [m]		
Abbreviations			
ANN	Artificial Neural Network	MFP	Mass flow parameter
DOE	Design of experiment	OP	Operating point
FEM	Finite-element-method	RPM	Revolutions per minute
LHS	Latin Hypercube Sampling		

ACKNOWLEDGEMENT

The authors gratefully acknowledge Imperial College London and RWTH Aachen for supporting a scientific exchange between both research centres. Furthermore, the authors would like to thankfully acknowledge the Centre for Low Carbon Transport (LoCARtic) at Universiti Teknologi Malaysia and the Natural History Museum in London for providing their experimental facilities required for conduction of underlying research. The responsibility for the content of this publication lies with the corresponding authors.

REFERENCES

[1] Pierret, S., Van den Braembussche, R.A., "Turbomachinery Blade Design Using a Navier-Stokes Solver and Artificial Neural Network," *Journal of Turbomachinery*, vol. 121, no. 2, pp. 326–332, 1999, DOI: 10.1115/1.2841318.

[2] Pierret, S., Demeulenaere, A., Gouverneur, B., Hirsch, C., Van den Braembussche, R., "Designing Turbomachinery Blades with the Function Approximation Concept and the Navier-Stokes Equations," *8th Symposium on Multidisciplinary Analysis and Optimization*, 06–08 September 2000, DOI: 10.2514/6.2000-4879.

[3] Mueller, L., Alsalihi, Z., Verstraete, T., "Multidisciplinary Optimization of a Turbocharger Radial Turbine," *Journal of Turbomachinery*, vol. 135, no. 2, pp. 021022 (9 pages), 2013, DOI: 10.1115/1.40075078.

[4] Winkler, N., Ångström, H. E., "Study of Measured and Model Based Generated Turbine Performance Maps within a 1D Model of a Heavy-Duty Diesel Engine Operated During Transient Conditions," *SAE World Congress & Exhibition*, 16–19 April 2007, DOI: 10.4271/2007-01-0491.

[5] Chen, T., Zhuge, W., Zheng, X., Zhang, Y., He, Y., "Turbocharger design for a 1.8 Liter Turbocharged Gasoline Engine Using an Integrated Method," *ASME Turbo Expo 2009*, 08–12 June 2009, DOI: 10.1115/GT2009-59951.

[6] Zhuge, W., Zhang, Y., Zheng, X., Yang, M., He, Y., "Development of an advanced turbocharger simulation method for cycle simulation of turbocharged internal combustion engines," *Journal of Automobile Engineering*, vol. 223, pp. 661–672, 2009, DOI: 10.1243/09544070JAUTO975.

[7] Pesiridis, A., Salim, W.S-I., Martinez-Botas, R.F, "Turbcharger Matching Methodology for Improved Exhaust Energy Recovery," *10th International Conference on Turbocharging and Turbochargers*, 15–16 May 2012, DOI: 10.1533/9780857096135.4a.203.

[8] Gugau, M., Roclawski, H., "On the Design and Matching of Turbocharger Single Scroll Turbines for Pass Car Gasoline Engines," *Journal of Engineering for Gas Turbines and Power*, vol. 136, no. 12, p. 122602 (10 pages), 2014 DOI: 10.1115/1.4027710.

[9] Halamek, M., Hohenberg, K. G, Maeda, K., Lafossas, F. A., Newton, P. J., Martinez-Botas, R. F., "Development and Application of Physical Turbine Model for Multi-Objective Design Optimization," *Aufladetechnische Konferenz*, 20–21 September 2018.

[10] Whitfield, A., Baines, N.C., "Design of Radial Turbomaschines," *Longman Scientific & Technical*, 1990, ISBN: 0582495016.

[11] Hohenberg, K., Luczynski, P., Wirsum, M., Martinez-Botas, R., "Analytic Approach for a Prediction of Radial Turbine Performance Based on the Experimental and Numerical investigation," *In accepted for: ASME Turbo Expo 2020*, 22–26 June 2020, GT2020–15698.

[12] Wasserbauer, C. A., Glassman, A. J., "FORTRAN program for predicting off-design performance of radial-inflow turbines," *NASA Technical Paper*, 1975.

[13] Glassman, A. J., "Enhanced analysis and users manual for radial inflow turbine conceptual design code RTD," *NASA Contractor Report 195454*, 1995.

[14] Rode, M., Suzuki, T., Iosifidis, G., Durbiano, L., Filsinger, D., Starke, A., Starzmann, J., Kasprzyk, N., Bamba, T., "Boosting the Future with IHI: a comparative evaluation of state-of-the-art TGDI turbo concepts," *24. Aufladetechnische Konferenz*, 26–27 September 2019.

[15] "High Altitude Observatory of the National Center for Atmospheric Research," [Online]. Available: https://www.hao.ucar.edu/modeling/pikaia/pikaia.php#sec2. [Accessed 23 02 2020].

Maximum turbocharger efficiency for an engine operating at 50% brake thermal efficiency

P. Ananthakrishnan

Cummins Turbo Technologies – Performance team, UK

S. Egan, J. Archer

Cummins Turbo Technologies – Advanced Engineering, UK

T. Shipp

Cummins Research and Technology – Systems integration, USA

ABSTRACT

With the announcement of a reduction in CO2 emissions by 15% by 2025 and a further planned 15% reduction by 2030 in Europe, commercial vehicle manufacturers have become increasingly focused on reducing fuel consumption by improving overall engine efficiency. As a result, commercial vehicle manufacturers are turning more towards optimising the internal combustion engine, focussing on highly efficient air handling systems, in which the turbocharger plays a key role.

This paper will aim to show how a well matched, optimised turbocharger can impact the engines Brake Thermal Efficiency (BTE). Carefully matching and optimising the turbocharger enables an increase in both open cycle and closed cycle efficiency which are key factors for engine performance. When combined, the optimisation of the turbocharger compressor, turbine and bearing system along with thermal energy management allow for increased overall turbocharger performance which is required to achieve high levels of BTE.

1   INTRODUCTION

The heavy-duty line haul truck market has come under increasing pressure from the introduction of stringent emissions legislation over the last decade. These emissions legislations are set to get even tougher over the coming years with the European union announcing a 15% reduction in CO2 emissions by 2025 with a further 15% reduction in 2030, as shown by Figure 1. Similarly, in the US, diesel engines are coming under increased scrutiny through greenhouse gas emissions (GHG) reduction and low NOx standards developed under the California Air Resources Board (CARB) (1).

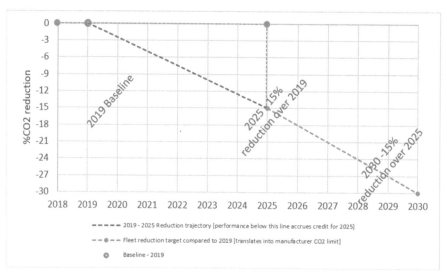

Figure 1. European Union CO$_2$ reduction guidelines for heavy duty truck market.

Despite the introduction of these legislations the turbocharged, diesel, internal combustion engine remains a critical resource for heavy the duty truck market and development must continue for the interim. Currently, a production heavy duty diesel engine has a brake thermal efficiency (BTE) between 42-46% (2); to achieve emissions targets the current BTE level must be improved upon. Cummins has been working in partnership with the United States Department of Energy (DOE) as part of a program called Supertruck II. Within the research program Cummins Research and Technology has committed to develop a new heavy-duty diesel engine targeted to achieve greater than 50% BTE at a target development point on the test cell without the use of waste heat recovery (WHR). This program will eventually aim to demonstrate 55% BTE at a single point in a vehicle system level with WHR, however, for this paper the 50% BTE demonstration point is in focus.

Cummins Research and Technology will focus on a system level optimisation where they will work closely with Cummins Turbo Technologies (CTT) to utilise a carefully matched, highly efficient turbocharger to help increase current BTE levels. The content of this paper will detail how matching and optimising the turbocharger while considering the turbocharger interaction with the engine can yield benefits at system level.

2   SYSTEM INTEGRATION

As discussed previously the target of 50% BTE was selected based on the project requirements. To achieve 50% BTE with an internal combustion I/C engine, system level optimization is necessary. The supporting systems must be mechanically capable and operate at peak or near peak efficiency. Brake thermal efficiency can be represented by:

$$\eta_{Thermal} = \eta_{Closed} \times \eta_{Open} \times \eta_{Mechanical}$$

These efficiencies can be used to evaluate the sub system performance and inform decisions. The combustion system was developed to reduce in-cylinder heat loss and achieve a high heat release rate. The lube oil system and base engine design have been optimized to increase mechanical efficiency while providing enough mechanical capability for enabling efficient operation. A highly efficient turbocharger was developed to provide the required airflow at a good open cycle efficiency and an exhaust manifold was developed to reduce heat loss, providing additional energy available to the turbine. A more detailed summary of the development work is documented in SAE paper 2019-01-0247.

A full break down of the engine efficiency is shown in Figure 2 below. This figure shows the factors affecting each contributor to the overall engine efficiency. In terms of turbocharging, it is fair to say that the turbocharger will have an impact on each contributor to engine efficiency which reinforces its importance at an overall system level.

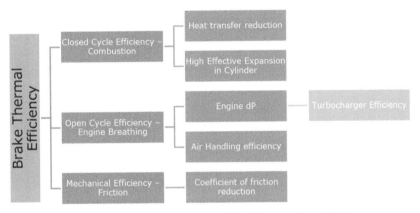

Figure 2. Breakdown of factors affecting engine efficiency.

2.1   Closed cycle efficiency

Closed cycle efficiency or combustion efficiency reflects the portion of the 4-stroke cycle when both the intake and exhaust valves are closed. Turbocharger efficiency affects the charge (air and EGR) delivered to the cylinder – a key enabler of high closed-cycle efficiency. Engine tests have shown good correlation between charge fuel ratio and close cycle efficiency. It is therefore important to ensure compressor and turbine stages are very efficient.

2.2   Open cycle efficiency

Open cycle or engine breathing efficiency is effectively a measure of how well the air will enter and exit the engine cylinder. It is in this area where a carefully matched and optimised turbocharger becomes a key enabler to help increase the engine BTE. For the turbocharger to aid open cycle efficiency, a balance must be struck between targeting a sufficiently high AFR while not increasing pumping work on the engine. In terms of stage efficiency, increasing the turbine stage efficiency will allow more exhaust gas energy to be extracted by the turbine which results in more power available for the compressor stage. Taking steps to reduce heat loss prior to the turbine will also increase the levels of energy in the exhaust gas at the turbine which will encourage a turbocharger work balance.

2.3   Mechanical efficiency

While the turbocharger will have minimal effect on system level mechanical efficiency, some options can be considered to allow for increased efficiency, for example reduction of oil flow through the turbocharger bearing system which is enabled through a variable lube pump on the engine.

3   INITIAL MATCHING

The selection of turbomachinery was dictated by the requirement for positive pumping on the engine – higher the overall machine efficiency, lower the pumping work done by the engine. The data for the 50% BTE demonstration point was used to select a compressor and turbine from the product catalogue. Consideration was also given to the other running points to ensure engine torque targets were achieved. The image below shows the map for the compressor selected from the product catalogue. The key demo point is in the heart of the map with an efficiency of ~81.5%.

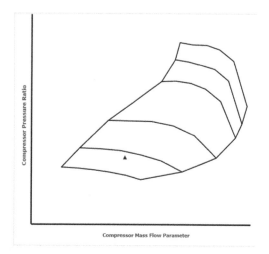

Figure 3. Running points on the compressor map.

The key drivers for engine pumping work when the exhaust valves are open are the turbine-stage and exhaust after-treatment back-pressure. The AFR target could be achieved using a relatively small turbine stage, however, the associated pumping work penalizes any gains in BTE from the closed-cycle (higher AFR). The over-riding requirement was, therefore, the delivery of air at the lowest restriction from the turbine stage, assuming a fixed back-pressure from the exhaust after-treatment at the key operating point. This fact also places a premium on the turbine stage efficiency.

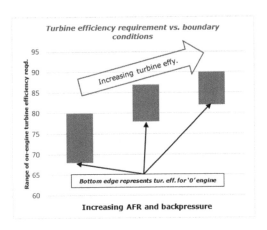

Figure 4. Turbine effy. requirement vs. boundary conditions.

Before carrying out the matching, the trade-off between turbine swallowing capacity, efficiency and engine backpressure were studied using a tool developed internally at CTT. This tool enables evaluation of 'what-if' scenarios associated with turbocharger boundary conditions such as after-treatment back-pressure, turbine inlet temperature etc. The tool iterates on both AFR and BSFC targets to derive an operating space or 'bubble' whose edges are defined by the boundary conditions at a specific engine running point. An example of the tool output for the 50% BTE point is shown below.

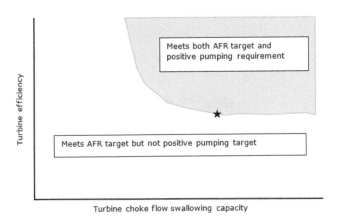

Figure 5. Example of a bubble plot.

The figure shows the turbine choked flow swallowing capacity on the X-axis and turbine efficiency on the Y-axis. The bottom edge of the bubble, in this case, is defined by the BSFC tolerance that is acceptable due to variation in turbine flow and efficiency. Critical boundary parameters show strong trends in the flow vs. efficiency space which helps with their identification. As EGR driveability was not a concern at the demo point, the region in the upper right corner in Figure 5 indicates the entitlement for turbine efficiency and flow. The lower edge of this boundary is defined by the '0 kPa' engine delta P line which represents the minimum for this parameter.

Engine delta P is notionally used as a measure of the pumping work in this case as there was insufficient data to quantify the change to PMEP of the engine. Based on the output from this tool, a few candidate turbine stages were evaluated using another internal tool to determine the turbo aero specification. The underlying principles of the matching tool are the power balance between compressor and turbine and convergence on an AFR target. This step resulted in the identification of 4 turbine stages that were evaluated in GT-Power to verify the performance at a system level.

The results from GT-Power showed that the best turbo match would require a 79mm compressor with a 74mm turbine from the product catalogue.

4   PATH TO TARGET

To plan out the path to target, careful analysis was carried out to consider the system level trade-offs. Predicted running conditions for the turbocharger were obtained by making educated assumptions as data for an engine of this size running over 50% BTE was not readily available.

4.1   Initial match

Figure 6. Turbocharger efficiency target for required AFR.

The initial turbocharger match had the expected on-engine performance, however the AFR did not meet the requirements to achieve 50% BTE. It was planned to take the catalogue turbocharger offering from CTT and apply efficiency improvements to achieve the AFR and PMEP targets required to achieve 50% BTE. The on-engine turbocharger performance confirmed the compressor and turbine match to be running near the peak efficiencies, which is a requirement to enable the engine BTE targets. The optimum solution was to increase turbine inlet temperature by minimizing heat loss between the engine cylinder and the turbocharger.

4.2 Heat transfer effects

During the matching phase, the turbocharger delta p required to achieve the required PMEP was predicted to be difficult to attain assuming reasonably high turbocharger efficiencies. Various system level options to reach the goal were considered. The optimum solution was to increase turbine inlet temperature by minimizing heat loss between the engine cylinder and the turbocharger. Analytical work comparing a non-insulated turbine housing versus an insulated turbine housing showed an apparent increase in compressor power for the insulated turbine housing.

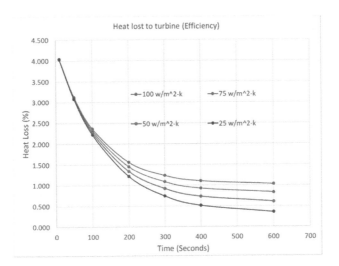

Figure 7. Decrease in heat loss from turbine housing with insulation.

The turbocharger turbine housing was designed to have low heat loss and low thermal inertia through careful design, material selection, and external insulation. A low heat transfer exhaust manifold has been designed to minimize heat loss between the engine and turbocharger.

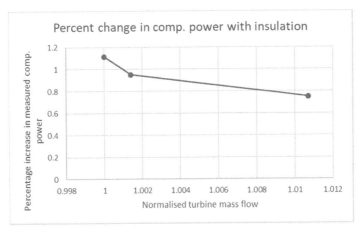

Figure 8. Increase in compressor power on gas stand with insulted turbine.

Back-to-back gas stand testing of a turbine stage with and without insulation showed an apparent increase in measured compressor work when the turbine stage was fully insulated. However, there was little confidence in the results due to measurement uncertainty in compressor outlet temperature due to ambient heat transfer on the gas stand. As the results from this test were directionally similar to the analytical work carried out earlier, the final prototype turbocharger would have an insulated turbine stage.

Figure 9. Sample thin-wall turbine housing without insulation.

5   TURBOCHARGER OPTIMISATION

To hit the turbocharger efficiency target, various technologies were employed across the turbomachine to boost its overall efficiency. As mentioned previously, boosting the turbine and compressor stage efficiency will allow the pumping work to be reduced but also the AFR target to be met, hence increasing the engines open and closed cycle efficiency; which are key factors for engine performance.

5.1   Compressor stage efficiency

From the product catalogue, a compressor stage was matched which achieved an ~81.5% efficiency value at the key demo point. To improve the compressor stage performance, the optimisation was mainly focused on compressor impeller tip clearance as well as surface friction improvements. To reduce the tip clearance an abradable coating was designed into the compressor housing whereby the impeller tip clearance was effectively reduced to zero. The abradable coating allows for cases of extreme rotor movement and avoids damage to the impeller in such conditions.

Regarding surface friction, the compressor volute surface finish was found to be one of the main parameters affecting surface friction in the compressor stage. Improvements were made to the volute surface finish through a fluid honing process whereby casting irregularities are removed from the volute. Previous work by Javed, A., and Kamphues, E. has illustrated the effects of improving surface finish on the compressor volute (3). A combination of both technologies resulted in an approximate 1.5-2% point efficiency delta over the baseline compressor stage, highlighted in the figure below.

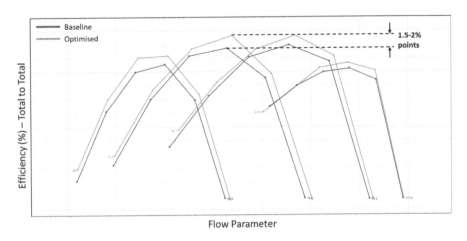

Figure 10. Compressor performance improvement on gas stand.

5.2 Bearing efficiency

While CTT's floating ring bearing (FRB) design offers high levels of efficiency, the util-isation of a rolling element bearing (REB) in the design of the turbomachine unlocks certain benefits, which would not be possible with an FRB, as well as a slight increase in turbomachine efficiency. The use of a REB contributes in total an approximate gain of 1% point turbomachine efficiency due to decreased bearing loss. Using a REB allows for a reduction in oil flow which has benefits for mechanical efficiency on the engine. An REB's range of radial movement is also less than an FRB system which has benefits for the turbine. The benefits of using rolling element bearings for turbo-chargers is widely documented in literature like (4)

5.3 Turbine stage efficiency

The turbine stage from the product catalogue achieved an efficiency value of ~80% at the key engine demo point. Like the compressor, the optimisation of the turbine stage was mainly focused around wheel tip clearance as well as surface friction improvements. To reduce the wheel tip clearance, the REB was utilised. Due to the low radial movement of the REB the clearance between the wheel and housing could be reduced significantly after considering rotor dynamics and thermal affects, hence reducing tip leakage.

The turbine housing volute surface finish proved to be a critical parameter in reducing the surface friction of the turbine stage. Like the compressor, improvements were made to the volute surface finish via a fluid honing process which removed any casting irregularities from the volute. A significant change in surface finish was noted between the production and newly fluid honed turbine housings. In total an increase of just over 4.5% points was noted on the turbine stage efficiency from the improvements and careful specification of machining tolerances.

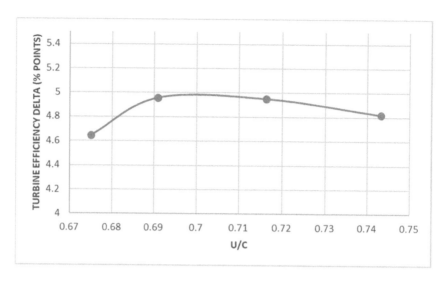

Figure 11. Turbine efficiency delta (from gas stand) above the baseline unit.

5.4 Cumulative turbocharger efficiency

With each of the individual turbocharger stages now optimised, a total increase in turbocharger efficiency could then be approximated. Shown below is an efficiency walk of the total increase in turbocharger efficiency from the added technologies:

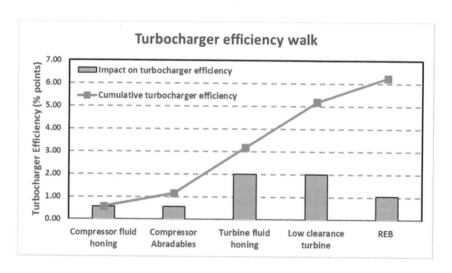

Figure 12. Turbocharger efficiency walk.

These results were used as an indication of what the turbocharger could offer the system in terms of BTE. It was estimated that the turbocharger efficiency increase would be ~6.24% points which would give a total turbocharger efficiency of ~69.7% points.

6   ENGINE EFFICIENCY

The on engine BTE comparison between the baseline and optimized turbocharger was not possible due to combustion system changes on the engine between the two evaluations. However, conclusions about the experimental turbocharger performance can be made as all parameters impacting engine breathing were kept constant. The optimized turbocharger demonstrated on-engine turbocharger efficiency of 69.5% at the 50% demonstration point. Figure 13 shows the relationship between turbocharger efficiency and AFR.

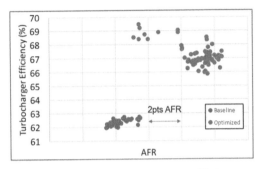

Figure 13. Experimental turbocharger efficiency vs. AFR.

Figure 14 presents a CCE to AFR relationships to assign a BTE improvement to the optimized turbocharger.

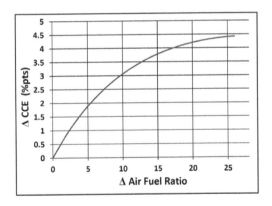

Figure 14. Delta Closed Cycle Efficiency (CCE) vs Delta Air Fuel Ratio (AFR).

The experimental AFR ran in a range on the relationship in Figure 14 where a 5 point improvement in AFR results in 0.5% point increase in closed cycle efficiency. The open cycle efficiency was kept almost constant between the two turbochargers. This is expected, as turbine swallowing capacity is kept constant, the turbocharger efficiency improvement materializes as increased airflow keeping open cycle efficiency constant. This is especially challenging given the fact that turbine inlet temperature decreases as AFR increases which makes a high turbine stage efficiency mandatory.

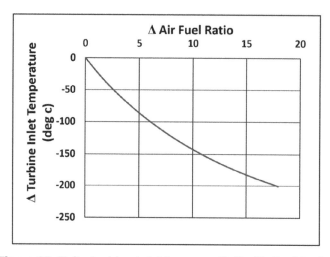

Figure 15. Delta turbine inlet temp. vs. Delta Air Fuel Ratio.

The engine BTE was improved by 0.32%pts with the addition of the high efficiency turbocharger as shown in Figure 15. Overall, the optimized exhaust manifold and the high efficiency turbocharger combined to achieve a 0.69%pts BTE improvement.(5).

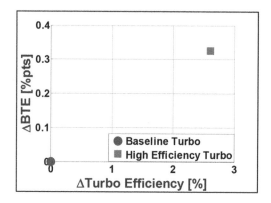

Figure 16. Experimental turbocharger efficiency vs. AFR.

7   CONCLUSION

From the process outlined in this paper system level interactions are key when matching and optimising a turbocharger for a specific BTE target. The effects of the air handling system on the contributors to engine efficiency are clearly understood from the outset, resulting in the selection of an optimal turbocharger for the demonstration point. The initial match ran on the peak efficiency islands for both the compressor and turbine stage and further optimisation of the turbocharger enabled the engine to meet its overall AFR target. The predictions for the estimated turbocharger efficiency are largely in line with the observed efficiency on the engine and indeed the expected levels of BTE increase.

REFERENCES

[1] *The Future of Heavy-Duty Vehicle Fuel Efficiency.* Khan, S. Chicago: 9th Integer Emissions Summit & DEF Forum, 2016.

[2] Thiruvengadam, Arvind, et al. *Heavy-Duty Vehicle Diesel Engine Efficiency Evaluation and Energy Audit.* s.l.: Center for Alternative Fuels, Engines & Emissions West Virginia University, 2014.

[3] *Evaluation of the Influence of Volute Roughness on Turbocharger Compressor Performance From a Manufacturing Perspective.* Javed, A., and Kamphues, E. Düsseldorf, Germany: ASME, 2014. Proceedings of the ASME Turbo Expo 2014: Turbine Technical Conference and Exposition. Volume 2D: Turbomachinery. V02DT42A040.

[4] *Rolling bearings in turbocharger applications.* Schweitzer, F., Adleff, K. s.l.: MTZ Worldw 67, 16–19, 2006.

[5] *The thermodynamic design, analysis and test of Cummins' Supertruck 2 50% brake thermal efficiency engine system.* Shipp, Timothy, Mohr, Daniel and Lu, Xueting. s.l.: SAE International, 2019. SAE 2019-01-0247.

Transfer of turbocharger shaft motion behavior from engine testing to hot gas component testing

F. Falke, A. Schloßhauer, H. Ruppert

Institute for Combustion Engines RWTH Aachen University, Germany

M. Stadermann, D. Lückmann, R. Aymanns

FEV Europe GmbH, Germany

ABSTRACT

Throughout the last decade, turbocharging the internal combustion engine has become the state-of-the-art technology when it comes to increasing power density and fuel economy of a given engine.

The investigation of the rotor dynamics is an important part in the turbocharger development process. This is especially true when it comes to advanced concepts such as new bearing technologies or matchings between turbine and compressor side as required for example for an electrically assisted turbocharger (eTC). Hence, the quantification and investigation of mechanical effects on the turbocharger (TC) and its components become more and more important in order to ensure long durability and high performance.

For the corresponding investigations, it is important to ensure realistic boundary conditions such as testing in combination with the original engine application on an engine test bench. However, in an early engine development phase, the latter might not yet be available or the desired measurement equipment can´t be installed due to limitations in available space. Before shifting such mechanical investigations on a component hot gas test bench, the impact of the engine boundary conditions on the results must be analysed.

Against this backdrop, the paper firstly covers the analysis of the TC shaft motion measured on-engine in different operating points. Thereto, the turbocharger of an 1.0 l three-cylinder spark ignited engine fuelled with compressed natural gas (CNG), has been equipped with shaft motion (displacement) sensors in order to quantify the relative movement of the turbocharger rotor assembly. While running the engine in most relevant operating points, possible effects of the exhaust pressure pulses upstream of the turbine on the rotor dynamics are analysed.

In a second step, the same investigations regarding the TC shaft movements are carried out on a component hot gas test bench using the identical turbocharger hardware. For these tests, a newly developed pulsation unit, the FEV PulseGen, is being installed between the stationary combustion chamber (exhaust gas generator) and the turbocharger itself. Upstream of the turbine, this unit enables the generation of exhaust gas pulsations that are realistic for real engine operation. For the first time a comparison of both measurement approaches is shown. This comparison has been carried out for the same engine like operating points which makes it possible to analyse the impact of the exhaust gas pulsations on the turbocharger shaft motion characteristics.

1   INTRODUCTION TO TURBOCHARGER ROTORDYNAMICS

The increasing electrification of powertrains has opened up new degrees of freedom for operating strategies of combustion engines. The optimum operating strategy in terms of efficiency and pollutant emissions can be determined based on driving cycle simulations [1]. In the next step, the development requirements for the turbocharger can be derived from its operating strategy. These requirements can generally be divided into three areas, which must be fulfilled with a suitable turbocharger design. First, there is the aerodynamic performance, i.e. the optimization to the highest possible turbocharger efficiency considering the interaction with the combustion engine. The heat flows in the turbocharger must also be covered. Important factors to consider here are the catalyst heating of the exhaust aftertreatment system and the temperature cycles in the bearing system.

Particularly in hybrid drive trains, there are very different requirements, so that the design must be questioned for each application. This paper deals with analysis possibilities for the third optimization area: the mechanical requirements regarding durability and NVH. Regarding this comprehensive task, the shaft motion of the TC rotor assembly will be in the focus.

From the ratio of unbalance and weight of the rotor, the properties of the rotor movement can be assumed. In the case of turbochargers for passenger car engines, due to the high speeds (up to n_{TC} = 300.000 rpm) and low rotor weight, the ratio is higher than a value of 100. From the dominance of the unbalance, it can be concluded that the rotor will move between the bearing center and the bearing wall primarily as a function of the rotor speed. In combination with the non-linear stiffness and damping behavior of the bearing as a function of the rotor eccentricity, a complex form of motion is produced. The investigations in this paper are performed with a fully floating ring bearing. Figure 1 gives a schematic overview of the bearing, highlighting the inner and outer oil films. The benefits in comparison with a semi floating bearing (locked floating ring) are the increased damping effects and the lower friction losses.

The following factors determine the shaft motion of a turbocharger rotor assembly with a fully floating ring bearing design:

- Not perfectly balanced rotor assembly: the remaining residual unbalance leads to centrifugal forces. These forces vary quadratically with the rotor speed. The resulting noise is also known as unbalancing whistle. This harmonic vibration has the same frequency as the rotor frequency (1st order)
- Self-excited vibration caused by the oil film in the turbocharger bearing: the oil whirl can lead to subsynchronous rotor vibrations with irrational frequency orders of rotor speed (0.25th to 0.7th order for the inner and 0.1th to 0.3th order for the outer oil film). The whirl of the outer oil film is responsible for the constant tone noise. It depends on many parameters of the rotor and bearing design as well as on the oil temperature [2].
- Vibration modes of the rotor assembly excited by the above-mentioned forces: the frequency depends primarily on the rotor geometry. For a small automotive TC, the 1st eigenmode (Conical mode) can be expected at 200-300 Hz. This is mostly excited by the outer lubricating film, recognizable by the low frequency at a low order. Frequency is then usually 0.5 times ring speed. The 2nd eigenmode is a purely cylindrical movement, but often superimposed with a bending at approx. 500 Hz.

205

Often excited by the inner lubricating film, because the excitation frequency is 0.5 times (TC Speed + Ring Speed). A pure bending is expected at approx. 3000 Hz and therefore excited by the 1st order. While for low TC speeds the ring speed plays an important role, for higher TC speeds the lateral bending vibration of the TC shaft becomes dominant [2].

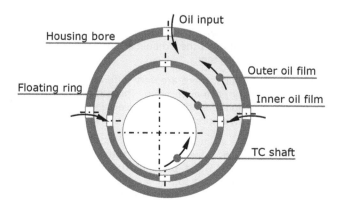

Figure 1. Schematic cross section of a fully floating ring bearing.

More detailed information on the subject can be found in [3] and [4].

2   FREQUENCY ANALYSIS TO EVALUATE ROTOR DYNAMICS

In order to investigate the TC shaft motion characteristics, the measurement signals of the shaft motion sensors have to be post-processed. In terms of oscillation phenomena, generally it can be distinguished between a time domain and a frequency domain. The link between the time domain and the frequency domain can be done by performing a Fast Fourier Transformation (FFT). Here it is important to carefully select the mathematical boundary conditions in order to maintain the data content. For the investigations within this publication the data post-processing has been done using the software ArtemiS from HEAD acoustics GmbH. With a sampling rate of 100 kHz, a spectrum size of $2^{14} = 16384$ has been realized (Hanning window with 81% overlap, $\Delta f = 6.1$ Hz, $\Delta t = 31$ ms). In Figure 2 the three previously mentioned main phenomena of the TC shaft movement (unbalance, inner & outer oil whirl) are theoretical described in the time domain (upper graph) and the frequency domain (lower graph). According to the literature, the highest amplitudes may be found at lower frequencies. Due to influential parameters such as oil pressure, oil temperature, oil type, kind of materials, clearances, geometry and bearing type, usually only frequency ranges are given in which the oscillation phenomena may occur. Since the amplitudes at different frequencies cannot just be summed up in order to calculate the maximum amplitude (the phasing has to be considered), the resulting maximum amplitude cannot be plotted into the graph. According to [5] usually less than 50% of the bearing clearance is utilized throughout operation.

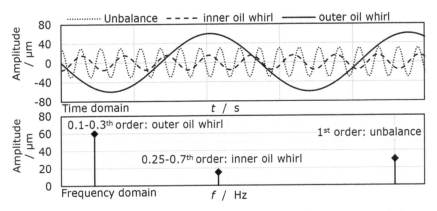

Figure 2. Frequency analysis of the TC shaft motion phenomena: unbalance, inner oil whirl and outer oil whirl (theoretical example).

3   EXPERIMENTAL SETUP

For the experimental investigations, the identical turbocharger hardware of a 1.0 l three-cylinder spark ignited engine (fuelled with CNG) has been used on all three test bench configurations. The TC is a series production sample with a fully floating bearing system. After installation of the shaft motion sensors as depicted in Figure 3, the compressor housing has not been moved to ensure best comparability. The shaft motion sensors have been calibrated prior to testing using the hardware's compressor nut. Throughout the measurements the sensor electronics have been synchronised in order to eliminate mutual interference. The sampling rate for all sensors has been set to 100 kHz for each operating point. For the measurements performed on the engine test bench, the original oil supply of the engine has been used. According to the oil boundary conditions of the engine test bench (oil pressure at 4 bar (abs) and oil temperature at 90°C), the same values have been controlled on the hot gas test bench and the setup with the pulsation unit (see chapter 3.3).

Figure 3. Installation of two shaft motion sensors in the compressor housing perpendicular to the shaft with 90° offset.

3.1 Engine test bench setup

The shaft motion measurements of the TC in combination with its original combustion engine has been carried out on a conventional engine test bench (see Figure 4). The engine speed n_E and load *BMEP* have been controlled in order to cover the most relevant operating points in the engine map like the full load curve and additionally load variations in $\Delta BMEP$ = 2.5 bar steps at constant engine speed n_E = 2000 rpm & 4000 rpm. All other parameters, especially related to the TC control, have not been manipulated to ensure most realistic engine boundary conditions for the TC and thus the shaft motion signals. In addition to the steady state operation points (temperatures used as convergence criteria) engine (respectively TC) speed sweeps at 100% load from n_E = 1000-6000 rpm have been performed.

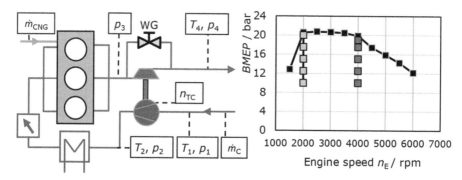

Figure 4. Setup to investigate the turbocharger shaft motion on the engine test bench with the original internal combustion engine.

Besides the transient data recording for the speed sweep for each operating point, the shaft motion signal has been captured for at least five seconds at a sampling rate of 100 kHz. As shown in Figure 4, temperature and pressure values for calculating the TC operating conditions have been monitored and recorded. Additionally, the mass flow rates of fresh air and fuel (CNG) are known. With these information, the compressor operating point can be calculated together with the TC shaft speed n_{TC}. These operating points in the corresponding compressor map will be controlled on the hot gas test bench in order to compare the resulting TC shaft motion.

3.2 Turbocharger hot gas test bench setup

As a second step, the same TC shaft motion investigations have been carried out on a stationary hot gas test bench for turbochargers. The setup of the applied test bench is shown in Figure 5.

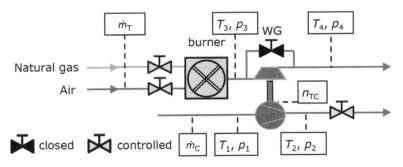

Figure 5. Setup to investigate the turbocharger shaft motion on the stationary hot gas test bench.

Here, the turbine of the TC is driven by the exhaust gases of a steady-state combustion chamber instead of the internal combustion engine. The combustion chamber burns natural gas in compressed fresh air and is connected to the turbine volute inlet with a measurement pipe used for temperature and pressure measurements. While the wastegate is closed, the exhaust mass flow through the turbine wheel is known by the amount of natural gas plus the amount of fresh air measured upstream of the combustion chamber. On the compressor side the compressor sucks in the air from the test bench environment. In the flow path upstream of the compressor, a mass flow meter and measurement pipe for temperature and pressure is installed. Downstream of the compressor is the measurement pipe for the outlet conditions of the compressor before a throttle valve which controls the backpressure and thereby the operating point of the turbocharger.

To find the corresponding operating points of the on-engine tests (see chapter 3.1), the turbocharger speed n_{TC}, the compressor pressure ratio Π_C and the mass flow rate $\dot{m}_C$ are calculated and then controlled on the hot gas test bench via turbine mass flow rate (for n_{TC}) and compressor back pressure valve (for Π_C and $\dot{m}_C$). The turbine inlet temperature has been kept constant at $T_3 = 600°C$. The impact of the wastegate position on the TC shaft motion is assumed to be negligible small. Therefore the wastegate is mechanically locked in closed position. For each operating point, the steady-state conditions of temperature, pressure and mass flow rate are checked before triggering the shaft motion measurements for at least five seconds at 100 kHz sampling rate.

3.3   Hot gas test bench setup with pulsation unit
In the last step of the investigations, a newly developed pulsation unit (*FEV PulseGen*) is installed between the steady-state combustion chamber and the turbine inlet (see Figure 6). This unit enables to transfer the constant pressure upstream of the turbine p_3 = const. into engine realistic exhaust gas pulsations $p_3 = f(t)$ comparable to the original engine application. Setup and design of this pulsation unit are covered in more detail in [6] and are not part of this publication.

However, one advantage of such a setup is the possibility to control the turbo-charger operating point such as speed n_{TC} or pressure ratio Π_C and mass flow capacity $\dot{m}_C$ independent of the pulsating turbine admission $p_3 = f(t)$. Besides the additional unit, the overall setup of the turbocharger and measurement equipment is the same as described in chapter 3.2.

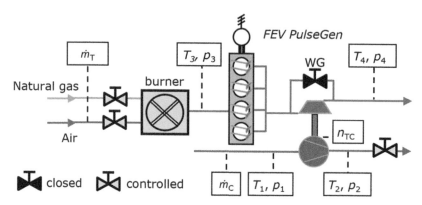

Figure 6. Setup to investigate the turbocharger shaft motion on a gas stand in combination with a pulsation unit.

4   EXPERIMENTAL RESULTS

The following chapter presents the results of the measurements conducted on the three different test bench setups.

4.1   Experimental results from engine & stationary hot gas test bench

The dynamic measurements with the motion sensors, see Figure 3, were performed at stationary conditions and for a transient TC speed range on the TC hot gas test bench and on the entire engine assembly. Since the amplitudes for both x and y directions were very similar, only the x axis is shown in the following graphs.

The linear wave motion amplitudes are shown in Figure 7 for five load levels from $BMEP = 10$ to 19 bar for $n_E = 4000$ rpm for the hot gas test bench and the engine test bench. Since the measurement results for load steps at $n_E = 2000$ rpm have been similar to the plotted results, the focus has been set to $n_E = 4000$ rpm only.

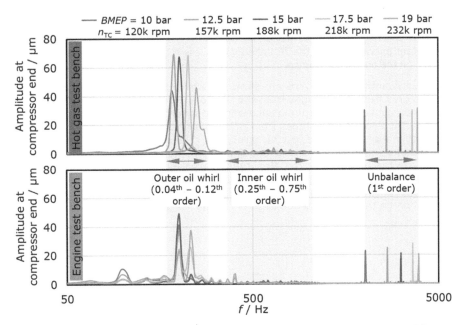

Figure 7. Narrow band amplitude of shaft motion at compressor end for steady state load variations from *BMEP* = 10-19 bar at n_E = 4000 rpm on stat. TC hot gas test bench and engine test bench in frequency domain.

Three characteristic order ranges can be distinguished in grey for the logarithmic frequency range shown from 50 to 5000 Hz:

Outer oil whirl: 0.04^{th} to 0.12^{th} order of n_{TC}
Inner oil whirl: 0.25^{th} to 0.75^{th} order of n_{TC}
Unbalance: 1st order of n_{TC}

For this specific TC with fully-floating bearing, both the outer oil whirl and the unbalance are the dominant phenomena. As shown in Figure 8, on both test rigs the amplitude is greatest by the outer oil whirl, followed by the unbalance and finally by the inner oil whirl.

The amplitudes of the outer oil whirl are smaller with ~30 to ~50 µm at the engine test bench than with ~40 to ~70 µm at the hot gas test bench. This difference can be explained by a theoretical comparison: on the hot gas test bench, the wave motion for the lower frequencies is determined exclusively by the self-excited outer oil whirl, which changes into very high amplitudes. In contrast to the TC hot gas test bench, on the engine test bench, additional structural vibration harmonics are introduced in the same frequency range by unbalanced free mass moments and ignition orders of the engine. This leads to a destructive influence on the self-excited outer oil whirl and thus to lower amplitudes on the engine test bench.

The inner oil whirl is hardly present for both setups and is therefore not part of further analyses.

The amplitudes of the unbalance order remain largely constant over the various operating conditions. Similar to the outer oil whirl, the amplitudes on the engine test bench appear to be lower. This is not the case, but is due to slight engine speed changes and thus TC speed changes. The frequency spectrum is slightly widened. However, for a constant operating point the TC speed varies less than 2 % even for the tests conducted on the engine test bench.

Detailed quantitative analyses are shown in Figure 8: the sum acceleration level for the grey frequency ranges from the previous Figure 9 is evaluated here. Both the amplitudes caused by unbalance and the inner oil whirl remain largely constant in the various load stages from $BMEP$ = 10 - 19 bar. Both effects are therefore not influenced by the test rig structures. The resulting energy level in both phenomena (inner oil whirl and unbalance) are not affected by the test bench setup. On the other hand, the amplitude caused by the outer oil whirl in the TC hot gas test bench is greatly increased compared to the engine test bench (~70 µm compared to ~50 µm). This can be explained by the undisturbed self-excited oscillation on the hot gas test bench as mentioned above.

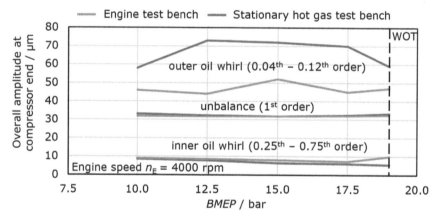

Figure 8. Broad band amplitude of shaft motion at compressor end for 0.04th to 0.12th order (outer oil whirl), 0.25th to 0.75th order (inner oil whirl) and 1st order (unbalance) for load variations $BMEP$ = 10 - 19 bar at constant engine speed n_E = 4000 rpm.

The effects of the motor arrangements on the shaft motion are further analysed with transient TC speed sweeps at the engine test bench and at the stationary hot gas test bench as shown in Figure 9. The TC speed ranges from n_{TC} = 180000 – 215000 rpm. The TC speed ramp on the engine test bench is realized by transient engine operation.

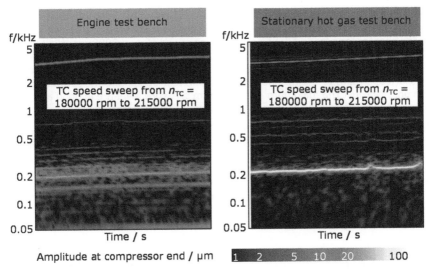

Figure 9. Amplitude of shaft motion at compressor end for transient TC speed sweep at full load from n_{TC} = 180000 rpm to 215000 rpm on engine test bench and on TC hot gas test bench.

The engine orders are clearly visible in the frequency spectrum on the left in Figure 9. Overall the agreement for the outer oil whirl frequency and amplitude is well. The first engine order below and above the outer oil whirl contributes significantly to the overall level in this frequency range. In comparison on the right, the frequency spectrum of the outer oil whirl on the TC hot gas test bench is very narrow and additionally more pronounced than on the left. As expected, there are slight differences in terms of frequency and magnitude, which can be explained by various effects:

- Due to the additional structural excitation by the engine vibrations the stiffness of the oil bearing can be altered leading to shifted frequencies of the outer-oil-whirl.
- Most importantly the presence of multiple structural engine order excitations disturb the outer oil whirl, because the exciting and the eigenfrequencies are close, but do not match exactly. Therefore the outer oil whirl mode is suppressed by the engine's unbalanced mass forces. Whether engine order excitations can lead to higher amplitudes still has to be tested with different test rig constructions and sufficiently strong vibration excitation: if an external vibration with the correct frequency (= eigenfrequency of the oil whirl) and the correct phase is used, the so called resonance catastrophe can occur.

However, further work is required to understand if this is a general phenomenon. In summary the agreement between both measurement setups is well in terms of frequency and magnitude. Altogether, the hot gas test bench seems to deliver a worst case scenario with regard to NVH and mechanical development with regard to the vortex phenomena of the outer oil.

4.2 First experimental results of tests with pulsation unit

In a last step, the shaft motion investigations have been performed for selected engine operating points on the hot gas test bench in combination with the newly developed pulsation unit. For the first time, a comparison between all three test bench configurations (see Figure 4, Figure 5 and Figure 6) is possible and shown in the following chapter for a load variation at an engine speed of n_E = 2000 rpm ($BMEP$ = 10 to 20 bar). To match the turbine pressure ratio with the results from the engine tests, an exhaust back pressure valve has been used. Figure 10 shows the comparison of the pressure upstream of the turbine for the engine test bench and the setup with pulsation unit. Not just the maximum and minimum pressure level are in good agreement, but also the rising and falling pressure gradients are well matching.

Considering the same compressor and turbine pressure ratios for the operating points on the hot gas test bench with pulsation unit and the engine tests itself, the axial thrust forces on the bearing system are assumed to be very much the same. This is beneficial for the transfer of investigations from the engine test bench on the component test bench with a pulsation unit. In contrast to that, the thrust forces on the stationary hot gas test bench without the engine-like exhaust pulses will differ to the investigations on the engine test bench [7].

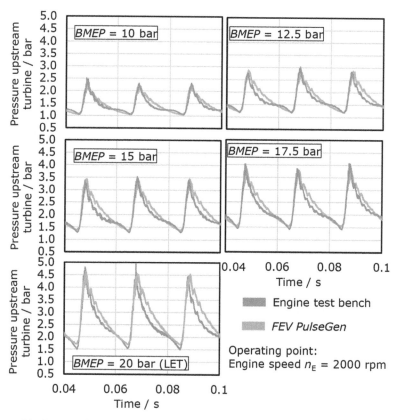

Figure 10. Comparison of the pressure pulse upstream of the turbine: engine test bench vs. pulsation unit at engine speed n_E = 2000 rpm.

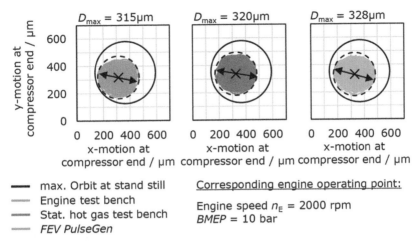

max. Orbit at stand still
Engine test bench
Stat. hot gas test bench
FEV PulseGen

Corresponding engine operating point:

Engine speed n_E = 2000 rpm
BMEP = 10 bar

Figure 11. Shaft motion at compressor end in *x*- and *y*-direction including max. orbit at standstill for *BMEP* = 10 bar at engine speed n_E = 2000 rpm on the engine test bench, hot gas test bench and with the pulsation unit.

In Figure 11, the orbit trajectories are shown as *x*- and *y*-motions of the compressor end throughout operation of the TC. Additionally, the trajectories at stand-still with washed bearing are plotted as reference for the maximum possible motion. The latter trajectories have been measured prior to all tests after removing the oil films within the bearing system with the help of petroleum.

On all three test benches, almost the same shift of the centre of rotation in south-west direction can be noticed. The trajectory itself shows the same overall shape which indicates similar oscillation modes on all three test benches. This behaviour will be further analysed for the three main oscillation phenomena (inner & outer oil whirl and unbalance) in the following section. Additionally to the dislocation of the center of the shaft and the overall shape, also the maximum distance between two measurement points D_{max} is comparable. With D_{max} = 315 µm, the maximum shaft motion is the smallest on the engine test bench which again might be explained by the stabilizing impact of the sub-excitations by the engine movement (crank shaft rotation and resulting free forces and moments) on the shaft motion amplitudes.

To gain a better understanding of the different oscillating modes of the TC shaft motion, the three phenomena inner & outer oil whirl and unbalance will be further analysed. Thereto, the corresponding amplitudes of the shaft motion at compressor end are shown in Figure 12 as overall level for the same frequency ranges as described in chapter 4.1.

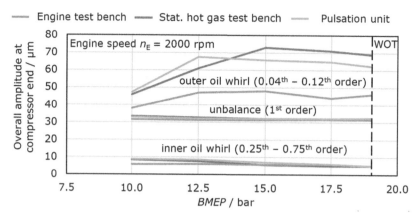

Figure 12. Amplitude of shaft motion at compressor end for 0.04th to 0.12th order (outer oil whirl), 0.25th to 0.75th order (inner oil whirl) and 1st order (unbalance) for load variations *BMEP* = 10 to 19 bar at engine speed n_E = 2000 rpm on the engine test bench, hot gas test bench and with the pulsation unit.

It becomes obvious that the amplitudes for the 1st order (unbalance) are very similar although this is mainly due to the matching TC speed n_{TC} on all three test benches. The TC speed varies less than 2 % for the tests with pulsation unit, which therefore shows the same behaviour as for the tests on the engine. Again, the inner oil whirl phenomena are of minor importance due to the small amplitudes, but still on a comparable level. Especially the results of the TC hot gas test bench and the test bench in combination with the pulsation unit are almost identical. The outer oil whirl phenomena again shows the largest amplitudes on all test benches, although the result on the engine test bench is the smallest. This also proofs the results of the maximum shaft motion on the engine test bench which has been shown in Figure 11 already. In general, the comparison between the test benches shows promising results in terms of transferability. For high loads, where the exhaust gas pulsations upstream of the turbine are assumed to play a more important role in the rotor dynamics, the comparability between engine tests and the pulsation unit is even higher than for low loads.

5   SUMMARY AND OUTLOOK

It has been shown that the shaft motion at the compressor end is dominated by the rotor dynamics such as 1st order unbalance and inner and outer oil whirls. The highest amplitudes can be found at lower frequencies which are characteristic for the outer oil whirl phenomena. Compared to the rotor dynamics, the shaft motion caused by the exhaust gas pulsations and the internal combustion engine movement do not play a significant role (approx. 20% of the max. amplitude of the outer oil whirl).

The measurements of the TC shaft motion on the hot gas test bench show discrete amplitude peaks for the outer oil whirl phenomena (0.04th to 0.12th order of TC speed). For the same operating points at engine speed n_E = 4000 rpm measured on the engine test bench, it has been shown that these amplitude peaks or no longer that distinctive but more synchronized with ambient frequencies. Additionally, the amplitudes are at a lower level on the engine test bench

216

(e.g. for *BMEP* = 17.5 bar, the amplitudes are reduced about approx. 35 % from 70 µm to 25 µm). The synchronization and the resulting lower amplitudes on the engine test bench have been explained by the additional sub-excitations caused by the engine movement (free mass moments and ignition orders). For a better separation of rotor dynamics and engine vibration, further work will be required with special test rigs (e.g. shaker tests). Assuming an operation beyond any resonance frequencies, the higher amplitudes on the hot gas test bench may be used to produce worst-case scenarios for the TC shaft motion. Besides higher amplitudes, it is also possible to have better separation of the different oscillation modes on the hot gas test bench without sub-excitations caused by the internal combustion engine.

Finally, this publication shows for the first time a comparison of TC shaft motion measurements on an engine test bench and on a hot gas test bench with and without the newly developed gas pulsation unit, the *FEV PulseGen*. The comparison between the different test setups shows promising results in terms of transferability. To reproduce worst-case scenarios with high amplitude TC shaft motion, shifting investigations from the engine test bench to a turbocharger hot gas test bench provides advantages in terms of costs, time and accessibility regarding sensor installation, while having the most realistic engine boundary conditions applicable at the TC at the same time.

The shown results of the phenomena of TC shaft motion are a good basis to start further research based on this publication, e.g. investigations with different bearing designs.

NOMENCLATURE

		Subscripts	
n	Rotational speed (rpm)	E	Engine
x/y	Direction of shaft motion at compressor end (µm)	C	Compressor
		T	Turbine
f	Frequency (Hz)	1	Compressor inlet
D_{max}	Max. shaft motion diameter	2	Compressor outlet
t	Time (s)	3	Turbine inlet
Π	Pressure ratio	4	Turbine outlet
BMEP	Break mean effective pressure (bar)	**Abbreviations**	
$\dot{m}$	Mass flow rate (kg/s)	WG	Wastegate
		TC	Turbocharger
		PG	Pulsation unit
		CNG	Compressed natural gas
		NVH	Noise, Vibration, Harshness
		FFT	Fast Fourier transformation

REFERENCES

[1] D. Lückmann, A. Schloßhauer, A. Müller, K. Kannan, R. Wohlberg, S. Yadlaa, A. Balazs, T. Uhlmann and M. Thewes, "Assessment of an Electrified Boosting System to Increase the Efficiency of a Hybrid Powertrain by Exhaust Energy Recovery," in *22. Aufladetechnische Konferenz*, Dresden, 2017.

[2] S. Pischinger, H. Stoffels, C. Steffens, R. Aymanns and R. Stohr, "Acoustics Development for Exhaust Gas Turbochargers," *MTZ*, vol. 69, 03/2008.

[3] H. Nguyen-Schäfer, Rotordynamics of Automotive Turbocharger, Stuttgart: Springer, 2012.

[4] L. San Andrés, A. Maruyama, K. Gjika and S. Xia, "Turbocharger Nonlinear Response With Engine-Induced Excitations Predictions and Test Data," in *Proceedings of the ASME Turbo Expo 2009: Power for Land, Sea, and Air. Volume 6: Structures and Dynamics, Parts A and B, pp. 637–647*, Orlando, Florida, 2009.

[5] S. Sahay and G. LaRue, "Turbocharger rotordynamics instability and control," in *Rotordynamic Instability Problems in High-Performance Turbomachinery*, Texas, 1996.

[6] A. Schloßhauer, M. Stadermann, J. Klütsch, F. Falke, D. Lückmann and R. Aymanns, "Virtual development of a mixed-sequential boosting system for small gasoline engines," in *24. Aufladetechnische Konferenz*, Dresden, 2019.

[7] B. Lüddecke, P. Nitschke, M. Dietrich, D. Filsinger and M. Bargende, "Unsteady Thrust Force Loading of a Turbocharger Rotor During Engine Operation," *J. Eng. Gas Turbines Power*, pp. GTP-15-1286, 2015.

Experimental investigation on the transient response of an automotive turbocharger coupled to an electrically assisted compressor

S. Marelli, V. Usai, M. Capobianco

University of Genoa, Italy

ABSTRACT

An experimental investigation on a turbocharger coupled to an electrically assisted boosting system was developed at the turbocharger test rig of the University of Genoa. In the paper, a description of the experimental set-up and the measuring equipment adopted to test the e-booster coupled to the main turbocharger is reported. The e-booster system (eSC) was maintained in an upstream configuration and the main instantaneous parameters (pressure, mass flow rate and turbocharger rotational speed) were measured using high frequency response transducers.

This layout allowed to analyse the transient response of the main turbocharger and to highlight the gain in turbo lag when the eSC is working. A specific experimental technique was developed in order to provide experimental information on the time requested from the turbocharger to achieve a compressor outlet pressure value with and without the eSC. The target was to properly control each variable (throttle valve, waste-gate valve and the by-pass valve of the eSC) to perform reproducible and repeatable tests in the standard and electrically assisted configuration.

The work presented in this paper was conducted with support from The H2020 UPGRADE (High efficient Particulate free Gasoline Engines) Project (Grant Agreement Number: 724036).

1   INTRODUCTION

The reduction of CO_2 and, more generally, GHG (Green House Gases) emissions imposed from the European Commission (EC) and the Environmental Protection Agency (EPA) for passenger cars has driven the automotive industry to develop technological solutions to limit exhaust emissions and fuel consumption without compromising vehicle performance and drivability (1).

Even if Electric Vehicles (2) and Fuel Cell propulsion systems (3) are considered as alternative solutions, the Well to Wheel energy balance is still now unfavourable (4). Besides, fossil fuels will continue to prevail as energy source for road transportation in the next decades, thanks to a more substantial adoption of advanced liquid and gaseous biofuel (5) to promote a progressive introduction of renewable energy sources. Therefore, the Internal Combustion Engine will play a fundamental role in the near future, thanks to the progressive hybridization of the powertrain (6), and development of integrated technologies.

Turbocharging (TC) is one of the most promising ways to achieve such targets, along with downsizing, Variable Valve Actuation systems, and Gasoline Direct Injection (7, 8, 9). In such rapidly changing and challenging scenario, it appears that, in order to remain competitive in financial and technical terms, the future development of electrically-driven and electrically-assisted boosting systems may also require adequate compatibility with the future "low carbon" vehicle. Ideally, the advanced boosting solution may offer multiple functional roles, where synergies with electrification solutions could be offered in addition to their conventional role of providing boost. To this aim, the possibility of coupling an electric drive to the turbocharger (Electric Turbo Compound, ETC) to recover the residual energy of the exhaust gas, as well as to extend boost range and reduce turbo lag, is becoming more and more attractive. Another possible solution to optimize the boosting technology is represented from the waste heat recovery using, for example, a low pressure turbine for electric turbocompounding. However, the improvement of transient response of the boosting system is a target that cannot be omitted. The problem of poor performance affecting the engine load or rotational speed conditions must be attributed to the nature of the energy transfer path between the engine and the turbocharging system. It requires time to fill the intake and the exhaust manifolds and to raise their pressure. Under accelerating conditions, only part of the energy exploited from the turbine is delivered to the compressor, due to turbocharger inertial effects, which delay the acceleration of its rotating components.

In this scenario, the adoption of an electrically driven compressor located in an upstream or downstream configuration could significantly improve the transient response of the main turbocharger in the region of low engine rotational speed. The electric machine operates independently from the engine exhaust gases and from the inertia of the turbine and friction losses in turbocharger bearings. Besides, in a scenario in which the Hybrid Vehicles will probably be the common solution to satisfy European Commission regulations, the electric energy can be supplied by a starter generator powered by the engine shaft or regenerative braking system (10).

In the paper, the main results of an experimental investigation carried out on a turbocharger coupled to an electrically driven compressor is presented. A description of the experimental set-up, the experimental technique and the control system adopted is reported. Then, the analysis concentrates on the transient response of the turbocharger and the gain in the turbo lag. The downstream installation probably allows faster transient responses compared to the upstream configuration characterized by a larger volume that requires pressurization. Due to engine layout reasons, the experimental activity is focusing on the upstream configuration. In a further step of the investigation, a comparison between the turbocharger transient response achieved in upstream and downstream configuration will be presented.

2   OVERVIEW OF TEST BENCH AND MEASURING EQUIPMENT

The experimental investigation was developed on the turbocharger test bench of the University of Genoa, fully described in previous papers (11,12). Dry clean air is delivered by three screw compressors, which can supply a total mass flow rate of 0.6 kg/s at a maximum pressure of 8 bar. The turbocharger performance is measured over an extended range by independently controlling the upstream pressure of the turbine and compressor feeding lines, and through a motorized valve. Experimental investigations can be performed under "cold" and "hot" conditions modulating the thermal power of an electric air heating station that allows to heat turbine inlet air up to 750 °C, depending on the size of tested component. Turbine and compressor maps can be also measured under unsteady flow conditions typically occurring in automotive application using rotating valves or a motored cylinder head.

The experimental activity here reported is referred to the specific test rig configuration shown in Figure 1.

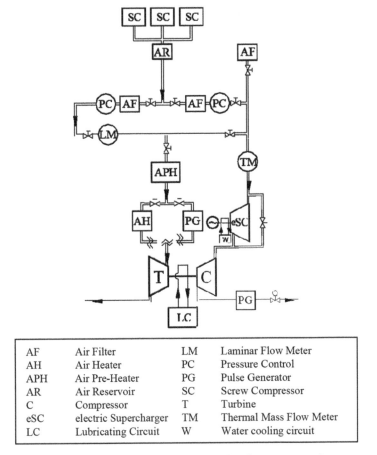

AF	Air Filter	LM	Laminar Flow Meter
AH	Air Heater	PC	Pressure Control
APH	Air Pre-Heater	PG	Pulse Generator
AR	Air Reservoir	SC	Screw Compressor
C	Compressor	T	Turbine
eSC	electric Supercharger	TM	Thermal Mass Flow Meter
LC	Lubricating Circuit	W	Water cooling circuit

Figure 1. University of Genoa turbocharger test rig.

The intake circuit consists of an air filter, the electrically-assisted compressor (called eSC) located upstream of the main turbocharger compressor and positioned in parallel to a by-pass valve to avoid the use of the eSC. This configuration was an OEM's requirement. A water cooling circuit was designed in order to cool the driving system of the eSC at a fixed temperature to avoid over temperature issues. Measurements of pressure, mass flow rate, temperature and rotational speed were performed upstream and downstream of the eSC, and of the turbocharger compressor. Average and instantaneous pressure levels were measured through piezoresistive transducers (with accuracy of ±0.15% of full scale). Platinum resistance thermometers (Pt 100 Ohm, with accuracy of ± 0.15 °C + 0.2% of measured value) were used to detect the mean temperature levels. The average compressor mass flow rate was measured through a thermal mass flow meter (with an accuracy of ±0.9% of measured value and ±0.05% of the full scale), while the instantaneous level at the eSC inlet and outlet section was detected using a hot-wire anemometric system through fiber-film probes. Turbocharger rotational speed measurement was performed through an eddy current probe (with a full-scale frequency precision of 0.009%), able to detect each blade passage.

Instantaneous measurements were performed using two synchronized data acquisition cards controlled from interactive procedures in LabVIEW® environment. During transient operation, a large number of experimental data has been acquired for a long period of time (about 10 seconds).

3   PROCEDURES FOR TESTING

The experimental investigation was developed on an electrically assisted compressor coupled to a small turbocharger for a downsized three cylinders Spark Ignition engine.

In order to provide the experimental information relative to the time requested by the turbocharger to reach a compressor outlet pressure with and without the eSC, a specific experimental technique was followed to maintain an equal increase of turbine power delivered to the compressor in both cases. The experimental goal was to adequately control each available variable to make tests performed with and without the eSC reproducible and repeatable.

In a typical pressure ratio steady flow map reported in Figure 2, the experimental investigation starts from a fixed level of corrected mass flow rate, pressure ratio and constant rotational speed to goes towards a final operating point characterized by different working parameters achieved thanks to an increase of the energy delivered from the turbine.

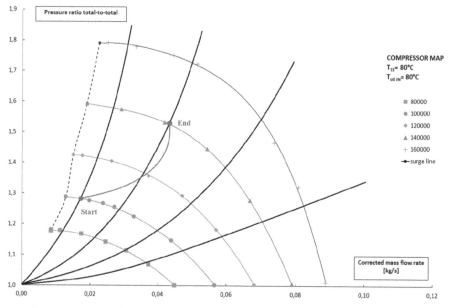

Figure 2. Example of the path followed on a pressure ratio map during transient operation.

In order to move along the path shown in Figure 2 as an example, it was decided to use different actuators available in the system (Figure 3): the by-pass valve (in parallel with the eSC), the waste-gate valve fitted to the turbocharger turbine, and an engine throttle valve located downstream the turbocharger compressor. These variables allow to perform investigation in a repeatable way in order to give a transferable information to be adopted in the simulation code. The experimental tests were conducted by maintaining the waste-gate valve fully open and the throttle valve in a fixed position (starting point), corresponding to a specific level of the energy transferred by the turbine. The final point was achieved by increasing the turbine energy through a partial closure of the waste-gate valve and by changing the position of the throttle valve, which causes an alteration of the external circuit characteristic curve (black curves in Figure 2). This experimental technique was then replicated with the eSC in idle (i.e., boosting system driven at 5000 rpm with the by-pass valve open) and when the eSC is working (i.e., rising the eSC rotational speed and closing the by-pass valve). In other terms, the variation of work transferred from the turbine to the compressor was kept unchanged in the presence and without the eSC, in order to highlight the benefit of the turbocharger transient response in two operating conditions characterized by the same energy delivered from the exhaust gases.

Since the engine and the ECU were not adopted, a control strategy developed in C++ using Arduino was set-up, maintaining linked all control variables of each system.

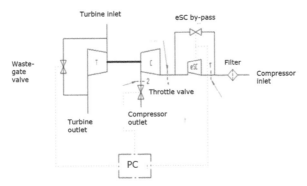

Figure 3. Schematic layout of the advanced boosting system and control variables.

The experimental investigation was performed for different corrected compressor rotational speed, mass flow rate and pressure ratio taking into account different engine operating points. Measurements performed with and without the eSC were synchronized to the start of transient operation.

4   EXPERIMENTAL RESULTS

As first step of the investigation, the steady flow curves of both compressors were measured at the test facility of the University of Genoa. With reference to the eSC compressor, the steady flow characterization required a specific driving system due to the limitation of the eSC working time imposed by over temperature problems. Therefore, a turbocharger turbine was selected as the compressor driving system, based on the required power levels. Thus, it was possible to measure the total-to-total compressor efficiency with good accuracy, starting from the measurements of the pressure ratio and the inlet and outlet temperature levels, which take time to stabilize.

For example, a transient operation represented by the instantaneous variation of the mass flow rate and by the overall static-to-static pressure ratio (red curve in Figure 4), is plotted on the turbocharger compressor steady flow map in Figure 4. The starting point corresponds in particular to an operating condition of 12 bar of BMEP (Brake Mean Effective Pressure) and 1750 rpm of the engine rotational speed. It should be noted that the overall static-to-static pressure ratio highlights the engine intake pressure.

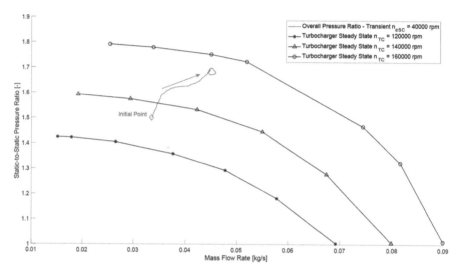

Figure 4. Example of a turbocharger transient operation in presence of the eSC.

As mentioned above, the analysis of the turbocharger transient operation was performed by properly checking the position of the waste-gate, the throttle and the by-pass valves. The path followed by the instantaneous static-to-static pressure ratio and the instantaneous mass flow rate is shown in red in the presence of the eSC driven at 40000 rpm.

The experimental investigation was performed without the eSC (here called idle condition) and varying the rotational speed of the eSC from 40000 to 65000 rpm. In Figure 5 the static-to-static pressure ratio measured through the eSC is plotted versus time during a transient operation performed by changing the position of the downstream throttle valve, partially closing the waste-gate valve and closing the by-pass valve when the eSC is operated. It is apparent that, when the eSC is activated, the pressure ratio through the compressor increases with an insignificant overshoot probably due to the logical control of the boosting system.

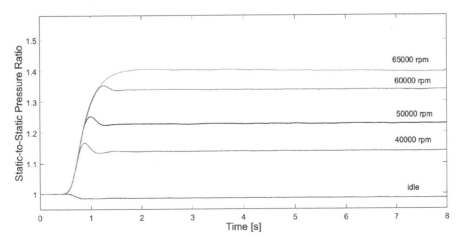

Figure 5. Static-to-static eSC pressure ratio during transient operation.

At the same time, the static-to-static pressure ratio of the turbocharger compressor decreases when the eSC is activated (Figure 6). This is probably due to the fact that, when the eSC works, the compressor is fed by a flow characterized by a higher density, which causes an increase in the mass flow rate and therefore a decrease in the pressure ratio. In other terms, the compressor does not reach the same pressure ratio with the same energy supplied by the turbine.

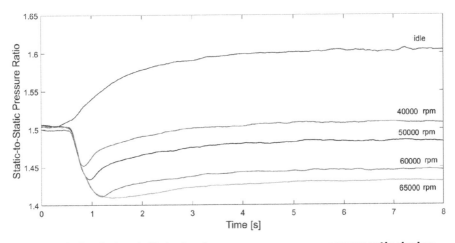

Figure 6. Static-to-static turbocharger compressor pressure ratio during transient operation.

Since the turbine power has been kept unchanged compared to the higher power required by the turbocharger compressor to process a higher mass flow rate, the rotational speed of the turbocharger decreases, as shown in Figure 7. This result is attributable to the logic of the experimental technique adopted.

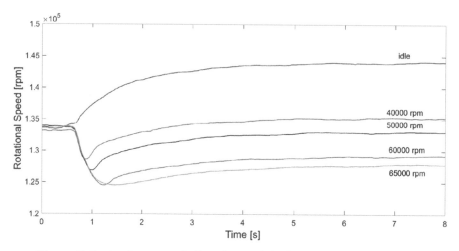

Figure 7. Turbocharger rotational speed during transient operation.

However, thanks to the availability of the eSC, the overall static-to-static pressure ratio during transient operation increases substantially when the eSC is in operation. The outlet pressure level (i.e., the engine intake pressure) is reached with different tendency in less time, with a consequent significant reduction of the turbo-lag as shown in Figure 8. For this specific operating condition, it seems that, independently from the eSC rotational speed, the target of intake pressure (red dashed line in Figure 8) is achieved in less time (about 3 seconds) than the idle condition.

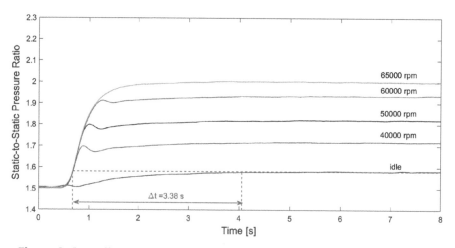

Figure 8. Overall static-to-static pressure ratio during transient operation (12 bar of BMEP and 1750 rpm).

To highlight the same improvement in terms of reduction of the turbo lag, Figure 9 shows the overall static-to-static pressure ratio during transient operation, with reference to a different starting point of the engine (10 bar of BMEP and 1500 rpm of the engine rotational speed).

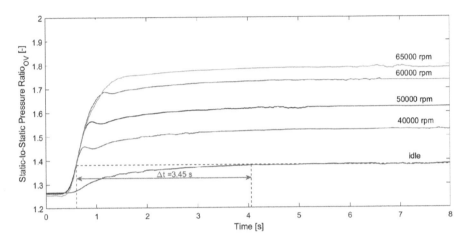

Figure 9. Overall static-to-static pressure ratio during transient operation (10 bar of BMEP and 1500 rpm).

In order to highlight the benefit of the eSC, the pressure signals recorded without the use of the eSC (continuous lines) and in the presence of the eSC dashed dotted lines) are shown in Figure 10. To reproduce the reported transient condition, the experimental test was performed starting from the waste-gate valve completely open and closing it completely to reach the final target of the engine intake pressure with the eSC in idle and the by-pass valve open. When the eSC is activated (dashed dotted lines), the waste-gate valve has been kept in a fully open position. This result underlined that the engine intake pressure target can be achieved with reduced engine back pressure in the presence of the eSC due to the fact that the waste-gate valve has been kept fully open. It is also possible to note that the turbo lag is reduced for each operating conditions, with a dependence on the rotational speed of the eSC (between 32500 to 41500 rpm).

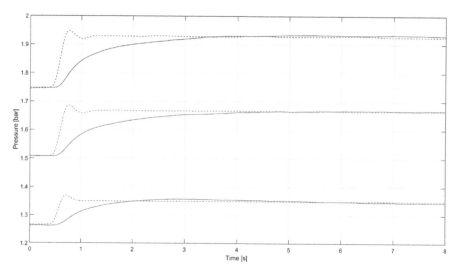

Figure 10. Instantaneous turbocharger outlet pressure without (continuous line) and with (dashed dotted lines) the eSC during transient operation.

5   CONCLUSIONS

The paper presents an advanced solution of the engine boosting system, considering the adoption of an e-booster system (eSC) coupled to a waste-gated turbo-charger (TC) typically adopted alone, in order to provide a reduced turbo-lag, i.e. an optimized transient response of the system. In order to highlight the behaviour of the eSC coupled to the TC, the first step of the investigation concerned the steady flow characterization of the eSC compressor. Due to over temperature problems, the working time of the eSC is limited, avoiding an accurate evaluation of the compressor efficiency, which is related to the upstream and downstream temperature measurements. In a second phase of the analysis, a specific experimental layout was set-up in order to study the interaction between eSC and TC in transient operations. This activity required a significant effort for the set-up of the circuit configuration, of the control systems, of the experimental technique to be adopted for the tests, and of the measuring equipment and data acquisition systems. Experimental activity has shown that the engine intake pressure can be reached in significantly less time, which can be reflected in the reduction of engine fuel consumption. The correct behaviour of the system (eSC coupled to the turbo-charger) and this configuration allows further analysis of the transient condition over an extended range.

Thanks to this previous investigation, it was possible to highlight the advantages of adopting an electrically assisted compressor. In the presence of eSC, when the boosting system is activated, the operating parameter of the turbocharger can be set differently from the standard application, for example by opening the waste-gate valve to reduce the engine back-pressure and therefore the fuel consumption, keeping the target of the intake pressure.

ACKNOWLEDGEMENTS

The work presented in this paper was conducted with support from The H2020 UPGRADE (High efficient Particulate free Gasoline Engines) Project (Grant Agreement Number: 724036). The authors would like to thank the team of Centro Ricerche Fiat and of VALEO for the technical support.

REFERENCES

[1] Bandel, W., Fraidl, G.K., Kapus, P.E., & Sikinger, H., "The turbocharged GDI engine: boosted synergies for high fuel economy plus ultra-low emission", SAE Technical Paper 2006-01-1266, doi: 10.4271/2006-01-1266, 2006.

[2] Guzzella, L. et al "Vehicle Propulsion Systems". Springer, 2013.

[3] Leal Filho, W. et al. "E-Mobility in Europe", Springer, 2015.

[4] Edwards, R. et al. Well-to-Wheels analysis of future automotive fuels and power-trains in the European context – Well-to-Wheels Report", Version 2c, 2007.

[5] Padula, A.D. et al. "Liquid Biofuels: Emergence, Development and Prospects", Springer-Verlag London, 2014.

[6] Onori, S. et al. "Hybrid Electric Vehicles", Springer-Verlag London, 2016

[7] Nikowitz, M., "Advanced Hybrid and Electric Vehicles", Springer International Publishing, 2016.

[8] Baines, N., "Encyclopedia of automotive engineering. Intake boosting". John Wiley & Sons, 2014.

[9] Bermúdez, V. et al., "Assessment of pollutants emission and aftertreatment efficiency in a GTDi engine including cooled LP-EGR system under different steady-state operating conditions", Applied Energy 2015.

[10] Tavcar, G., Bizja, F., & Katrasnik, N., "Methods for improving transient response of diesel engines – influences of different electrically assisted turbocharging topologies", Trans. Journal of Automobile Engineering, vol 255, no.9, pp. 1167–1185, 2011.

[11] Marelli, S., & Capobianco, M., "Measurement of Instantaneous Fluid Dynamic Parameters in Automotive Turbocharging Circuit", SAE Technical paper 2009-24-0124, 10.4271/2009-24-0124, 2009.

[12] Marelli, S., Marmorato G., & Capobianco, M., "Evaluation of heat transfer effects in small turbochargers by theoretical model and its experimental validation", Energy, Elsevier, Volume 112, 264–272, doi: 10.1016/j.energy.2016.06.067, 2016.

Steady state and transient tuning of a driven turbocharger for commercial diesel engines

T. Waldron, R. Sherrill, J. Brin

SuperTurbo™ Technologies, USA

ABSTRACT

Use of a driven turbocharger on a commercial diesel engine allows for direct control over the speed of the turbomachinery. Directly controlling the turbomachinery speed makes it possible to supply the necessary charge air independent of the energy available in the exhaust for the turbine to collect. This ability removes some of the constraints that exist in conventional turbochargers and allows for better optimization of emissions and efficiency during transient conditions as well as steady state conditions.

OEM testing of the SuperTurbo has shown that improvements in efficiency and emissions are possible. Driven turbos can carefully balance PM, NOx, EGR and AFR. Results show that driven turbochargers can balance NOx and PM and BSFC through engine transients with direct control of the speed of the turbomachinery. Precise control over AFR at all times, coupled with turbo-compounding from the driven turbocharger combine to improve engine efficiency. The use of a driven turbocharger allows additional turbine and compressor design freedom and many system enabling effects not typically possible with traditional turbochargers.

Actual engine OEM test data in this paper will demonstrate how driven turbochargers can be applied in the commercial diesel engine industry to help meet current and future emission regulations and efficiency targets. Test data from internal, third party, and customer test cells is presented to support claims.

1   INTRODUCTION

Driven turbochargers represent a new technology that can be adapted specifically for efficiency improvement and emission reduction strategies. Driven turbos can be used to control the speed of the turbomachinery independently of the engine's exhaust flow and vary the relative ratio between engine speed and turbo speed. This is done by utilizing either mechanical or electrical components to either add or subtract power from the turbocharger shaft. Mechanically driven turbos utilize a speed reduction fixed ratio planetary drive coupled with a continuously variable transmission. Electric turbos can apply motor/generators between the turbine and compressor, or extended on the common shaft, both of which operate at the speed of the turbo. Electric turbos can also utilize a planetary drive system to relocate the motor/generator and reduce its speed. The amount of power that can be exchanged with the turbocharger will correlate to the power specification of an individual driven turbo.

Driven turbos are utilized for several reasons, including performance, efficiency, and emissions. They can be largely considered as an 'on-demand' air device. The engine can now prescribe precise and variable air flow across different operating conditions, thus providing more options for engine control and tuning. During transient operation, a driven turbo will behave like a supercharger and draw mechanical or electrical energy to accelerate the turbomachinery for improved engine response. Unlike a traditional supercharger, a driven turbo also receives transient power from its turbine. The net effect is both a fast transient response and a more efficient power draw for supercharging. Driven turbochargers can also utilize different turbine and compressor designs that benefit from eliminating a conventional turbo's direct balance between turbine and compressor power. The turbines can be designed with reduced constraints related to inertia and overspeed. At higher power engine operating conditions, a driven turbo will have more turbine power than the compressor requires. In this case the driven turbo will return mechanical or electrical power to the engine in the form of turbo-compounding, and thus be recovering extra exhaust power to improve efficiency. These attributes allow a driven turbo to perform all the functions of a supercharger, turbocharger, and turbo-compounder.

2   DESIGN

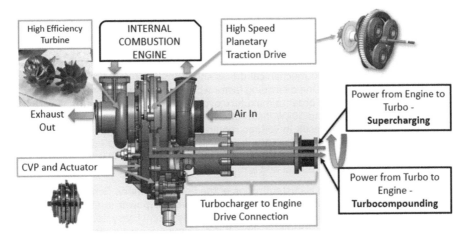

Figure 1. SuperTurbo™ architecture.

This paper will focus on the design and subsequent testing results for a specific mechanically driven turbocharger, otherwise identified as The SuperTurbo™. The name of this device is derived from, but not limited to, the benefits provided from both superchargers and turbochargers (1). Figure 1 outlines the basic architecture and components that will be discussed.

Fundamental to the design is the ability to forgo any need to have a variable ratio component operating at the speed of a turbocharger, which typically exceeds 100,000 rpm. In order for this to be possible, the SuperTurbo incorporates a fixed ratio high speed planetary drive, which is situated between the compressor and turbine. This planetary drive utilizes traction rollers that capture and position the turbo shaft both axially and radially. There is thus no requirement for fluid, ball, or thrust bearings on the main turbo shaft. The traction drive is preferred over a gear based system due to the high surface velocity at the interface between the main shaft (sun) and the rollers (planets). Traction rollers are more cost effective than high speed/precision gears and likewise provide improved NVH characteristics. The balance between torque capacity and drive efficiency/life relates directly to the amount of normal force exerted between the traction surfaces. The Super-Turbo planetary utilizes a bi-directional ball ramp positioned in the torque path which translates system torque into the optimal amount of normal force. In this way power/

torque specification is purposely designed to balance with optimized efficiency and the expected durability of heavy-duty diesel turbo systems.

With the speed of the system reduced by the planetary drive, a variable component can be interfaced at speeds below 8,000 rpm. The current SuperTurbo utilizes a throughput continuously variable transmission. This CVT(P) uses a tilted axis ball configuration with traction drive input and output contacts along the orbit of the spinning balls. A small amount of force is applied via an electric actuator which tilts the balls in different directions, thus changing the relative position of the input and output races as they relate to the center of rotation. This varies respective input and output speeds, thus providing overall controllability between the engine speed and the turbomachinery speed. The CVT, like the high-speed drive, also utilizes torque based loading to balance power, efficiency, and life.

The SuperTurbo can replace the CVT with an electric motor/generator to create an e-SuperTurbo for hybridized CV applications, although the results discussed in this paper will focus on the mechanical system. Most current e-turbos place the motor/generator on the turbo shaft, requiring the electric components to operate at high speed and in proximity to high temperature components. The e-SuperTurbo utilizes a more off-the-shelf motor/generator that can operate below 10,000 rpm and be positioned further from the turbine and manifold. Evaluations of e-turbo motor/generator efficiencies versus the SuperTurbo mechanical drives show similar efficiencies with variations based on speed and load. One overriding factor when applying e-turbos to CV engines is the power requirement. In order to maximize overall benefits, the SuperTurbo will operate at 30kW continuous and 50kW instantaneous power for larger vehicles. This power level from a e-turbo can result in voltage requirements in the 280V to 400V range (2). The high voltage requirement for CV e-turbo will need to be matched with the power electronics and battery storage of a hybridized vehicle, whereas the mechanical SuperTurbo forgoes such a requirement and is more agnostic to vehicle architecture.

Key to driven turbo design is customization of the aero machinery. Conventional turbochargers always have turbine power equal to compressor power in steady operation. This can be controlled to some extent through variable geometry systems, waste gates, or simply the fundamental design of the wheels. Conventional turbochargers are also designed with an eye towards minimizing inertia and correlating turbo-lag. Conventional turbos also must consider overspeed control. Driven turbos can use engine or electrical power to overcome system inertia and speed the system up like a supercharger. This results in less concern over turbine inertia and size. Driven turbos inherently will not overspeed, and instead translate any extra turbine power into mechanical power returned back to the engine through compounding. Driven turbos can also have more control over the relationship between turbine blade speed and exhaust flow. All of these factors allow for more freedom of design for a driven turbo turbine. Driven turbo designs will often have turbines that are high inertia, large diameter (even larger than the compressor), with full back walls and optimized blade shapes. These turbines are designed to maximize efficiency and the full capture of available exhaust energy. Heavy duty diesel engines that utilize high pressure EGR have disproportionately more exhaust power available than the compressor requires. The energy available from the pressure drop related to driving high pressure EGR can be captured by the turbine and utilized. All these factors allow for driven turbo aero designs to be higher efficiency and different than conventional turbos. The SuperTurbo generally utilizes a fixed geometry turbine with split entry designed to provide specific customer requirements for target efficiency points. In a traditional turbocharger, this kind of turbine design would create unacceptable turbo-lag and likely overspeed without a waste gate. However, in a driven turbo, it is preferred and necessary for the best energy recovery. To prove this fundamental difference, the SuperTurbo turbine has been gas-stand tested against normal turbines and can be seen in Figure 2.

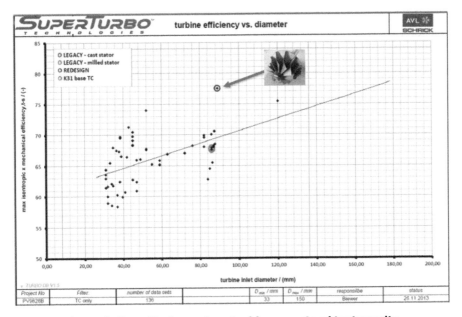

Figure 2. SuperTurbo custom turbine gas stand test results.

The SuperTurbo also needs a mechanical interface with the engine. The engine interface includes a clutch, which allows the driven turbo to be disengaged during idle or other unbeneficial operating ranges. There is also a torsional damper included in the system, although traction drives provide a degree of damping as they transmit torque through a shear based fluid film. Additionally, the SuperTurbo can be configured differently based on the engine/vehicle requirements. It can interface with the engine PTO or be configured to transmit power via belt drive at the front of the engine.

3   EFFICIENCY IMPROVEMENTS

3.1   Steady state efficiency

Using a driven turbo to tune an engine for steady state efficiency is a balance of several factors, of which the engine OEM now has more direct control. At any given steady state condition, it is possible to use the SuperTurbo CVT to manually prescribe a compressor speed and desired boost. If the compressor speed is increased, the mass air flow and correlating AFR rises, while the amount of compounding power decreases. For each operating condition there is an optimal position where the in-cylinder combustion efficiency is balanced with the amount of exhaust energy converted into mechanical compounding. Sweeping the turbo speed manually will yield a total efficiency curve that clearly shows the most efficient point. If the turbo speed is too low, the compounding power will be high, but the pumping losses will increase and the AFR will be too low. Likewise, if the turbo speed is too high, the compounding power will decrease too far and AFR will be too high.

For engines running high pressure EGR, there is an extra factor included in the tuning. Depending on the emissions goal and existing aftertreatment, engine tuning will target a specific level of engine-out NOx. So, while manually sweeping the turbo speed, EGR valve position must also be considered. Manipulating the EGR valve will directly affect the pressure delta across the turbine and thus the amount of compounding power collected. Of course, the EGR rate will fundamentally affect both the in-cylinder efficiency and the engine out NOx. Unlike Conventional turbochargers, moving the EGR valve will not change the speed of the turbo or the charge air flow. This is discussed in more depth in Section 4.1 and Figure 6.

The design of experiments for steady state tuning are used to create a control map for the driven turbo. Each operating point can now precisely balance: boost pressure, mass air flow, AFR, compounding power, EGR rate, engine out NOx, and total BSFC. There will be some operating conditions where the best efficiency does not include compounding power. At lower engine speed/load conditions, the best efficiency can sometimes be found by using a small amount of supercharging power. When in supercharging mode, the driven turbo is feeding power to the compressor from both the engine and the turbine. This is a more efficient form of supercharging than traditional superchargers, which take all their power from the engine or electrical system. At these lower engine operating conditions, it can be beneficial for a driven turbo to borrow some power for supercharging and achieve overall system efficiency by increasing the AFR for more optimal in-cylinder combustion.

This steady state tuning process was carried out on an Isuzu 7.8L engine with several generations of SuperTurbos over the last few years (3). Before the turbine and compressor are designed or selected, its necessary to run a through engine modeling project, usually utilizing GT Power. During this process the OEM can identify which operating points are prioritized in the design. Engines targeting high power/load efficiency will have turbines designed to maximize full power compounding, while sacrificing some lower engine power efficiency. In the case of the Isuzu program, the targeted engine efficiency was placed around the mid-rpm range and around 75% load. The turbine design was thus developed to collect compounding power in that part of the operating map. That turbine was thus more restrictive, and at full power had the necessary tradeoff of increased pumping losses. There can be a point where the pumping losses grow faster than the increasing compounding power, at which time a pressure based waste gate can be applied to balance that area of operation. The final result of the Isuzu steady state tuning effort can be seen in Figure 3, which shows the BSFC delta between the SuperTurbo and the baseline two stage turbo.

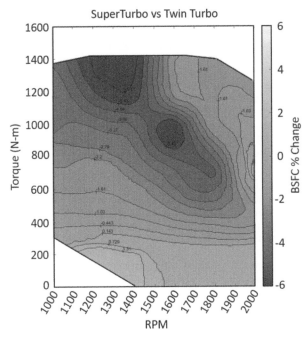

Figure 3. Engine efficiency comparison of SuperTurbo vs two stage turbo on 7.8L diesel engine.

Driven turbo integration and the space required in a vehicle have definable constraints. Figure 4 below shows the Isuzu 7.8L engine configured for testing with both the two-stage turbo and with the SuperTurbo. In this case, the SuperTurbo requires less overall space. While driven turbos will never be smaller than traditional turbos, their overall capability must be considered. Packaging a turbocharger, supercharger, and turbo-compounder would be far more challenging.

Figure 4. Installation of two stage turbo vs gen5 SuperTurbo.

3.2   Transient and drivecycle efficiency

Future regulations are focusing on drivecycle evaluations for CO2 and other criteria emissions. Embedded in this new direction for the industry is the requirement to look beyond steady state optimization and focus on the transient portions of full drive-cycles. Many heavy-duty diesel engines have inherent inefficiencies while performing transients. With today's turbochargers an engine transient is controlled by the smoke limiter. After the naturally aspirated torque rise, the engine is basically waiting on the turbocharger to provide enough air to accomplish the transient rise, all the while turning fuel into PM with inefficient combustion. In many cases the EGR valve is also closed during the transient in order to provide more turbine power, and thus increasing NOx in exchange for turbine power.

A driven turbo can be used during transients to change the status quo. As discussed previously, a driven turbo uses efficient supercharging by combining power from the engine (or battery) and the turbine to perform a quick rise in boost pressure. The engine can now perform each transient rise without using a smoke limiter. Optimizing the AFR through the transient provides more efficient combustion, with less parasitic loss than previously available from superchargers or e-boosters. Figure 5 below shows the Super-Turbo air-fuel ratios throughout the JE05 drivecycle. The collective efficiency gain from improved transient combustion along with compounding power has shown a 6% efficiency gain through the cycle, which can be extended to 9% with a downspeed strategy.

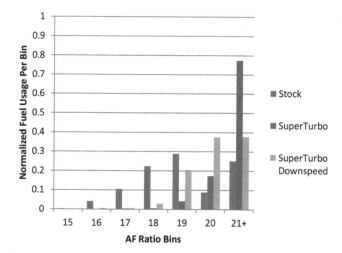

Figure 5. Histogram showing fuel usage over the drive cycle split into air fuel ratio bins.

Key to successfully implementing a downspeed is the ability to run fast transients. Meeting the drivecycle transient engine power requirements at lower engine speeds requires fast response from the boosting system. Being able to do this without extremely low AFR is an added benefit for a driven turbo.

Of course, there are always tradeoffs with how a transient is executed. Running high AFR through a transient can have the negative effect of increased NOx production, inability to flow EGR, and lower temperatures for the aftertreatment. Luckily, if you have a driven turbocharger, the exact ramp rate of the compressor can be controlled and even continually modified during engine operation. The transient ramp rate can be slowed down if the goal is to push EGR, lower transient NOx, and have good temperature rises for the after-treatment. This strategy has been applied with the SuperTurbo during cold start operation. For the best BSFC and CO2 the driven turbo can be controlled to ramp more quickly and improve transient combustion. The key is using this new controllability to optimize the full cycle based on the individual OEM requirement.

4   EMISSION REDUCTION

Heavy duty diesel emission reduction using a driven turbo is largely enabled by the enhanced controllability of the entire engine system. There are inherent combustion based tradeoffs between PM, NOx, and CO2. The benefit of implementing a driven turbo is that it provides an extra level of control. The ability to precisely control air flow, even through transients, gives the engine tuner new methods for manipulating the engine's emission profile. Driven turbos also have some unique enabling capabilities to push most emissions to new lower levels.

4.1   NOx emissions

With future regulation focusing on drivecycle based emissions (4), applying a driven turbo can improve EGR control, reduce transient engine out NOx, and improve aftertreatment temperatures during cold start and throughout the cycle. A current diesel engine running HP EGR and a VGT has compromises in controlling the balance between EGR fraction and charge flow. If the EGR valve is actuated, the resulting pressure change directly affects the turbo speed. Likewise, if the VGT vane position is changed, the result is a pressure change that affects the EGR fraction. A driven turbo utilizing a fixed geometry turbine changes this relationship because the speed of the turbo is now being directly controlled. If the EGR valve is actuated, it will have no effect on the turbo speed or charge flow. Likewise, without the need for variable vanes, the turbo speed can be changed without a large effect on EGR fraction. This now allows independent and more linear optimization of both air flow and the EGR rate.

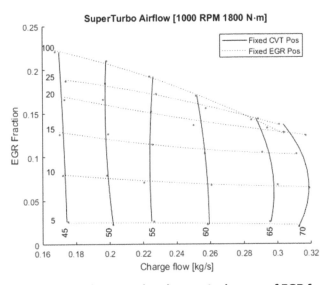

Figure 6. 15L diesel engine test showing control space of EGR fraction and charge flow using a driven turbo.

237

Similarly, a driven turbo doesn't rely solely on exhaust power to perform transient accelerations. Many traditional turbochargers close the EGR valve during transient operation in order to accelerate the turbo, minimize time spent on the smoke limiter, and achieve faster transients. Driven turbos can keep the EGR valve open during the transient and vary the turbo ramp rate to balance EGR flow (EO NOx) and correlating air flow/engine power increase.

Driven turbochargers can also be used to facilitate improved performance of diesel aftertreatment. Key to aftertreatment performance is minimizing the effects of cold start operation and low load cycles as they relate to SCR conversion percentages. There are several technologies, beyond driven turbos, that can be used to add heat when necessary for aftertreatment (5). Electrically heated catalysts, heated urea dosing, cylinder deactivation, close coupled light off SCR, turbo bypass, EGR cooler bypass, and mini burners are all viable options. Of primary concern is the correct combination of both heat and flow. The large aftertreatment sizes for heavy duty diesel vehicles may benefit more from the high-quality heat and flow provided directly from engine exhaust and accessed best through a full bypass of the turbine. The charts below in Figure 7 show test results from a SuperTurbo demonstration conducted in part with the CARB Phase III Low NOx program at SwRI.

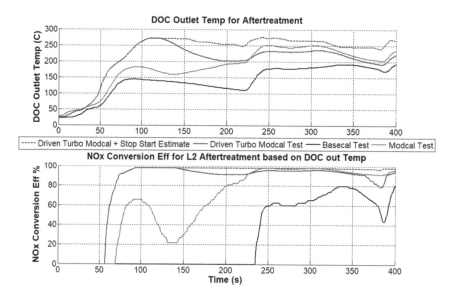

Figure 7. FTP cold start test showing AT temperature and NOx conversion for a driven turbo with bypass.

In the above case, the driven turbo was outfitted with a bypass valve that could fully direct all exhaust flow around the turbine and directly to the aftertreatment. This removed the heat sink of the turbine scroll and the loss of temperature from the pressure drop across the turbine. Driven turbos are unique in this capability in that they can fully bypass the turbine and still provide the necessary boost pressure to the engine through supercharging. This is dependent on the power specification of the driven turbo, which in the case of the 15L X15 used in this test, required up to 50kW instantaneous supercharging power during early portions of the cycle. In this test the bypass was only utilized during the initial FTP transient section in order to quickly achieve SCR operating temperatures and then balance CO2 versus exhaust temperature. The bypass can likewise be used across a LLC to provide quick temperature rises when the SCR requires it, which aligns with studies showing SCR temperature (6). The quick temperature rise provided by a driven turbo bypass pairs well with start-stop strategies. Main AT temperature drops were observed around 2°C/min, during shutdown once the aftertreatment was hot. Thus start-stop can maintain AT temperature during idle periods if a bypass can push initial temperatures above 250°C, and additionally it can reduce CO2 production compared to idling the engine.

In the first 400 seconds of the FTP, based on the conversion efficiency for the latest aftertreatment, the bypassed driven turbo combined with idle start-stop can reduce the cumulative tailpipe NOx from 2.9g to 1.4g, achieving a 50% reduction. The quick rise in temperature also can allow the engine to return to its more efficient base calibration early in the cycle. While the bypassed driven turbo does consume power while supercharging, the combined effect of the quick temperature rise, utilizing start stop, late cycle compounding, and resuming normal calibration can limit the overall CO2 penalty to <1% on the cold FTP cycle, while providing the best NOx conversion early in the cycle.

4.2 PM emissions
The discussion on transient tuning in Section 3.2 correlates directly with particulate emissions. Figure 8 below shows the SuperTurbo testing on a 13L on-highway diesel engine. The comparison in this figure is between a current production VGT and the SuperTurbo, showing first the difference in torque rise. It's clear where the engine with the VGT finishes the naturally aspirated rise and then is controlled by the smoke limiter. The SuperTurbo, utilizing efficient supercharging, has a much faster torque rise. This test also measured opacity in the exhaust, which can be seen on the right side. While running the faster transient, the SuperTurbo also kept the AFR at higher levels and thus the amount of smoke created was significantly reduced. Not only does this show better transient efficiency, but there are also benefits for the aftertreatment. The collection of soot on the DPF throughout the cycle will be reduced, and the amount of time and fuel spent in the DPF regeneration process will also decrease.

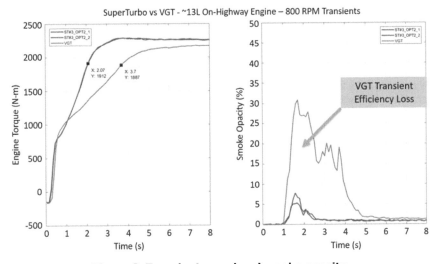

Figure 8. Transient speed and smoke opacity.

5   FUTURE TESTING

The SuperTurbo has been through several years of testing on engines varying in displacement from 7L to 15L. With 90% part commonality between SuperTurbo units, only the aero machinery (turbine/compressor) and the mechanical integration are currently changed for different engines. Steady state and drivecycle dyno testing will continue, but the focus is shifting more to vehicle testing. Figure 9 below shows a SuperTurbo equipped class 8 truck testing with a 15L engine. This test vehicle is showing the expected 3-4% steady state highway efficiency gains and 6% dynamic drivecycle gains (7). The vehicle is also used for high altitude testing, system durability, and controls development.

Figure 9. SuperTurbo vehicle testing.

Future testing will also employ improvements to the SuperTurbo bypass system for improved control of aftertreatment temperatures during cold start and low load cycles. The new bypass system will be fast actuating based on SCR temperature feedback and have the added benefit of controllable back pressure for dynamic EGR control.

6   SUMMARY

Driven turbos, either electric or mechanical, are demonstrating new benefits that are well suited to respond to the regulatory challenges facing the entire ICE industry. The controllable flexibility of on-demand air and exhaust energy recovery, along with all the enabling effects covered in this paper, demonstrate how OEMs can apply this new technology to reach the next level of CO_2 and NOx reduction.

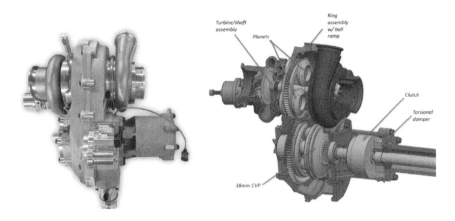

Figure 10. Current generation SuperTurbo unit.

REFERENCES

[1] Sherrill, R., Brown, J., and Waldron, T., "Design and Testing of a Mechanically Driven Turbocharger for Improved Efficiency and Drivability," in 13th International Conference on Turbochargers and Turbocharging, London, 2018.

[2] Koegeler, Hans-Michael, et al. "Final Report on Fuel Consumption and Emission Reduction Evaluation." 2019, http://www.imperium-project.eu/fileadmin/user_ upload/IMPERIUM_D6.2_Final_report_on_FC_and_Emission_reduction_evalua tion.pdf. Accessed 21 Feb 2020.

[3] Suelter, B., Itou, T., Waldron, T., and Brin, J., "Optimizing Steady State Diesel Effi-ciency and Emissions Using a SuperTurbo™ on an Isuzu 7.8L Engine," SAE Tech-nical Paper 2019-01-0318, 2019, https://doi.org/10.4271/2019-01-0318.

[4] Boriboonsomsin, K., Johnson, K., Scora, G., Sandex, D. et al., Collection of Activity Data from On-Road Heavy-Duty Diesel Vehicles. Final Report, California Air Resources Board and the California Environmental Protection Agency, May 2017.

[5] California Air Resources Board (CARB), "California Air Resources Board Staff Cur-rent Assessment of the Technical Feasibility of Lower NOx Standards and Associ-ated Test Procedures for 2022 and Subsequent Model Year Medium-Duty and Heavy-Duty Diesel Engines," White Paper, April 2019. https://ww3.arb.ca.gov/ msprog/hdlownox/white_paper_04182019a.pdf

[6] Manufacturers of Emission Controls Association (MECA), "Technology Feasibility for Heavy-Duty Diesel Trucks in Achieving 90% Lower NOx Standards in 2027," White Paper, February 2020. http://www.meca.org/resources/MECA_2027_Low_ NOx_White_Paper_FINAL.pdf

[7] Brown, J., and Waldron, T., "Drivecycle Benefits of Controlling Airflow with the SuperTurbo™," SAE Technical Paper 2018-01-0970, 2018, https://doi-org/ 10.4271/2018–01–0970.

Development of new generation MET turbocharger

Y. ONO, Y. ITO

Mitsubishi Heavy Industries Marine Machinery and Equipment Co., Ltd, Nagasaki, Japan

ABSTRACT

MHI-MME (Mitsubishi Heavy Industries Marine Machinery and Equipment) developed new generation turbochargers named MET-MBII and MET-ER to respond to the environmental regulations and the market needs. MET-MBII is axial turbine turbocharger for mainly 2 stroke main engine. MET-ER is radial turbine turbocharger for mainly 4 stroke generator engine. This paper describes the development history and the details of the new MET turbochargers.

1 INTRODUCTION

In recent years, the consciousness to the environmental problem is increasing worldwide, and it has become likely to tighten up gradually the environmental regulation about the air-pollution substance and carbon dioxide which are emitted by ships.

Especially in the ECA in North America or Europe, in addition to the 3rd regulation (TierIII) published in 2016, SOx regulation which regulates the sulfur content concentration in fuel is also carried out. It has become likely to regulate the upper limit of sulfur content concentration also in the general ocean area excluding ECA from 2020. Moreover, decreasing of GHG containing carbon dioxide emissions is required. EEDI, used as the index of GHG, will be more stringent gradually and therefore fuel consumption reduction is always argued.

On the other hand, LNG emits zero SOx and much less air-pollutants and carbon dioxide. Due to LNG's superior environmental performance, more and more ships are using DF engines, which can run on both heavy fuel oil and LNG.

Because turbochargers play a very important role in engine performance, MHI-MME newly developed new type turbochargers MET-MBII and MET-ER to respond to above background.

2 MET TURBOCHARGERS

Turbochargers designed by MHI-MME for marine diesel engines have been known as MET turbochargers. MHI-MME started production of axial turbine turbochargers in 1960 and introduced first MET turbocharger with no-cooling in 1965. Since then, MHI-MME has continued developing new technologies and expanding new series since then in response to various requirements from engine manufacturers, and is deploying the latest axial model, the new MET-MBII series (Figure 1).

The MET-SR series, which is the first radial turbine MET turbocharger, was developed based on the axial turbine turbocharger MET-SC series in 1988, then the MET-SRII series, which was developed by improving the MET-SR series, was released in 2002, and the current model on the market is the MET-SRC series developed in 2005. The cumulative production number of the radial turbine MET turbochargers of the MET-SR series, SRII series, and SRC series has reached 15,000 units. Especially the current MET-SRC series has obtained a good reputation in the market for easy overhaul. The compressor pressure ratio was increased as new series were developed, and has been raised to 6.0 at maximum for the new MET-ER series (Figure 2).

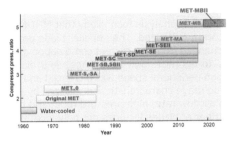

Figure 1. History of axial turbine MET turbocharger.

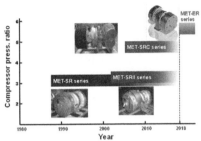

Figure 2. History of radial turbine MET turbocharger.

3   MARKET REQUIREMENT AND DEVELOPMENT CONCEPT

Conventionally the reduction of CAPEX and OPEX has continuously been required by the market. Recently, a newly built ship is subjected to evaluation of its greenhouse gas emission performance with EEDI (Energy Efficiency Design Index).

Environmental regulation of NOx emissions (Tier III) was also focused. Engine manufacturers are planning to abide by this regulation using technologies such as EGR, SCR and DF. We, as a turbocharger manufacturer, have to provide turbochargers that are optimized for these technologies. For example, for HP (High Pressure) EGR, wide range turbocharger for which the efficiency reduction with respect to changes in the pressure and flow rate is small is necessary for stable engine operation because the engine operating point moves greatly when EGR is switched ON/OFF. In the case of DF engine which used premixed gas and air for meeting Tier III regulations, the air-fuel ratio control in the gas mode is very important. So the turbocharger, which plays the role of supplying the air to the engine, needs to have high efficiency at high load to match engine performance needs. Especially for 4 stroke engines, the Miller cycle, which advances the closing timing of the intake valve, has been increasingly adopted as a technique for reducing NOx emissions. Turbochargers are required to respond to the increase of the pressure ratio to compensate for the air amount decrease caused by the early valve closing. Table 1 shows summary of market requirements and development items.

Table 1. Market requirements and development items.

Main market requirements	T/C development items
Low cost (Reduction of CAPEX and OPEX)	T/C Downsizing by large capacity compressor and turbine Easy maintainability and low cost service parts High efficiency turbocharger for fuel reduction
NOx reduction (EGR application)	Wide range compressor for stable engine operation
NOx reduction (SCR application)	Axial T/C gas inlet casing for engine and turbocharger piping layout with SCR system
NOx reduction	High pressure ratio compressor, two-stage turbocharging
Dual fuel tuning	Optimization of air amount for gas mode by T/C efficiency tuning at high load

4 PERFORMANCE AND CONSTRUCTION

4.1 Development of impeller

Based on our market research, large air flow rate and wide operation range impeller is required for axial type turbocharger. Especially for DF engine, high load optimized impeller is also required.

To increase the air flow rate, based on the latest flow analysis technology that MHI-MME cultivated, the impeller and the related parts were optimized. The impeller blade profile was reviewed and the number of blades was changed from the current combination of 11 whole and 11 splitter blades to 7+7 blades. To optimize high load range, 9+9 blades impeller is also prepared. In addition, the design peripheral speed was increased and the blade angle distribution was adjusted. The wide range characteristic was realized by utilizing flow analysis while maintaining the high air flow rate, high pressure ratio, and high efficiency. The advantage of reducing the number of blades includes not only increasing the air flow rate but also improving the machinability and reduction of the processing cost. Figure 3 shows photographs of the MET-MB and MET-MBII impellers. In terms of the strength, the blade thickness was revised so as to maintain the same size as at present even when the blade profile is reviewed, the number of blades is reduced, and the design peripheral speed is increased. Due to the above method, the air flow rate is increased by approximately 16% compared with the existing impeller. Figure 4 shows compressor maps of the MET-MB and MET-MBII impellers. As mentioned in chapter 3, new compressor has wide range and high pressure characteristic as shown in Figure 4, and it is suitable for EGR, DF tuning engines.

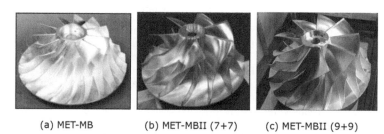

(a) MET-MB   (b) MET-MBII (7+7)   (c) MET-MBII (9+9)

Figure 3. Impellers.

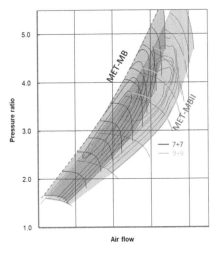

Figure 4. Compressor map.

For the radial type turbocharger, high pressure ratio impeller is required. Figure 5 shows a newly developed impeller with optimized backward angle and larger outlet width to reach 6.0 pressure ratio. In order to ensure the range at high pressure ratio, the number of diffuser blades was reduced with keeping throat area, and stall characteristics were improved. By using the parametric optimized design method and latest flow analysis technology, compressor pressure ratio was increased compared to conventional type and flow separation in the diffuser was disappeared as shown in Figure 6. As mentioned in chapter 3, high pressure ratio compressor is suitable for Miller-cycle engine which cycle is one of technology for reducing NOx emissions.

Figure 5. MET-ER impeller.

245

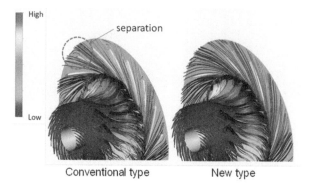

Figure 6. Comparison of compressor pressure ratio.

4.2 Development of turbine

New turbine blade has been developed for axial turbine turbocharger. To be balanced with the high flow rate impeller, the necessary turbine capacity was studied with reference to the planned values of each engine manufacturer and development to increase the turbine capacity was carried out. First, in response to the increase in the design peripheral speed, the turbine blade shape and the number of turbine blades were examined so as to maintain the same strength as the existing turbine blade. As a result, the number of blades increased from 35 in the existing turbine to 42. To increase the capacity of a turbine blade, it is necessary to increase the turbine blade throat distribution. However, if the throat distribution is increased too much, the dynamic pressure component (leaving loss) at the outlet of the turbine increases and the turbine efficiency will decrease. Therefore, the optimum turbine blade shape was examined using the latest flow analysis technology that MHI-MME cultivated. By optimizing the blade shape, as shown in Figure 7, exhaust chamber pressure recovery performance was improved and resulted in the turbine efficiency improvement. Figure 8 shows the shapes and Figure 9 shows the efficiencies of MET-MB and MET-MBII turbine blades. Turbine efficiency development was also successful. While turbine capacity was increased, turbine efficiency drop was also minimized.

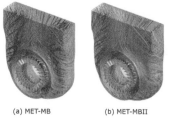

(a) MET-MB   (b) MET-MBII

Figure 7. Exhaust chamber pressure recovery.

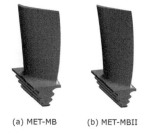

(a) MET-MB   (b) MET-MBII

Figure 8. Turbine blades.

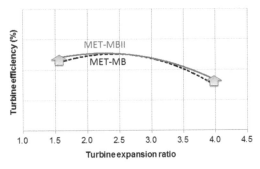

Figure 9. Turbine efficiency.

For the radial turbine turbocharger, high response is important based on market requirement. The response was improved by about 25% compared with the existing MET-SRC series by developing a new-type turbine wheel as shown in Figure 10 that reduces the inertia of the turbine by making its diameter smaller. High response contributes to engine high-speed starting up performance and reduction of smoke at engine low-load range.

To ensure turbine capacity and improve matching characteristic of exhaust diffuser and duct, turbine wheel throat area and distribution were adjusted and incidence at leading edge hub side was improved to reduce flow loss in the turbine wheel as shown in Figure 11.

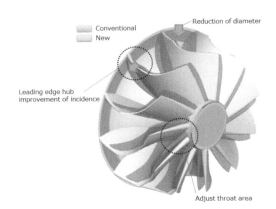

Figure 10. New turbine wheel.

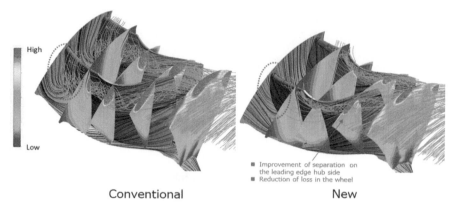

High

Low

■ Improvement of separation on
  the leading edge hub side
■ Reduction of loss in the wheel

Conventional   New

Figure 11. Distribution of loss in the wheel.

4.3   Construction

For the axial turbine turbocharger, as the turbocharger rotation speed increases, the bearing peripheral speed increases, so does the loss on the bearings. To reduce the bearing loss, the bearing diameter and the rotor shaft diameter were made smaller to reduce the bearing temperature. As a result, the bearing temperature was decreased and stayed well under the design limit at even the increased speed. When the diameter of the rotor shaft is reduced, it is necessary to consider the stability of the shaft. Rotor dynamics analysis was carried out in advance to confirm that it was within tolerance even if the diameter was reduced. Shaft vibration measurement test was conducted to confirm that the shaft vibration and the unstable vibration were below our criteria.

Axial turbine MET turbocharger has maintained the thrust balance using the seal air pressure passing through the inside of the bearing pedestal (see Figure 12). The seal air uses part of the compressed air at the impeller outlet. As the pressure ratio is increased, the amount of seal air increases and the thrust balance changes, which may cause an increase of the force working on the thrust bearing on the compressor side. To reduce this thrust force, a structure that increases the thrust force working on the turbine side was considered. By changing the shape of the gas labyrinth on the turbine side so that the area where the seal air led to the turbine side pushing the turbine disk is increased, the thrust force working on the compressor side was reduced. Figure 13 shows the gas labyrinth structure applied for MET-MBII.

Figure 12. Seal air passage.

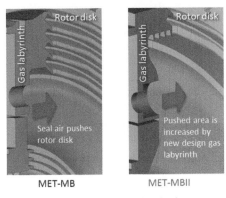

Figure 13. Gas labyrinth.

For the radial turbine turbocharger, compactness and maintainability are also required from the market. Based on these technological developments, further compactness as compared with the existing MET-SRC series was attained and each frame size was optimized. A model suitable for 1,000-kW class engines is the MET18SRC in the case of the existing MET-SRC series, and the MET16ER in the case of the new MET-ER series. As shown in Figure 14, when comparing the external shapes of these turbochargers, the size of the MET16ER is reduced by about 40% from the MET18SRC. With respect to the maintainability, the number of parts was reduced by about 30% compared to the existing MET-SRC series while maintaining the conventional concepts of "Easy overhaul", "Crew-maintainable design", and "Condition-based maintenance".

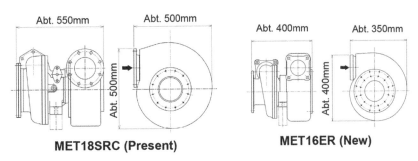

Figure 14. Comparison of external shape of turbochargers.

4.4 Turbocharger overall efficiency

MET33MBII axial turbine turbocharger prototype was prepared and performance test was carried out. Figure 15 is photograph of MET33MBII prototype. The performance measurement tests were carried out at 16% higher air flow rate of the operation line of MET-MB turbochargers. As a result, 16% higher air flow increased is confirmed while keeping the same efficiency as the MET-MB series. Verification test of comparison of Blade numbers of 7+7 and 9+9 was also carried out. As a result, optimization of high load area is confirmed with 9+9 blade numbers impeller as shown in Figure 16.

Figure 15. MET33MBII prototype.

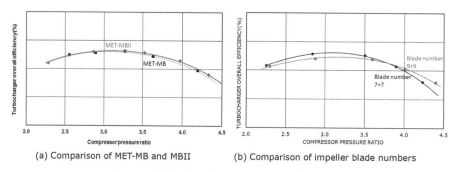

(a) Comparison of MET-MB and MBII   (b) Comparison of impeller blade numbers

Figure 16. Turbocharger overall efficiency.

MET16ER radial turbine turbocharger prototype was prepared and performance test was carried out. Figure 17 is photograph of MET16ER prototype. Figure 18 compares the performance of the MET16ER and the existing MET18SRC in a bench test. When comparing them for the same compressor and turbine specifications, it was confirmed that the turbocharger efficiency of the MET16ER was wholly improved by 1 to 1.5 points as compared with the MET18SRC. Figure 19 compares the performance of a new type compressor for a high-pressure ratio and a conventional type compressor. It was confirmed that the efficiency peak was shifted toward the high load side by applying the new-type compressor. In this way, the MET-ER series can be optimum-tuned according to engine requirements. Figure 20 shows example cases. In tuning case 1, the turbocharger efficiency is improved over all regions compared to the existing MET-SRC series, aiming to reduce the fuel consumption of the engine. In tuning case 2, a high-pressure ratio is attained by improving the peak turbocharger efficiency to the high load region while maintaining the same level low load turbocharger efficiency as the MET-SRC series.

Figure 17. MET16ER prototype.

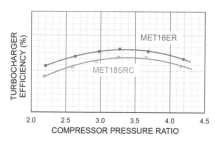

Figure 18. MET16ER bench test result.

250

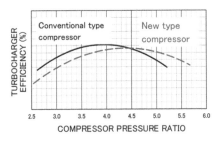

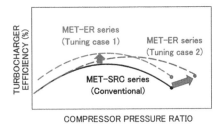

Figure 19. New compressor test result.

Figure 20. MET-ER series performance tuning case examples.

5   CONCLUSION

With the start of the SOx regulations in general sea areas in 2020, many ship owners are obliged to use expensive low sulfur fuel, and the fuel cost is anticipated to continue to be on an increasing trend over the long term, so ship owners are seeking ways to reduce fuel consumption. Furthermore, regulations based on the CO_2 emissions index (EEDI) introduced for ships built in 2013 and later will be more stringent in stages in the future. Since a turbocharger has great influence on the fuel combustion, the turbocharger plays a significant role in solving the above-mentioned problem. In addition, customers constantly seek to reduce CAPEX and OPEX. To meet the above market requirement, MHI-MME developed MET-MBII and MET-ER turbochargers.

The MET-MBII development focus was on the rotating parts and it adopts casings from the existing MET-MB. This development made 16% higher air flow rate and wide operation range than the MET-MB turbocharger while maintaining the turbocharger efficiency, the concept of easy maintenance, which is a trademark feature of the MET turbochargers, and reduction in size of the turbocharger is realized. MET-MBII turbochargers are a great choice even for special cases such as EGR, SCR, and DF engines.

The MET-ER contributes to meeting regulations on diesel engine emissions and to fuel saving, due to the high-pressure ratio and the high efficiency. In addition, the initial cost and the maintenance cost can be lowered by making the size smaller and reducing the number of parts by approximately 30% while keeping the great maintainability of the existing model, which can offer economic merits to customers.

Through turbochargers, MHI-MME aims to contribute to global environmental conservation and will continue technology innovation and product improvement to meet customer needs that will diversify further in the future.

REFERENCES

[1] Y Ono, "Development of New radial Turbocharger MET-ER Series", CIMAC Congress 2019
[2] Y Ito, "Development of New Generation MET Turbocharger", CIMAC Congress 2019

An approach to thermo-mechanical fatigue life prediction for turbine housings in gasoline engine application

H. Nakai, M. Takanashi

Structural Strength Gr., Technology Platform Center, Technology & Intelligence Integration, IHI Corporation, Japan

H. Kojima, K. Ito

Development Department, Engineering Center, Vehicular Turbocharger Business Unit, IHI Corporation, Japan

ABSTRACT

The prevention of thermo-mechanical fatigue (TMF) cracking in turbine housings for gasoline application has been one of crucial issues in automotive industry from a perspective of structural integrity. In the present study, an approach to thermo-mechanical fatigue life prediction for austenitic cast steel that enables to accelerate design loops was developed. The present approach has two major features as follows: (i) elastic-analysis-based approach and (ii) fatigue design curve determined by iso-thermal fatigue data. This paper describes the details of the approach and its application to actual turbine housing.

1   INTRODUCTION

Turbine housings of a turbo charger are under cyclic thermo-mechanical loadings, which leads to thermo-mechanical fatigue (TMF) cracking. The prevention of TMF cracking in turbine housings for gasoline application has been one of crucial issues in automotive industry from a perspective of structural integrity. Recently, the risk of TMF cracking in turbine housings for gasoline application is increasing due to increasing exhaust gas temperature to extend lambda 1 operation to deal with future stringent emission regulations. Therefore, accurate TMF life prediction method is demanded more than before. Additionally, because the demand to shorten design loop has been increasing to achieve cost reduction and to keep market superiority, simple but accurate TMF life prediction method that accelerate design loop is strongly demanded.

TMF is essentially complicated phenomenon, where fatigue, creep and oxidation are mutually related [1, 2]. Also, phase difference between temperature fluctuations and strain fluctuations is known to affect TMF life [3]. Hence, a number of researchers have been developing TMF life evaluation methods where complicated behavior of TMF is considered. Most popular approach is to evaluate accumulative damage during TMF loadings, which consists of fatigue damage, creep damage and oxidation damage [4, 5, 6]. Other approaches to correlate plastic strain range or inelastic plastic strain range with TMF life have been also widely developed [7, 8]. An important point in common between these two approaches is being based on elasto-plastic FE stress analysis that might take long time and high CPU cost. As stated above, to shorten design loops is strongly demanded. Therefore, more simplified approach based on elastic FE stress analysis has a noteworthy value as an engineering tool.

In the present study, the authors proposed an approach to thermo-mechanical fatigue life prediction for austenitic cast steel that enables to accelerate design loops. The present approach has two major features as follows: (i) elastic-analysis-based approach and (ii)

fatigue design curve determined by isothermal fatigue data. First, mechanical strain range, which actually occurs in turbine housings, is predicted by elastic thermal stress analysis, not by elasto-plastic analysis; mechanical strain range obtained from elastic thermal stress analysis is multiplied by strain amplification factor that is conservatively determined by a series of analyses in advance. Secondary, fatigue design curve for austenitic cast steels is determined by isothermal fatigue data, not by thermo- mechanical fatigue data; both of many isothermal fatigue tests by changing test temperature and a limited number of thermo-mechanical fatigue tests were conducted in order to determine fatigue design curve based on isothermal data, which enables to estimate TMF life conservatively. This paper describes the details of the approach and its application to actual turbine housing below.

2   OVERVIEW OF THE PRESENT APPROACH

Figure 1 shows the schematic flowchart of the present approach. First step to calculate elasto-plastic based mechanical strain range in turbine housings is conducting CFD (Computational Fluid Dynamics) analysis and heat transfer analysis to obtain thermal boundary conditions in elastic FE thermal stress analysis. Although details of these analyses are not described in this paper, CFD analysis is based on steady-state analysis, and heat transfer analysis is based on unsteady-state analysis. CHT (Conjugate heat transfer) analysis is desirable from the perspective of prediction accuracy. However, CFD analysis and heat transfer analysis are conducted separately because CHT analysis takes considerable high CPU cost. Subsequently, elastic FE stress analysis is conducted based on temperature boundary conditions obtained from heat transfer analysis. Finally, elastic mechanical strain range obtained by elastic FE stress analysis is converted to elasto-plastic based mechanical strain range using strain amplification factor. The detail of the conversion method is described in section 6.

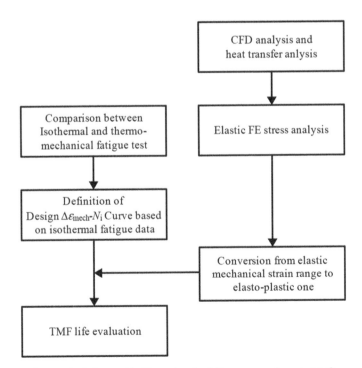

Figure 1. Schematic flowchart of the present approach.

On the other hand, TMF design curve is necessary for evaluating TMF life. Both of iso-thermal fatigue tests and TMF tests are conducted to compare their result and to define TMF design curve based on isothermal data. The detail of fatigue tests is described in section 4, and how to determine TMF design curve is explained in section 5.

3   ISOTHERMAL AND THERMO-MECHANICAL FATIGUE (TMF) TESTS

Isothermal fatigue tests and TMF tests were carried out on round bar of austenitic cast steel using servo-hydraulic test machine. Isothermal fatigue tests were conducted according to ASTM E606 [9], and TMF tests were conducted according to ISO 12111 [10]. Gauge length and diameter of all specimens were 22.5mm and 10.0mm, respect-ively. All tests were under strain-controlled, and strain ratio was set as -1. Table 1 shows the test matrix. Test temperatures of isothermal fatigue test were 650°C, 750°C, 850°C and 950°C. In the TMF test, test temperature ranged from 100°C to 950°C. Moreover, phase difference between temperature and mechanical strain was set as 180 degrees throughout TMF tests, which is known as out-of-phase TMF tests, because areas in turbine housings, where TMF crack might occur, should be under almost out-of-phase TMF loadings.

Figure 2 shows test results of all specimens. Crack initiation life was determined as number of cycles when tensile peak stress drops by 5% or specimen separation hap-pens. Tensile load drop criterion is defined in ISO 12106 [11]. Arrow marks in Figure 2 represent run-out data; both of isothermal fatigue tests and TMF tests were inter-rupted at specified cycles. For isothermal fatigue tests, data at 650°C had longer fatigue life than data at other test temperatures in the lower strain range. On the other hand, test temperature effect on fatigue life was not found in the higher strain range. Moreover, data at test temperature higher than 750°C has almost the same trend; there is a clear dividing line between 650°C and 750°C possibly due to tem-perature-dependent deformation behavior and fracture mechanisms. For TMF tests, data showed shorter fatigue life than isothermal data in the higher strain range, whereas data had longer fatigue life than isothermal data especially for data at tem-perature higher than 750°C in the lower strain range.

Table 1. Test matrix for isothermal fatigue and TMF tests.

Isothermal or TMF	Temperature °C	Strain ratio	Number of specimens	Remarks
Isothermal	650		8	Strain rate:0.1%/sec Heating method: electric furnace
	750		8	
	850		7	
	950	-1	5	
TMF Out-Of-Phase	100-950		5	Temperature rate: 2°C/sec Heating method: induction heating system

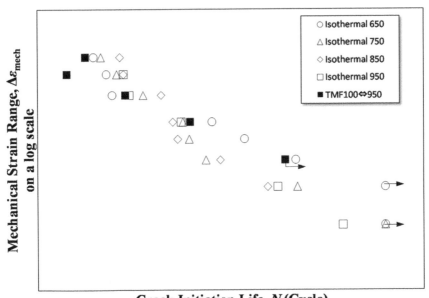

Figure 2. Test results of isothermal fatigue tests and TMF tests.

Figure 3 shows examples of photos of fractured TMF specimens. All TMF specimens except for unbroken one fractured near the edge of gauge section, showing bulging at the middle of gauge section. On the other hand, almost all isothermal specimens fractured within gauge section, showing no bulging. This difference might be mainly due to different heating method; induction heating system inevitably generates temperature distribution in the axial direction, then creep possibly occurs at the middle of gauge section with highest temperature when maximum compression stress happens at lowest test temperature, which leads to bulging. Electric furnace does generate no temperature distribution in the axial direction. Moreover, for isothermal fatigue tests, amplitude of creep deformation should be almost the same when heating and cooling. Hence, no bulging was observed for isothermal fatigue tests.

To investigate temperature-dependent fracture mechanisms for isothermal fatigue tests, SEM (Scanning electron microscope) – EDX (Energy Dispersive X-ray spectrometry) observation was carried out. Surfaces of specimens just below the main fracture surface were observed, where residual sub cracks are expected to be found. Figure 4 shows SEM – EDX observation results. For isothermal fatigue specimen with lower test temperature, residual sub cracks were found at areas with high content of Nb, which follows that fatigue cracks were likely to be initiated at niobium inclusions such as niobium carbide at lower test temperature. On the other hand, for isothermal fatigue specimen with higher test temperature, residual sub cracks were found at areas with high content of O, which follows that fatigue cracks were likely to be initiated at oxidation layer at higher test temperature. This temperature-dependent fracture mechanism is one of the reasons why there is a clear dividing line for isothermal fatigue life between 650°C and 750°C.

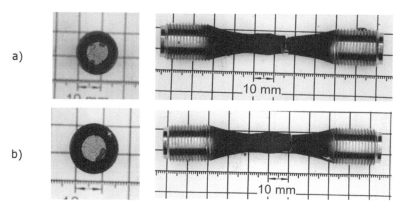

Figure 3. Examples of fractured TMF specimens; a) higher strain range condition, b) lower strain range condition.

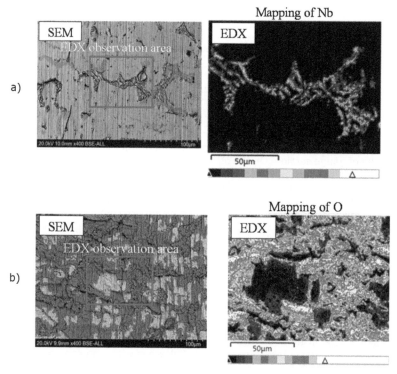

Figure 4. Examples of SEM-EDX observation results for isothermal fatigue specimens; a) lower temperature condition, b) higher temperature condition.

4   THERMO MECHANICAL FATIGUE DESIGN CURVE BASED ON ISOTHERMAL DATA

The major goal for TMF life evaluation is to check in advance whether or not an interested turbine housing could pass endurance tests conducted by automobile companies, where severe heat cycles simulating motoring and idling are applied to turbine housings. In the endurance tests, target cycles are set as 3,000cycles at most. Therefore, TMF fatigue design curve shall be defined conservatively, especially at thousands of cycles.

As stated in section 4, isothermal data can be divided into two types: date at 650°C and data at temperature higher than 750°C. Figure 5 shows isothermal mean curve for each type in black and red solid lines. If isothermal mean curve for date at temperature at higher than 750°C (red solid line) can estimate fatigue life conservatively as compared with TMF data at around target cycles, it is reasonable that a curve with 0.5% fracture probability, which is obtained based on isothermal mean curve (750°C-950°C), is used for TMF life evaluation as design curve. Therefore, TMF design curve can be determined as follows;

➢ TMF design curve is
  ✓ a curve with 0.5% fracture probability based on isothermal mean curve (750°C-950°C) if $650°C \leq T_{max} \leq 950°C$
  ✓ a curve with 0.5% fracture probability based on isothermal mean curve (650°C) if $T_{max} < 650°C$

where T_{max} is maximum temperature during TMF loadings. In this approach, it is assumed that only maximum temperature of TMF has a correlation with TMF life, then maximum temperature of TMF determines reference temperature of isothermal fatigue, at which shorter fatigue life could be obtained as compared with TMF life. It is clear that minimum temperature of TMF also affect TMF life. To investigate the minimum temperature effect is remaining issue.

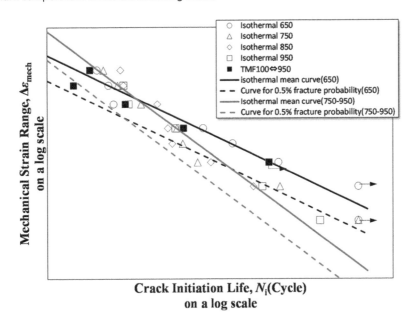

Figure 5. TMF design curve based on isothermal data.

5 CONVERSION FROM ELASTIC-BASED STRAIN TO ELASTO-PLASTIC-BASED STRAIN

In this research, elasto-plastic mechanical strain range is calculated by multiplying elastic mechanical range obtained from elastic FE stress analysis by strain amplification factor (SAF). This approach is quite similar to "simplified elastic-plastic analysis" in ASME BPVC. Sec. III. 1. NB-2015 (hereinafter called ASME code) [12]. In the ASME code, SAF has a maximum limit value of 3.3 for austenitic stainless steel; SAF increases with increasing elastic strain, and reaches maximum limit if elastic strain reaches two times yield strain or uniform strain, whichever is smaller. ASME code defines some requirements for the approach, of which is that temperature should not exceed 425°C. Considering turbine housings of turbochargers, maximum temperature reaches 1000°C or more, recently. Moreover, considerable larger plastic strain does exist in turbine housings than that in nuclear facility components to which ASME code is applied. Therefore, direct application of ASME code to turbine housings is not appropriate.

To determine appropriate SAF in turbine housings conservatively, both of elastic FE stress analysis and elasto-plastic FE stress analysis were conducted on actual turbine housing FE model with various constitutive laws, temperature conditions and constrain conditions. Figure 6 shows three constitutive laws, three temperature conditions and two constrain conditions which were applied to FE stress analysis;

➢ Constitutive laws
  ✓ a: original one obtained by round tensile test
  ✓ b: 1.5times yield stress with the original work hardening behavior
  ✓ c: the original yield stress with different work hardening behavior

➢ Temperature conditions
  ✓ t1: Temperature distribution 10 sec. after heating
  ✓ t2: Temperature distribution 30 sec. after heating
  ✓ t3: Temperature distribution 280 sec. after heating

➢ Constrain conditions
  ✓ i: No out-of-plain displacement on inlet surface
  ✓ ii: No out-of-plain displacement on both of inlet and outlet surface

FE stress analyses were carried out for all combinations of various conditions. Temperature conditions were determined based on results of unsteady-state heat transfer analysis whose boundary conditions were set based on steady-state CFD analysis. Conditions of unsteady-state heat transfer analysis and steady-state CFD analysis were based on representative endurance test. For elastic FE stress analysis, a constant constitutive law was applied; Young's modulus was assumed to be temperature-dependent, and Poisson's ratio was assumed to be temperature-independent. For elasto-plastic FE stress analysis, temperature dependency of constitutive laws was considered based on hot round tensile tests at various temperature although Figure 6 show only constitutive laws at room temperature. For FE stress analysis, temperature of all nodes was set as room temperature at initial condition, then thermal stress/strain was calculated at each temperature condition. Histories of temperature between initial condition and each temperature condition were not considered in elasto-plastic FE stress analysis.

SAF at all nodes were calculated using the following equations;

$$\varepsilon_{\mathrm{el,eq}} = \frac{\sigma_{\mathrm{el,eq}}}{E(T)} \tag{1}$$

$$\varepsilon_{\mathrm{pl,\ eq}} = \frac{\sigma_{\mathrm{pl,\ eq}}}{E(T)} + \frac{\sqrt{2}}{3}\left[\left(\varepsilon_{\mathrm{p1}} - \varepsilon_{\mathrm{p2}}\right)^2 + \left(\varepsilon_{\mathrm{p2}} - \varepsilon_{\mathrm{p3}}\right)^2 + \left(\varepsilon_{\mathrm{p3}} - \varepsilon_{\mathrm{p1}}\right)^2\right]^{0.5} \tag{2}$$

$$\mathrm{SAF} = \frac{\varepsilon_{\mathrm{pl,\ eq}}}{\varepsilon_{\mathrm{el,\ eq}}} \tag{3}$$

where $\varepsilon_{\mathrm{el,\ eq}}$, $\varepsilon_{\mathrm{pl,\ eq}}$, $\sigma_{\mathrm{el,\ eq}}$ and $\sigma_{\mathrm{pl,\ eq}}$ are equivalent strain and equivalent stress for elastic and elasto-plastic stress analysis, respectively. $E(T)$ is temperature-dependent Young's modulus. $\varepsilon_{\mathrm{p1}}$, $\varepsilon_{\mathrm{p2}}$ and $\varepsilon_{\mathrm{p3}}$ are principal plastic strain components, respectively.

Figure 7 shows an example of maximum SAF in classified area. Figure 7 also shows classification of areas in turbine housings. For area G of tongue part, SAF was higher than other areas. While value of SAF was dependent on various analytical conditions, it is possible to determine conservative SAF as a value larger than maximum SAF among all areas and all set of analyses with various analytical conditions. In this research, elasto-plastic based mechanical strain range at all nodes is calculated by multiplying elastic mechanical strain range by constant SAF determined in the way as stated above.

Although SAF might be affected by geometry of turbine housings, it is found that SAF did not show clear dependencies on stress triaxial factor, node temperature, strain amplitude itself, and so on. Hence, it can be assumed that geometry of turbine housings does not have a strong effect on SAF. It follows that conservative value of SAF determined by these analyses could be applied to turbine housings with different geometry.

6   APPLICATION OF THE PRESENT APPROACH

The present approach was applied to a turbine housing under TMF loadings with inlet gas temperature range between 100°C and 950°C simulating a severe endurance test. Steady-state CFD analysis and unsteady-state heat transfer analysis were conducted to obtain temperature boundary condition of elastic FE stress analysis. Elasto-plastic based mechanical strain range was calculated for all nodes by multiplying elastic mechanical strain range by SAF that was determined conservatively in the way as described in section 6. Finally, TMF life was evaluated for all nodes using TMF design curve based on isothermal data, which is described in section 5.

Figure 8 shows an example of TMF life contour calculated by the present approach. Red-colored areas, where fatigue life is short, are consistent with empirically critical area against TMF. Although the approach is obviously needed to be validated through comparison between the TMF life contour and actual endurance test results or other approaches that are introduced in section 2, the present approach has a potential to enable to accelerate design loops mainly because elastic FE stress analysis and TMF evaluation on turbine housing model with more than 1 million elements took only about half a day at the longest.

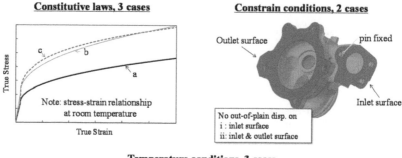

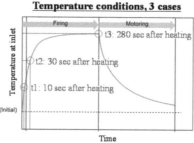

Figure 6. Various conditions for elastic/elasto-plastic FE stress to determine strain amplification factor.

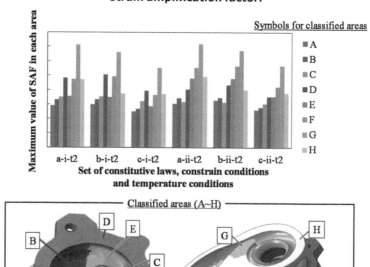

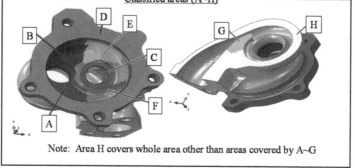

Figure 7. An example of maximum strain amplification factor in clasified area.

TMF life

Long

Shor

Figure 8. An example of TMF life contour calculated by the present approach.

7   CONCLUSION

This paper describes an approach to thermo-mechanical fatigue life prediction for turbine housings in gasoline engine application. To determine TMF design curve, both of isothermal fatigue and TMF tests were conducted. By comparing results of both fatigue tests, TMF design curve was determined as temperature-dependent 0.5% fracture probability curve based on isothermal data. On the other hand, elasto-plastic based mechanical strain range in turbine housings was calculated by multiplying elastic mechanical strain range calculated by elastic FE stress analysis by strain amplification factor (SAF). SAF was determined conservatively through series of elastic/elasto-plastic FE stress analysis on actual turbine housing FE model with various constitutive laws, constrain conditions and temperature boundary conditions. Finally, a result of application of the present approach is introduced, and its potential as an engineering tool to accelerate design loops is clarified.

NOMENCLATURE

TMF	Thermo-mechanical fatigue	$\Delta\varepsilon_{mech}$	Mechanical strain range
FE	Finite element	$\varepsilon_{el,\,eq}$	Equivalent strain obtained by elastic stress analysis
CPU	Central processing unit	$\varepsilon_{pl,\,eq}$	Equivalent strain obtained by elasto-plastic stress analysis
CFD	Computational fluid dynamics	$\sigma_{el,\,eq}$	Equivalent stress obtained by elastic stress analysis
CHT	Conjugate heat transfer	$\sigma_{pl,\,eq}$	Equivalent stress obtained by elasto-plastic stress analysis
SEM	Scanning electron microscope	ε_{pi}	Principal plastic strain components ($i=1,2,3$)

(Continued)

(Continued)

EDX	Energy dispersive X-ray spectrometry	$E(T)$	Temperature-dependent Young's modulus
SAF	Strain amplification factor	T	Temperature
N_i	Crack initiation life	T_{max}	Maximum temperature during TMF loadings

REFERENCES

[1] Neu, R. W. and Sehitoglu, H., "Thermomechanical fatigue, oxidation and creep: Part I. Damage mechanisms", Metallurgical transactions A, Volume 20A, pp. 1755–1767, 1989.

[2] Neu, R. W. and Sehitoglu, H., "Thermomechanical fatigue, oxidation and creep: Part II. Life prediction", Metallurgical transactions A, Volume 20A, pp. 1769–1783, 1989.

[3] Guth, S. and Lang, K.-H., "An approach to life time prediction for a wrought Ni-base alloy under thermo-mechanical fatigue with various phase angle between temperature and mechanical strain", International journal of fatigue, Volume 99, pp. 286–294, 2017.

[4] Laengler, F., Mao, T. and Scholz, A., "Investigation of application-specfic phenomena to improve the lifetime assessment for turbine housings of turbochargers", Procedia engineering, Volume 10, pp. 1163–1169, 2011.

[5] Nagode, M., Laengler, F. and Hack, M., "Damage operator based lifetime calculation under thermo-mechanical fatigue for application on Ni-resist D-5S turbine housings of turbocharger", Engineering failure analysis, Volume 18, pp. 1565–1575, 2011.

[6] Voese, F., Becker, M., Fischersworring-Bunk, A. and Hackenberg, H.-P., "An approach to life prediction for a nickel-base superalloy under isothermal and thermo-mechanical loading conditions", International journal of fatigue, Volume 53, pp. 49–57, 2013.

[7] Ohmenhaeuser, F., Schwarz, C., Thalmair, S. and Evirgen, H.S., "Constitutive modeling of thermo-mechanical fatigue and lifetime behavior of the cast steel 1.4849", Materials and design, Volume 64, pp. 631–639, 2014.

[8] Rajkumar, G., Kannusamy, R., Kulkarni, A. and Kumar, A., "Development of high temperature material constitutive model for thermo-mechanical fatigue (TMF) loadings in turbochargers", IMechE, London, pp. 217–230, 2018.

[9] ASTM international, "ASTM E606-12: Standard test method for strain-controlled fatigue testing", 2012.

[10] International organization for standardization, "ISO 12111-2011: Metallic materials –fatigue testing- strain-controlled thermomechanical fatigue testing method", 2011.

[11] International organization for standardization, "ISO 12106-2017: Metallic materials –fatigue testing- axial-strain-controlled method", 2017.

[12] The American society of mechanical engineers, "ASME Boiler & Pressure Vessel Code 2015 edition, Section III, Division 1-Subsection NB, Rules for construction of nuclear facility components", 2015.

Multidisciplinary and multi-point optimisation of radial and mixed-inflow turbines for turbochargers using 3D inverse design method

J. Zhang

Advanced Design Technology Ltd, UK

M. Zangeneh

Department of Mechanical Engineering, University College London, UK

ABSTRACT

The radial and mixed-inflow turbines have been widely used for the turbocharger application. The design of a turbocharger turbine with good performance still presents a lot of challenges. Apart from the traditional requirements such as high efficiency and low stress, the turbine blade is also required to achieve certain performance targets at multiple operating points, high unsteady efficiency under pulsating flow condition, reduced moment of inertia (MOI) and high vibration characteristic. To meet these challenges, it is important to optimise the radial and mixed-inflow turbines for the aerodynamic performance at multiple operating points and the structural performance subject to MOI, stress and vibration constraints. In this paper we propose an approach based on 3D inverse design method that makes such a design optimisation strategy possible under industrial timescales. Using the inverse design method, the turbine blade geometry is computed iteratively based on the prescribed blade loading distribution. The radial filament blading is always applied by the conventional design method to reduce the stress, while the inverse designed blade is three-dimensional (3D). A radial filament modification method is proposed to control the stress level of 3D blades. The turbine's aerodynamic and mechanical performance is evaluated using CFD (5 operating points) and Finite Element Analysis (FEA). A linear regression is performed based on the results of the linear DOE study. The number of design parameters is reduced based on a sensitivity analysis of the linear polynomial coefficients. A more detailed DOE with around 60 designs is generated and Kriging is used to construct a response surface model (RSM). Multi-objective genetic algorithm (MOGA) is then used to search the optimal designs which meet multiple constraints and objectives on the Kriging response surface. The performance of the final optimal design is evaluated in both the aerodynamic and mechanical aspects based on CFD and FEA simulations. The numerical results show that the optimal design leads to better performance in almost all aspects including improved efficiency in the design point and high U/C_{is} (velocity ratio), similar maximum stress, reduced MOI and increased vibration frequencies.

1   INTRODUCTION

The main challenge for the multidisciplinary and multi-objective optimisation of turbo-machinery blades are the time-consuming meshing, CFD, static structural and modal analysis which require a tremendous amount of computational resources (CPU time and computer memory). To accelerate and improve the optimisation process, surro-gate models have been widely used. The terms surrogate model, approximation model, response surface and metamodel are used as synonym in the literature. The surrogate model is constructed based on data from known designs (usually from DOE) and provides fast approximation and evaluation of objectives for different design parameters at new design points. The most commonly used surrogate models are polynomial approximation [1-5], artificial neural network (ANN) or radial basis func-tion (RBF) [6-11] and Kriging [12-13]. A detailed review of these methods can be found in Queipo et al [14].

In this paper, first-order (linear) polynomial, Kriging approximation and inverse design method will be used to optimise the aerodynamic and mechanical performance of a turbocharger turbine.

2   OPTIMISATION METHODOLOGY

The flowchart of the optimisation methodology used in this paper is shown in Figure 1.

To generate a blade geometry using the inverse design method, the meridional geom-etry, the thickness distribution and the blade loading distribution are necessary inputs. The parametrisation of all these outputs and their ranges of the variation have to be specified first during the optimisation process. The output parameters including the aerodynamic performance parameters and the mechanical performance param-eters of any designs in the optimisation will be evaluated using CFD and FEA.

A linear DOE and RSM model are generated based on the design parameters and per-formance parameters using first-order polynomial regression. The number of the design parameters is reduced based on the sensitivity analysis which compares the normalised coefficients of the linear polynomial and the most significant design parameters are selected whose variation have a larger effect on the performance parameters.

A new DOE with more designs is then generated for the new selected design param-eters and their performance parameters are evaluated using CFD and FEA simulations. The Kriging approximation is used to build the Kriging RSM based on the new DOE results.

Finally a Pareto front is generated through searching the optimal designs on the Kri-ging RSM quickly using MOGA and several optimal designs can be selected from the Pareto front. The performance parameters of these optimal designs are validated against CFD and FEA calculations.

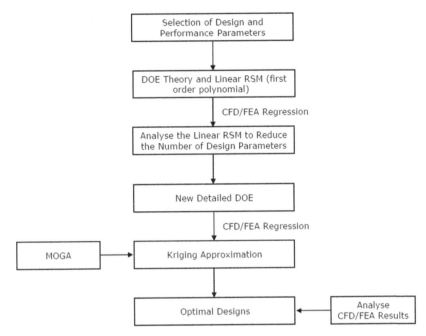

Figure 1. Flowchart of the optimisation methodology.

3   BLADE PARAMETERISATION

The design parameters consist of 6 meridional geometry parameters, 10 blade loading parameters and 1 thickness parameter.

It is shown in Figure 2 that the five control points A, B, C, D and E are used to create the hub curve and the shroud curve is created similarly using A', B', C', D' and E'. Both hub and shroud curves are created by the cubic spline method. The radial coordinate of point B' (maximum tip radius) is fixed (38 mm) while the axial coordinate of point B is also fixed (0 mm). Point D is fixed in both axial and radial directions to make sure all the blades have the same blade length and shaft radius as the baseline. The 6 design parameters used to define the blade meridional geometry are the inducer width W_1, the exducer width W_2, the LE angle α_1, the TE angle α_2, the hub and shroud control points Y_{hub} and Y_{shr}.

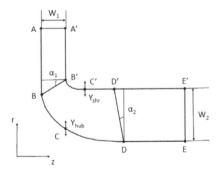

Figure 2. Meridional plane parameters.

The blade shape is computed iteratively based on prescribed blade loading using inverse design method. The blade loading is defined by LE/TE rV_θ and the derivative of rV_θ along the meridional direction ($\partial rV_\theta/\partial m$). The streamwise blade loading ($\partial rV_\theta/\partial m$) is defined by a three-segment method (more details can be found in [15]). The 10 blade loading parameters are $rV_{\theta TE,hub}$, $rV_{\theta TE,shr}$, NC_{hub}, NC_{shr}, ND_{hub}, ND_{shr}, $SLOPE_{hub}$, $SLOPE_{shr}$, $DRVT_{LE,hub}$ and $DRVT_{LE,shr}$.

The blade thickness is controlled by one non-dimensional factor called thickness parameters which is greater than 0.9 and less than 1.2. The shroud thickness of the blade remains the same and the hub thickness is multiplied by the thickness parameter. The thickness between the hub and the shroud sections is recalculated through linear interpolation.

3D radial turbine blades designed using the inverse design method show 2-3% higher efficiency than the conventional radial fibre design (validated against numerical and experimental results by Zangeneh-Kazemi [16]). However, their stress values are much higher than the material strength. To reduce the stress level of the 3D blades, a blade modification method called Radial Filament Modification method 1 (RFM1) is introduced. Basically, the wrap angle distribution at the LE and shroud of the blade remains the same and are mapped radially to all the other part of the blades. By doing this, the stress level of 3D blades will be reduced significantly.

For an optimisation process, it is desirable to explore the design space as much as possible, this requires a wide variation of all the design parameters in order to increase the change that the global optimal design can be found. However, if the range of design parameters is too big, a large number of poor designs have to be evaluated which will significantly increase the complexity and the cost of the optimisation process. Therefore, the range of all the 17 design parameters are carefully selected and are shown in Table 1.

Table 1. Variation ranges of design parameters.

Design parameter	Min	Max
W_1 (mm)	7	11
W_2 (mm)	15	24
σ_1	0°	40°
σ_2	0°	10°
Y_{hub} (mm)	16.5	21
Y_{shr}	0.2	0.4
$rV_{\theta TE,hub}$	0	0.04
$rV_{\theta TE,shr}$	0.06	0.1
NC_{hub}	0.05	0.2
NC_{shr}	0.05	0.4
ND_{hub}	0.6	0.85
ND_{shr}	0.6	0.85
$Slope_{hub}$	1	2.5
$Slope_{shr}$	-5	-1
$DRVT_{LE,hub}$	-1	-0.1
$DRVT_{LE,shr}$	-1	-0.1
Thickness	0.9	1.2

4   STEADY CFD ANALYSIS

The computational domain used for the steady CFD simulation is shown in Figure 3. The nozzle mesh is unstructured and generated using ANSYS Meshing. The inflation layers are applied on all the nozzle walls with a near wall element distance of 0.001 mm to capture the boundary layer effects. The rotor mesh is structured (hexahedron) and generated using ANSYS TurboGrid. The first element offset is also 0.001 mm. There are 20 layers of elements in the shroud clearance whose value is 0.5 mm. The total number of elements is around 2,200,000.

The nozzle domain is stationary and the rotor domain is rotating with a constant speed. Inlet boundary conditions are total pressure and total temperature. Inlet absolute flow angle is 40° from the tangential direction which is determined by a given volute geometry. Outlet boundary condition is atmospheric static pressure (1.0 bar). Rotational periodical boundary conditions are applied on all the periodic surfaces of the nozzle and rotor domains. The Stage (or mixing plane) method is used for the interface between the stator and the rotor. The Stage model performs a circumferential averaging of the fluxes through the interface and passes it to the component downstream. The turbulence model used is the shear stress transport (SST) k-ω. The working fluid is assumed to be ideal gas with gamma = 1.4. RANS equations are solved iteratively to obtain the whole flow field. T-S Efficiency (η_{50k}, η_{60k}, η_{70k}, η_{80k} and η_{90k}) and turbine mass flow parameter (MFP_{50k}, MFP_{60k}, MFP_{70k}, MFP_{80k}, MFP_{90k}) for five different RPM are used to evaluate the turbine aerodynamics performance and flow capacity.

Figure 3. CFD computational domain.

5   STATIC STRUCTURAL AND MODAL ANALYSIS

The blade geometry used in the steady CFD simulation is just a single blade which is not enough for the static structural and the modal analysis of the turbine wheel. To get accurate evaluation of the stress value and vibration characteristics during the turbine's rotation it is necessary to create the whole turbine wheel geometry from the single turbine blade by using Pro/ENGINEER which is a commercial CAD software now known as PTC Creo. Variable radius fillet is generated between the blade root and the hub to reduce the stress concentration. The minimum fillet radius has to be greater than 1mm due to the manufacture restrictions and the adjacent fillets must not contact each other. To reduce the MOI of the rotor the back face is scalloped by removing metal between the blades in the inducer region, which will reduce the turbine efficiency by ~1-3%.

The mesh is generated by using ANSYS Meshing as shown in Figure 4. And the total number of unstructured elements is around 150,000. Only one blade mesh is refined (element size = 0.6 mm) to save computational resources and time since the whole wheel geometry is axisymmetric. The mesh in the hub fillet and the blade trailing edge is refined further (element size = 0.3 mm) since they are locations where the maximum stress occurs.

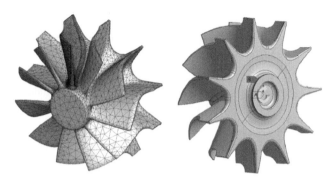

Figure 4. Mesh and boundary conditions for FEA.

The material used to manufacture the turbine is Inconel 713C. The wheel MOI can be obtained directly through ANSYS Mechanical once the geometry is imported. For the static structural analysis, the boundary conditions applied are the rotational velocity (A in Figure 4) and the cylindrical support (B in Figure 4) provided by the shaft connected to the compressor. The rotating speed is 130,000 rev/min which is the maximum working speed of the turbine. For the model analysis, the boundary condition applied is only the cylindrical support without pre-stress consideration.

The turbine mechanical performance parameters including the maximum principle stress, 1st and 2nd model natural frequencies and MOI will be obtained.

6   LINEAR DOE AND SENSITIVITY STUDY

The 25 design points are generated for the 17 design parameters with 25 different levels (the ranges specified in Table 1) using OLHS and allow several designs to diverge and fail to create geometries. Finally 19 designs converge and generate blade geometries using the inverse design method which is sufficient for the linear regression. The design matrix including the design parameters and the performance parameters of all 19 designs is called the linear DOE.

To reduce the number of design parameters (n = 17), a sensitivity analysis is performed by comparing the value of normalised polynomial coefficient $(\hat{a}_i^j)$ which is shown in Equation (1). Where $\hat{x}_i$ is the ith design parameter (normalised to 0-1) and $\hat{y}_j$ is the jth performance parameter (normalised to 0-1).

$$\hat{y}_j = \hat{a}_0^j + \sum_{i=1}^{17} \hat{a}_i^j \hat{x}_i \tag{1}$$

The normalised coefficients $\tilde{A}_i^j$ is defined by Equation (2). The range of $\tilde{A}_i^j$ is from -100 to 100. For a particular performance parameter y_j (j is constant), the greater the absolute value of $\tilde{A}_i^j$ is, the more significant the corresponding design parameter x_i is.

$$\tilde{A}_i^j = \frac{\hat{a}_i^j}{\max\left(\left|\hat{a}_i^j\right|\right)} \times 100 \tag{2}$$

The most significant design parameters are selected based on the summation of all the absolute values of $\tilde{A}_i^j$. The number of the significant parameters selected is directly related to the size (or the dimension) of the design space and the computational cost. The larger this number is, the more likely the optimal design can be found while more sampling points and computational resource are needed. Therefore, the 8 most significant design parameters are selected which are W_2, W_1, a_1, NC_{hub}, Y_{hub}, ND_{hub}, $SLOPE_{shr}$ and $rV_{\theta TE, shr}$. The variation of these 8 design parameters have much larger effect on the performance parameters compared to the other design parameters. Different weighting numbers can be applied for different performance parameters during this summation process and this will result in different collections of most significant design parameters. In this study the weighting numbers for different performance parameters are assumed to be identical.

7 KRIGING APPROXIMATION AND MOGA OPTIMISATION

In the previous section, the number of design parameters has been successfully reduced from 17 to 8 by a sensitivity analysis based on the linear DOE results. A more accurate approximation method (Kriging) will be used for these new 8 design parameters and MOGA will be used to search the design space to obtain the optimal design which meets multiple objectives and constraints.

The accuracy of the Kriging RSM is directly related to the number of the sampling points and the sampling method. The more points are used to build the Kriging RSM, the more accurate the model will be. In total 60 designs are generated by the OLHS method for the 8 new selected design parameters. The values of other design parameters are set as medial value or same as the baseline value since they have little effect on the performance parameters. The 53 of 60 designs converge and generate blade geometries using the inverse design method. Radial Filament Modification method 1 is performed for these 53 blade geometries to get RFM1 blades. CFD and FEA calculations are run for these 53 new RFM1 blades. A Kriging RSM then can be constructed using the design and performance parameters of these 53 designs. The performance parameters of any new designs in the optimisation can be evaluated quickly through the Kriging RSM instead of the expensive CFD and FEA simulations.

The constraints used in the optimisation are summarised in Table 2. The objectives are to maximise η_{70k} and minimise Stress. The flow chart of MOGA optimisation based on RSM is illustrated in Figure 5. NSGA-II is used to search the design space based on the constraints and objectives specified above. The performance parameters are evaluated through the Kriging approximation model which is much faster compared to the time consuming CFD and FEA simulations. The population size is set as 100 and the number of generations is set as 120. In total 12,000 designs are generated and their aerodynamic and mechanical performance values can be evaluated in 10 minutes.

Table 2. Constraints used in the optimisation.

	Constraints
η_{50k}	> 0.599
η_{60k}	> 0.665
η_{80k}	> 0.633
η_{90k}	> 0.547
MFP_{50k}	> 23.3
1st freq	> 7479
2nd freq	> 13535
MOI	$< 8.8342 \times 10^{-5}$

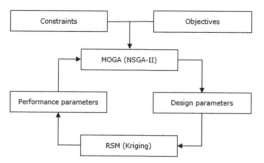

Figure 5. The flow chart of MOGA optimisation based on RSM.

A Pareto front is plotted in Figure 6. The Stress value is normalised by a constant value. Three optimal designs (design 5571, 7222 and 10535) are selected from the Pareto Font and their performance parameters are validated by CFD and FEA simulations. The comparison of performance improvements for design 5571, design 7222 and design 10535 compared to the baseline is shown in Figure 7. As it can be seen that design 5571 has the best efficiency but wor.st mechanical performance. Design 7222 has the best mechanical performance but worst efficiency. A clear trade-off between the aerodynamic performance and the mechanical performance is demonstrated. The error between the prediction value and CFD/FEA validation value for most of the performance parameters are between 0.8 and 4.4%. Design 10535 is selected and further analysis will be performed in the following section.

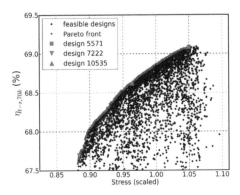

Figure 6. 2D scatter plot of η $_{70k}$ versus Stress (scaled) for Kriging approximation.

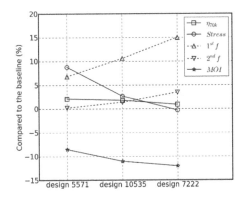

Figure 7. Comparison of performance improvements for design 5571, design 7222 and design 10535.

8   RESULTS

8.1   Comparison of meridional geometry and performance maps

The meridional geometry comparison between the baseline and design 10535 is shown in Figure 8. As it can be seen that the most obvious differences for design 10535 are the increased a_1 and reduced W_2 which are helpful to reduce MOI. The reduced blade exducer height W_2 of design 10535 is helpful to increase the blade stiffness. The MFP and $\eta_{t\text{-}s}$ comparison of the baseline and design 10535 are shown in Figure 9. The MFP of design 10535 is slightly higher at $U/C_{is} < 0.6$ and slightly lower at $U/C_{is} > 0.6$ compared to the baseline. The $\eta_{t\text{-}s}$ of design 10535 at $U/C_{is} < 0.64$ keeps almost the same as the baseline and is much higher (up to 5 percentage points) at $U/C_{is} > 0.64$.

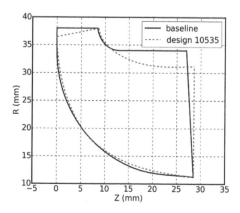

Figure 8. Comparison of meridional geometries.

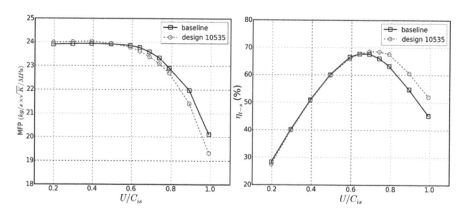

Figure 9. Comparison of MFP and t-s efficiency.

8.2 Comparison of internal flow field

In this subsection, the internal flow field details of the baseline and design 10535 at RPM = 80k (U/C_{is} = 0.79) where the efficiency improvement is much higher than that at design point (RPM = 70k).

The blade surface (suction side) streamlines comparison for the two designs is shown in Figure 10. One can see that design 10535 has a much better streamline distribution attached on the blade surface, since it has less secondary flow whose direction is from the hub to the shroud compared to the baseline.

The streamlines for the tip leakage flow for these two designs are compared in Figure 11. It can be seen that most of the tip leakage flow starts from the blade LE pressure side. The flow direction is from the pressure side to the suction side along the whole chord locations from the LE to the TE. A small leakage vortex is generated near the LE suction side and this vortex grows and mixes with any new leakage flow from the pressure side along the meridional direction. Design 10535 has a better leakage flow structure since the strength of the leakage vortex and its entropy generation for design 10535 is smaller than the baseline which can be seen in Figure 12.

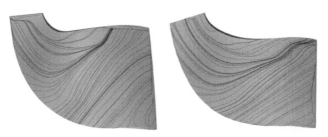

Figure 10. Comparison of blade surface streamlines on the suction side @ RPM = 80k (left – baseline, right – design 10535).

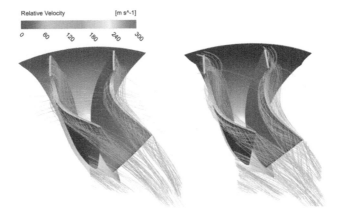

Figure 11. Comparison of streamlines across the tip leakage @ RPM = 80k (left – baseline, right – design 10535).

The comparison of static entropy contours for the two designs at three different streamwise locations is shown in Figure 12 and the last section is located at the TE. As it can be seen that most of the entropy is accumulated near the blade tip suction side where the tip leakage vortex locates. At the same streamwise location, the baseline has higher entropy than design 10535. Especially, in the TE, the baseline has two high entropy regions near the tip while 10535 only has one.

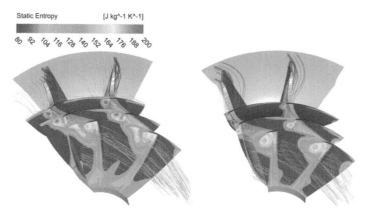

Figure 12. Comparison of streamlines associated with static entropy contours at different streamwise locations @ RPM = 80k (left – baseline, right – design 10535).

8.3   Comparison of static structural and modal analysis results

The comparison of stress contours for the two designs is shown in Figure 13. The stress distribution on the blade surface are very similar for these two designs. The stress level in the hub fillet of design 10535 is reduced compared to the baseline. The maximum stress occurs in the same location which is in the TE hub region. The maximum stress of design 10535 is 2.3% higher than the baseline and this can be easily reduced by increasing the fillet radius slightly near the TE. The frequency for 1st vibration mode of design 10535 is 10.6% higher than the baseline and the frequency for 2nd vibration mode of design 10535 is 1.4% higher than the baseline.

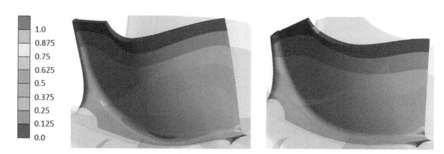

Figure 13. Comparison of maximum principle stress (scaled) contours on the suction surface (left – baseline, right – design 10535).

9   CONCLUSIONS

A systematic optimisation methodology using the inverse design method, DOE, RFM1, Kriging approximation and MOGA is presented in this paper. The inverse design method is used to generate the 3D blade geometry and Radial Filament Modification method 1 is used to modify the 3D blade shape to reduce the maximum stress. The number of design parameters is reduced from 17 to 8 through a sensitivity analysis based on the linear DOE results. The Kriging is used to construct a more accurate response surface for the new selected design parameters. An optimal design, design 10535, is obtained by searching on the Kriging RSM using MOGA with multiple constraints and objectives. Design 10535 shows better aerodynamic and mechanical performance compared to the baseline design, especially the efficiency at high U/C_{is} and MOI (-11.0%). The improved performance of design 10535 is confirmed by detailed CFD and FEA analysis.

The inlet to a turbocharger turbine encounters highly unsteady flow with varying pressure and temperature due to the pulsating nature of the exhaust gas from the internal combustion engine. It is very important to improve the cycle-averaged t-s efficiency which enables turbines to extract more energy from the exhaust gas during one pulse cycle. The pulsating engine exhaust gas with high pressure and temperature (low U/C_{is} region) carries more energy than high U/C_{is} region. Therefore, it is suggested that future work should involve improving the turbine efficiency at low U/C_{is} region while maintaining low MOI and stress.

ACKNOWLEDGEMENTS

This research is sponsored by Cummins Turbo Technologies and EPSRC. The author would like to thank their financial support.

NOMENCLATURE

ANN	artificial neural network
CFD	computational fluid dynamics
DOE	design of experiment
DRVT	streamwise blade loading parameter
FEA	finite element analysis
LE	leading edge
MFP	mass flow parameter
MOGA	multi-objective genetic algorithm
MOI	moment of inertia
NC	streamwise blade loading parameter
ND	streamwise blade loading parameter
OLHS	optimal Latin hypercube sampling
RANS	Reynolds-averaged Navier–Stokes
RBF	radial basis function
RFM1	radial filament modification method 1
RSM	response surface model
SLOPE	streamwise blade loading parameter

(Continued)

(Continued)

SST	shear stress transport
TE	trailing edge
a	polynomial coefficient
m	meridional coordinate
rV_θ	swirl velocity
U/C_{is}	velocity ratio
W_1	inducer width
W_2	exducer width
x:	design parameter
y:	performance parameter
Y_{hub}	hub control point
Y_{shr}	shroud control point
α_1	LE angle
α_2	TE angle
η	efficiency

REFERENCES

[1] Dornberger, R., Buche, D. & Stoll, P. 2000. *Multidisciplinary Optimization in Turbomachinery Design*. European Congress on Computational Methods in Applied Sciences and Engineering.

[2] Lian, Y. & Liou, M. 2005. *Multiobjective Optimization Using Coupled Response Surface Model and Evolutionary Algorithm*. AIAA Journal.

[3] Goel, T., Vaidyanatham, R., Haftka, R. T., Shyy, W., Queipo, N. V. & Tucker, K. 2007. *Response Surface Approximation of Pareto Optimal Front in Multi-objective Optimization*. Computer Methods in Applied Mechanics and Engineering, 196:879–893.

[4] Bonataki, E. & Zangeneh, M. 2009. *On the Coupling of Inverse Design and Optimization Techniques for the Multi-objective, Multipoint Design of Turbomachinery Blades*. Journal of Turbomachinery, 131(2):021014.

[5] Kim, J-H, Choi, J-H, Husain, A. & Kim, K-Y. 2010. *Performance Enhancement of Axial fan Blade through Multi-objective Optimisation Techniques*. Journal of Mechanical Science and Technology, 24(10):2059–2066.

[6] Verstraete, T., Alsalihi, Z. & Van den Braembussche, R. A. 2007. *Multidisciplinary Optimization of A Radial Compressor for Micro Gas Turbine Applications*. Journal of Turbomachinery, 132(3):031004.

[7] Pierret, P. 2005. *Multi-objective Optimization of Three Dimensional Turbomachinery Blades*. European Conference for Aerospace Sciences.

[8] Pierret, P., Filomeno Coelho, R. & Kato, H. 2007. *Multidisciplinary and Multiple Operating Points Shape Optimization of Three-dimensional Compressor Blades*. Structural and Multidisciplinary Optimization, 33(1):61–70.

[9] Roclawski, H., Bohle, M. & Gugau, M. 2012. *Multidisciplinary Design Optimization of A Mixed Flow Turbine Wheel*. ASME Turbo Expo 2012, GT2012-68233:499-509.

[10] Chahine, C., Seume, J. R. & Verstraete, T. 2012. *The Influence of Metamodeling Techniques on the Multidisciplinary Design Optimization of A Radial Compressor Impeller.* ASME Turbo Expo 2012, GT2012-68358:1951–1964.

[11] Mueller, L., Alsalihi, Z. & Verstraete, T. 2013. *Multidisciplinary Optimization of A Turbocharger Radial Turbine.* Journal of Turbomachinery, 135(2):021022.

[12] Chung, H-S & Alonso, J. J. 2004. *Multiobjective Optimization Using Approximation Model-based Genetic Algorithms.* 10th AIAA/ISSMO Symposium on Multidisciplinary Analysis and Optimization.

[13] Siller, U., Vob, C. & Nicke, E. 2009. *Automated Multidisciplinary Optimization of A Transonic Axial Compressor.* 47th AIAA Aerospace Sciences Meeting including the New Horizons Forum and Aerospace Exposition.

[14] Queipo, N. V., Haftka, R. T., Shyy, W., Goel, T., Vaidyanathan, R. & Tucker, P. K. 2005. *Surrogate-based Analysis and Optimization.* Progress in Aerospace Sciences, 41(1):1–28.

[15] Zhang, J., Zangeneh, M. & Eynon, P. 2014. *A 3D Inverse Design based Multidisciplinary Optimization on the Radial and Mixed-inflow Turbines for Turbochargers.* IMechE 11th International Conference on Turbochargers and Turbocharging, pages 399–410.

[16] Zangeneh-Kazemi, M. 1986. *Three-Dimensional Design of Radial-Inflow Turbines.* PhD thesis, Cambridge University, Engineering Department.

14th International Conference on Turbochargers and Turbocharging
Institution of Mechanical Engineers, ISBN: 978-0-367-67645-2

Experimental and computational analysis of the flow passing through each branch of a twin-entry turbine

J. Galindo, J.R. Serrano, L.M. García-Cuevas, N. Medina

CMT-Motores Térmicos. Universitat Politècnica de València, Camino de Vera s/n. València, Spain

ABSTRACT

This paper presents an experimental and computational analysis of the flow behaviour of twin-entry turbines. The experimental data were measured in a gas stand specially designed for this kind of turbines, controlling the flow characteristics passing through each volute. The computational analysis is carried out by means of CFD simulations that were globally validated against experimental data. In full and unequal admission conditions, the flow passing through each volute does not fully mix with the other within the rotor, which can be exploited in reduced order turbine models, simulating them as two single entry turbines working in parallel.

1   INTRODUCTION

Most land transportation vehicles use a reciprocating internal combustion engine (ICE) as powertrain system. It can be either as unique engine or as combined with an electric motor in the case of hybrid cars. However, these engines generate greenhouse gases such as CO_2 and other pollutant gases such as NO_x or unburnt HC.

Turbocharging has enabled downsizing in ICE which has been one of the most efficient techniques to reduce gaseous emissions of pollutants and CO2 [1]. During the last thirty years, the matching between turbocharger and ICE in operational points of turbocharger design has been well established. In part due to the ability of fast and reliable 1D and 0D models, as shown in Payri et al. [2] and Baines [3]. However, at turbocharger off-design conditions it is more difficult to ensure an optimal prediction of turbocharged engine operation.

In the last twenty years, computational power has increased exponentially. Many researchers have taken advantage of it and they have not only used CFD to simulate turbines in steady-state conditions, but also in pulsating flow operation, such as Palfreyman et al. [4] and Galindo et al. [5-7]. In real operation, the turbocharger flow is highly pulsating, becoming more important to understand turbocharger unsteady performance. Turbochargers sometimes work in extreme off-design conditions where some interesting phenomena arise, but they are difficult to extrapolate from design conditions. These conditions have caused a lot of interest, as can be seen in Serrano et al. [8], Walkingshaw et al. [9] and Binder et al. [10].

Nevertheless, 3D CFD simulations could not be reliable in their own and they may provide invalid results if they are not correctly carried out. To ensure that these results are valid, steady simulations should be compared with experimental data first. If both simulations and experimental data provide the same results, the 3D CFD simulations may be considered valid. Then, unsteady and pulsating conditions simulations can be performed with more confidence in the quality of their results.

Due to the typical operating pressures, mass flows and temperatures of automobile internal combustion engines, centripetal turbines tend to be the optimum configuration. The usual configuration to get the flow coming from the cylinders to the rotor used to be a single volute. Single volute turbines have some issues in multiple cylinder engines: pulsating flow cause interference between cylinders and impose limits to the opening and closing angles of the exhaust valves, reducing the volumetric efficiency. Twin-entry turbines could solve some of these problems.

Twin-entry turbines are becoming the standard in multiple cylinders, spark-ignition engines. Cylinders of adjacent firing order are connected to different volutes, isolating their pulses and reducing their interference. These turbines present some phenomena that do not appear in single volute turbines due to the union of both volutes before the rotor. These phenomena must be studied and analysed in order to calculate their main performance parameters correctly. To understand the flow characteristics of these turbines, different authors have characterised them with different approaches, as in Brinkert et al. [11], Romagnoli et al. [12] and Aghaali et al. [13].

Recently, some initial CFD simulations were carried out in order to better understand some phenomena, such as in Cerdoun et al. [14], Fürst et al. [15] and Ghenaiet et al. [16]. The flow at the rotor inlet of twin-entry turbines has also been studied as in Cerdoun et al. [17], Yokoyama et al. [18] and Hajilouy-Benisi et al. [19].

This work presents a twin-entry turbines flow behaviour analysis based on both experimental data and CFD simulations that allows to better understand the phenomena occurring in this kind of turbines. The flow of each volute has been tracked not only in the stator and in the vaneless space, but also along the turbine wheel. The purpose has been using 3D CFD to gain insight into a novel approach that treats the turbine as two different single entry and variable geometry turbines (VGT) working in parallel [20]. Tracking the degree of mixing of the flow from each volute along the rotor one can learn up to what extent the two VGTs hypothesis is valid and what is the typo of sub-models that can be proposed for a fast modelling of this type of twin-entry turbines.

2   EXPERIMENTAL METHOD

Twin volute turbines must be measured in a gas stand in order to obtain reliable experimental data. Then, these experimental data could be used for validating the results obtained in 3D CFD simulations.

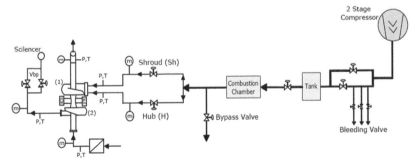

(1) Turbine, (2) Compressor, ICS-Independent cooling system
ILS-Independent lubrication system, m-Mass flow meter
T-Thermocouple, P-Pressure sensors, Vbp- Back-pressure valve

Figure 1. Experimental Setup Scheme.

The test bench used for this work is a special gas stand designed for twin and double volute turbines using compressed air, as detailed in Serrano et al. [20] and shown in Figure 1. The compressed air splits into two separated sets of pipe-work which connect with each turbine entry. There are also two control valves that allow controlling the mass flow rate independently in each branch.

The gas-supplier is a two-stage, oil-free, centrifugal compressor. It is powered by a 450-kW electric motor. The gas stand also has an independent lubricating system and an independent coolant system, which deliver oil and coolant with adjustable mass flow. All pipes and the turbocharger are insulated to ensure almost adiabatic operation. In order to measure the parameters such as turbocharger speed, pressure, temperature and mass flow, there are sensors at the essential sections, such as the turbine inlets and outlet. The sensors shown in Figure 1 are listed in Table 1.

There are two independent air flow lines arriving at the turbine inlet and the mass flow admission conditions can be adjusted and controlled with precision by highly linear control valves. This is done in order to assess the impact of each entry on the overall performance in steady state conditions. The possible admission conditions are divided into three different conditions.

Table 1. Gas stand measurement equipment and precision.

Variable	Sensor Type	Range	Typical uncertainty
Gas Mass Flow	V-cone, Thermal and Vortex	45 -1230 kg/h	<1%
Gas Pressure	Piezoresistive	0 - 5 bar	12.5 mbar
Gas/Metal Temperature	K-type thermocouple	273 - 1500 K	1.5 K
Oil/coolant Pressure	Piezoresistive	0 - 5 bar	12.5 mbar
Oil/coolant Temperature	RTD	173 - 723 K	< 0.5 K
Oil/coolant Mass Flow	Coriolis and Magnetic	Few tens g/s	2%
Turbocharger Speed	Inductive sensor	<300 krpm	<500 rpm

In full admission conditions, the mass flow rate is the same in both entries. In partial admission conditions, the mass flow only passes through one entry. In unequal admission conditions, the mass flow passes through both entries, but in different rates. Unequal conditions are the most common in real operation.

The relation of the mass flow passing through each volute should be characterised in order to describe the different mass flow admission conditions. From now on, the volute nearer the turbine hub will be called Hub Branch and the volute nearer the turbine shroud, Shroud Branch. The Mass Flow Ratio parameter (MFR) relates the mass flow passing through the Shroud Branch and the total mass flow, Eq. (1).

$$MFR = \frac{\dot{m}_s}{\dot{m}_s + \dot{m}_h} \qquad (1)$$

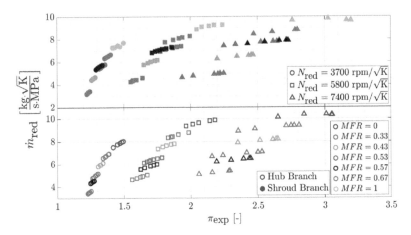

Figure 2. Experimental Map.

Other authors prefer using the Reduced Mass Flow Ratio parameter (RMFR) instead of the MFR, but to use the MFR eases the desired experimental data acquisition. Moreover, as will be shown in section 5, there are some parameters that relate directly to the MFR. Turbine maps can also be represented as a function of this MFR parameter, as it is typically done for variable geometry turbines with the position of the nozzle vanes, the detailed discussion can be read in [20]. Figure 2 shows the reduced mass flow versus expansion ratio map for shroud and hub branch with different MFR. The typical uncertainty of the reduced mass flow is less than 2%.

3 CFD SIMULATIONS

Reliable 3D CFD simulations should have the same exact geometry than the experiment. Therefore, the measured twin turbine has been digitised as a CAD (Computer-Aided Design) model.

All geometry parts have been digitally scanned. The rotor, the inlet ducts and the outlet duct are easy to scan since they are accessible to a conventional 3D scanner. However, the volutes and the plenum at the rotor outlet could not be scanned directly. Thus, an inverse non-destructive process is used for digitising these parts. They are filled with liquid silicone and, when cured, it is extracted. These silicone moulds are the negative part of the parts wanted to scan. Nevertheless, these parts should be pre-processed computationally because there could be some imperfections such as air bubbles and fissures in the moulds. After this pre-process, all parts can be united in a unique CAD model as shown in Figure 3.

With the geometry digitised and properly prepared, the next step is to create the mesh. It should be sufficiently fine in order to capture correct results independently of the cell size. Moreover, regions near the wall and regions with more complicated phenomena as the rotor tips should be meshed finer than other regions.

Table 2 shows the mesh independence study carried out plotting temperature and pressure at the rotor inlet section as a function of the number of cells for a case with MFR 0.53 and reduced speed of 3700 rpm/√K.

The results are dependent on the mesh size until 5.5 million cells. The case with more than 8 million cells gives almost the same results than at 5.5 million cells. Thus, the case with 5.5 million cells is considered sufficiently fine and all simulations carried out have used this mesh. Figure 3 shows a zoom into a radial view of the final mesh.

Table 2. Mesh Independence study. Pressure and temperature values at the rotor inlet section.

Number of Cells ($\cdot 10^6$)	Total Pressure [bar]	Total Temperature [K]
0.97	1.354	356.87
1.59	1.359	357.31
2.53	1.361	357.68
4.46	1.364	358.00
5.54	1.365	358.07
8.28	1.365	358.08

The simulations are performed using 3D unsteady Reynolds-Averaged Navier Stokes equations (URANS). The turbulence model used is the k-ω SST. The equations are solved using a second order, upwind coupled flow solver. To differentiate the air passing through each volute, a multi-component gas is selected, one component for each turbine entry. Both gases are modelled as non-reactive, ideal gases. The boundary conditions are stationary and extracted from the experimental data, as stagnation inlet (total pressure, total temperature, composition, and turbulence description) for each inlet and pressure outlet (static pressure, as well as static temperature, composition, and turbulence description for backflow) for the outlet.

In the beginning, the simulations are carried out using a steady solver, using moving reference frames for simulating the motion of the rotor. Their converged solutions are used as initial conditions for the URANS simulations. For them, a second order, implicit and constant time-step unsteady solver is used. The movement of the wheel in the URANS simulations is modelled using a rigid body motion, rotating the mesh of the rotor each time-step. While the inlet and outlet boundary conditions are still stationary, this moving mesh induces local unsteadiness in the rotor.

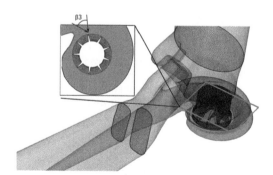

Figure 3. CAD model with final mesh.

In order to set up the time step of each case [21], the time spent in rotating an entire revolution is divided into 360 parts. Thus, it rotates one degree for each time step. Moreover, each time-step is solved iteratively, taking 20 inner iterations to ensure that the residuals of the different equations were low enough to ensure convergence.

Once the simulation is converged, the relevant information as pressure, temperature or mass flow ratio at different sections as volute inlet or rotor outlet is extracted from monitors. Then, these values can be post-processed in order to calculate other important parameters such as expansion ratio, efficiency or reduced mass flow and compare them with the experimental data.

On the other hand, it is also interesting to study the behaviour of the flow inside the turbine. With the multi-component option, two different airs are defined with the same properties and each one passes through each volute. Then, the flow passing through each volute can be differentiated and followed along all the geometry by plotting the Mass Fraction.

4   CFD SIMULATIONS VALIDATION

The CFD simulations are not reliable on their own, they need to be validated with experimental data. For this reason, the boundary conditions of the simulations carried out are the same that some of the points measured experimentally. One case was simulated per MFR and reduced speed.

The concordance between simulations and experiments can be represented through simulated versus measured turbine maps. Figure 4 shows the reduced mass flow from simulations and experimental data. Dotted lines represent ± 3% discrepancy, while the solid line represents perfect concordance.

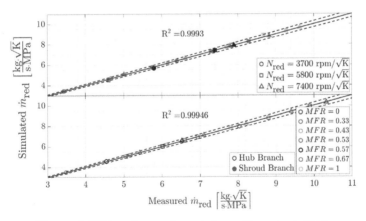

Figure 4. CFD and experimental results comparison.

The CFD simulations at low rotational speed are very close to the experimental data. When the rotational speed is higher, there are some cases with discrepancies, but they still are similar to the experimental data. All simulated cases have lower than 3% error and there is a very high R^2 value in both branches, which indicates a high concordance between both simulated and measured results. Thus, the results could be considered reliable for every MFR and rotational speed simulated.

5   ANALYSIS OF THE RESULTS

One of the main goals is to analyse the behaviour of the flow passing through each volute along the turbine and, more specifically, inside the rotor. With both flows differentiated and plotting the Mass Fraction of one of them, the flows can be followed inside the rotor. Figure 5 shows the mass fraction of air passing through the shroud branch along a rotor canal in the case of MFR 0.57 and rotational speed 3700 rpm/√K.

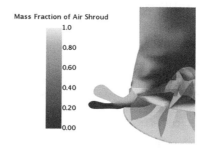

Figure 5. Mass fraction of air through shroud branch along a rotor passage.

It can be observed that the flow passing through each volute does not mix with the other along the rotor in a significant way. The air passing through the hub branch is mainly maintained near the rotor hub and the air passing through the shroud branch is mainly maintained near the rotor shroud inside the rotor. It can be seen that almost all the mixing is produced after the rotor. This fact could be exploited in reduced-order simulations such as in one-dimensional engine models, as it hints that the turbine may be modelled as two separated single volute turbines working in parallel. This behaviour occurs in all the rotor, as shown in the radial plane in Figure 5.

5.1   Effect on the rotor inlet and outlet areas

The behaviour for different MFR and rotational speed is the same in all cases. However, with MFR variations, the volume of each flow inside the rotor changes. This change in volume affects from the rotor inlet until the rotor outlet and it can be studied measuring the area occupied by each branch flow in both rotor inlet and rotor outlet sections in the CFD simulations, as shown in Figure 6. The numerical uncertainty of these measurements, as well as those shown in Figure 7, is given by the mesh error described in Table 2.

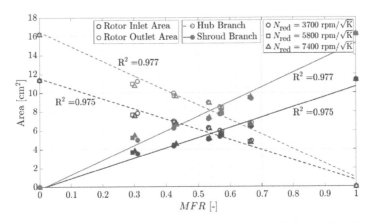

Figure 6. MFR and rotational speed influence on rotor inlet and outlet area occupied by each branch flow.

The area variation is mostly affected by the change in MFR both at rotor inlet and rotor outlet. The change in rotational speed (from 3700 to 7400 rpm/√K) has a very low effect in this area variation. Moreover, as shown in Figure 6, this variation is nearly linear with the change in MFR. This flow behaviour against the MFR in the rotor may be important for developing 1D meanline codes to calculate twin-entry turbines with the model of 2 VGTs in parallel.

5.2   Effect on the rotor inlet and outlet flow angle

The rotor inlet incidence angle is an important parameter. It allows calculating the incidence pressure losses at the rotor inlet. This parameter is commonly considered constant along all the rotor inlet section, as in Galindo et al. [5] and Chiong et al. [22]. In the case of twin-entry turbines, there will be two different rotor inlet incidence angles, one for each branch. One way to ensure that these angles are nearly constant along all the rotor inlet section for the same boundary conditions is to have a constant total pressure along the volutes. In all simulated cases, the total pressure is constant along both volutes, except the tongue. However, the typical deviation is low and it is confined in a small zone. Therefore, the rotor inlet incidence angle could be considered constant for each simulation.

Considering constant the rotor inlet incidence angle for each simulation, the influence of the MFR and the rotational speed on this parameter can also be studied. Figure 7 shows the effect of the MFR and the rotational speed on the rotor inlet incidence angle for hub and shroud branch. The value of this angle is computed from the radial axis (β_3 from Figure 3). The rotational speed has a low influence on the rotor inlet incidence angle, giving almost the same angle for three different rotational speeds (and different expansion ratio, as shown in Figure 2). However, the MFR has a higher influence on this parameter, increasing its value when the MFR increases for the shroud branch and vice versa for the hub branch. As in the rotor inlet and outlet areas previous study, this flow behaviour against the MFR may result important for 1D meanline codes.

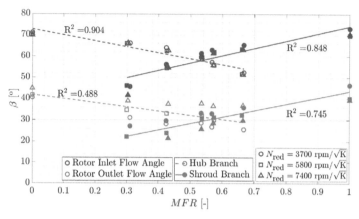

Figure 7. MFR and rotational speed influence on rotor inlet and outlet flow angle.

The same considerations could be done at the rotor outlet considering that all rotor channels have the same geometry and the rotor moves as a rigid solid. Figure 7 also shows the effect of the MFR and the rotational speed on the rotor outlet flow angle for the hub and shroud branch. The value of this angle is computed in this case from the axial axis. The rotational speed has a higher influence on the rotor outlet flow angle than on the rotor inlet incidence angle, increasing its value when the rotational speed increases for the shroud branch and vice versa for the hub branch. Moreover, the MFR also has a high influence on this parameter, decreasing its value when the MFR increases for the shroud branch and vice versa for the hub branch. But it is less than in the rotor inlet flow angle. Therefore, both rotational speed and MFR have a similar influence on the rotor outlet flow angle.

6   CONCLUSIONS

This paper presents an improved method for analysing the twin-entry turbine flow behaviour in terms of MFR and rotational speed using 3D CFD simulations. Moreover, with the experimental data available, the results obtained in the CFD simulations could be validated in spite of having only a few simulations due to their high computational cost in time.

CFD simulations have shown good agreement with the experimental data, with discrepancies in the reduced mass flow always lower than 3% and comparable to the experimental uncertainty. The simulations with extreme MFR values have a bit poorer agreement. It may be due to some flow recirculation at the volute outlet where, in these extreme cases, there is a sudden expansion. RANS simulations are not able to capture these phenomena correctly, and higher fidelity simulations may be needed.

The 3D CFD results show that the flow passing through each branch does not fully mix with the other until the rotor outlet.

It means that the flow is separated along the turbine. This information could be useful in 1D calculation models where twin-entry turbines could be studied as two separated single volute VGTs.

The rotor inlet area occupied by each branch flow depends mainly on the MFR, increasing its value when the MFR increases for the shroud branch and vice versa for the hub branch. Moreover, it could be approximately a lineal variation with the MFR.

The rotor inlet and outlet flow angles have also been studied. On one hand, the influence of the reduced speed in the rotor inlet incidence angle is low. It depends mainly on the MFR, increasing its value when the MFR increases for the shroud branch and vice versa for the hub branch. On the other hand, the rotor outlet flow angle is more influenced by the rotational speed and less by the MFR.

Generally, the main conclusion is that 3D CFD simulations are very useful to better understand the flow behaviour along twin-entry turbines. They provide information that is not intuitive a priori and they help to describe better the phenomena occurring inside twin-entry turbines, where it is more difficult to take measurements.

REFERENCES

[1] Fraser N., Blaxill H., Lumsden G., and Bassett M., «Challenges for increased efficiency through gasoline engine downsizing», SAE Int. J. Engines, vol. 2, pp. 991–1008, 2009.

[2] Payri F., Serrano J.R., Fajardo P., Reyes-Belmonte M.A. and Gozalbo-Belles R., «A physically based methodology to extrapolate performance maps of radial turbines», Energy Conversion and Management, vol. 55, pp. 149–163, 2012.

[3] Baines N., «Turbocharger turbine pulse flow performance and modelling – 25 years on», 9th International conference on turbochargers and turbocharging, pp. 347–362, 2010.

[4] Palfreyman D. and Martinez-Botas R., «The Pulsating Flow Field in a Mixed Flow Turbocharger Turbine: An Experimental and Computational Study», Journal of Turbomachinery, vol. 127, No. 1, p. 144, 2005.

[5] Galindo J., Tiseira A., Fajardo P. and García-Cuevas L.M., «Development and validation of a radial variable geometry turbine model for transient pulsating flow applications», Energy Conversion and Management, vol. 85, pp. 190–203, 9 2014.

[6] Galindo J., Climent H., Tiseira A. and García-Cuevas L., «Effect of the numerical scheme resolution on quasi-2D simulation of an automotive radial turbine under highly pulsating flow», Journal of Computational and Applied Mathematics, vol. 291, pp. 112–126, 1 2016.

[7] Galindo J., Fajardo P., Navarro R. and Garcia-Cuevas L.M., «Characterization of a radial turbocharger turbine in pulsating flow by means of CFD and its application to engine modelling», Applied Energy, vol. 103, pp. 116–127, 3 2013.

[8] Serrano J.R., Tiseira A., García-Cuevas L.M., Inhestern L.B. and Tartoussi H., «Radial turbine performance measurement under extreme off-design conditions», Energy, vol. 125, pp. 72–84, 4 2017.

[9] Walkingshaw J., Spence S., Ehrhard J. and Thornhill D., « An investigation into improving off-design performance in a turbocharger turbine utilizing non-radial blading», ASME Turbo Expo 2011, GT2011, pp. 2023–2032.

[10] Binder N., Carbonneau X. and Chassaing P., « Off-design considerations through the properties of some pressure-ratio line of radial inflow turbines», International Journal of Rotating Machinery, vol. 2008, pp. 1–8.

[11] Brinkert N., Sumser S., Schulz A., Weber S., Fieweger K. and Bauer H.J., «Understanding the Twin Scroll Turbine: Flow Similarity», de Volume 7: Turbomachinery, Parts A, B, and C, 2011.

[12] Romagnoli A., Martinez-Botas R.F. and Rajoo S., «Steady state performance evaluation of variable geometry twin-entry turbine», International Journal of Heat and Fluid Flow, vol. 32, No. 2, pp. 477–489, 4 2011.

[13] Aghaali H. and Hajilouy-Benisi A., «Experimental and theoretical investigation of twin-entry radial inflow gas turbine with unsymmetrical volute under full and partial admission conditions», ASME Turbo Expo 2007, GT2007–27807.

[14] Cerdoun M. and Ghenaiet A., «Characterization of a twin-entry radial turbine under pulsatile flow condition», International Journal of Rotating Machinery, vol. 2016, pp. 1–15, 12 6 2016.

[15] Fürst J. and Žák Z., «CFD analysis of a twin scroll radial turbine», EPJ Web of Conferences, vol. 180, p. 02028, 4 6 2018.

[16] Ghenaiet A. and Cerdoune M., «Simulations of steady and unsteady flows through a twin-entry radial turbine», ASME Turbo Expo 2014, GT2014–25764.

[17] Cerdoun M. and Ghenaiet A., «Unsteady behaviour of a twin entry radial turbine under engine like inlet flow conditions», Applied Thermal Engineering, vol. 130, pp. 93–111, 5 2 2018.

[18] Yokoyama T., Hoshi T., Yoshida T. and Wakashima K., «Development of twin-entry scroll radial turbine for automotive turbochargers using unsteady numerical simulation», de 11th International Conference on Turbochargers and Turbocharging, Elsevier, 2014, pp. 471–478.

[19] Hajilouy-Benisi A., Rad M. and Shahhosseini M.R., «Flow and performance characteristics of twin-entry radial turbine under full and extreme partial admission conditions», Archive of Applied Mechanics, vol. 79, No. 12, pp. 1127–1143, 12 2009.

[20] Serrano J.R., Arnau F.J., García-Cuevas L.M., Samala V. and Smith L., «Experimental approach for the characterization and performance analysis of twin entry radial-inflow turbines in a gas stand and with different flow admission conditions», Applied Thermal Engineering, vol: 159, 2019.

[21] Galindo J., Hoyas S., Fajardo P. and Navarro R., «Set-up analysis and optimization of CFD simulations for radial turbines», Engineering Applications of Computational Fluid Mechanics, vol: 7 pp: 441–460, 2013.

[22] Chiong M., Rajoo S., Romagnoli A., Costall A. and Martinez-Botas R., «Integration of meanline and one-dimensional methods for prediction of pulsating performance of a turbocharger turbine», Energy Conversion and Management, vol: 81 pp: 270–281, 2014.

Design and modelling of circular volutes for centrifugal compressors

H.R. Hazby*
PCA Engineers Limited, UK

R. O'Donoghue
PCA Engineers Limited, UK

C.J. Robinson
PCA Engineers Limited, UK

ABSTRACT

The current paper investigates the application of 1D performance models for external circular volutes. The aim is to substitute the volute in computationally expensive CFD calculations with lower order models during intermediate design iterations. The CFD predicted volute performance, available to the authors from designs of various stages including both vaned and vaneless diffusers, have been used as benchmarks. The conventional low order models have been examined and, where appropriate, further developments have been made. The final set of the models proposed in the paper are applicable to stages with both vanes and vaneless diffusers.

1   INTRODUCTION

The flow at the discharge of centrifugal compressor stages is typically collected by a volute or scroll, consisting of a spiral duct with increasing flow area around the circumference, followed by a diffusing cone downstream of the throat. In most cases, a significant diffusion has already taken place in the vaned or vaneless diffuser section upstream of the volute. The resultant low inlet dynamic head makes the overall performance of the stage less sensitive to the performance of the volute compared to the performance of the impeller and the diffuser. Nevertheless, inappropriate sizing of the volute can result in several percentage points loss in stage efficiency at the design point. Including the volute in a CFD analyses is therefore essential if realistic performance estimates are to be achieved.

The volute cross-sections are often 'one-sided' and can be external or overhung, depending on the available radial space. They also take a variety of shapes including circular, elliptical, rectangular, D-type, etc. The volute is normally modelled in CFD using an unstructured mesh due to its complex shape which requires high mesh density. Including this in stage CFD calculations significantly increases the computational overhead and consequently the turnaround time in assessing design iterations. This is particularly problematic when an entire compressor map must be predicted to determine the performance at multiple target operating points. The elapsed time of the design process can therefore be significantly reduced if the volute performance, notably its pressure recovery and total pressure loss, can be represented by lower order models.

The aim of this work is to investigate the applicability and improve the accuracy of 1D volute performance models specifically for external circular volute designs, commonly used by the authors, so that including the volute geometry in CFD calculations can be

---

* Currently at Mercedes AMG High Performance
Powertrains Ltd.

avoided in the intermediate design iterations. The 1D models are set out by Japikse [1], while some adaptions and additions have been suggested in [2] and [3]. A more recent summary of the 1D and 2D loss prediction methods is provided in [4]. Various combinations of these loss models have been used in the development and analysis of full compressor 1D modeling ([5-7]).

In the present paper, the results from existing 1D models have been compared with CFD calculations and then, where necessary, corrections have been proposed. First, the test cases used in the current study are discussed along with the numerical procedures used to calculate stage performance. This is followed by a brief discussion on volute performance and sizing parameters. Then, considering applications for both vaned and vaneless diffusers, various loss terms involved in volute performance modelling are examined and compared to CFD results when appropriate.

2   TEST CASES AND NUMERICAL METHOD

The volute geometries used in this study are of the constant inner radius type with circular cross sections (Figure 1). All the investigated cases were generated using a parametric setup in ANSYS Workbench, ensuring consistency of geometrical features. The volutes belong to compressors designed for a range of applications including turbochargers, air compressors and fuel cell compressors, representing a wide range of situations encountered in industry.

Figure 1. Geometrical shape of the investigated volutes.

The numerical calculations have been performed using ANSYS CFX 19.2 [8]. The meshes in the impeller and diffuser passages comprise structured grids with about 400,000 and 250,000 nodes respectively, generated using ATM topology in ANSYS Turbogrid. The volute domain was discretised using unstructured mesh with about 500,000 tetrahedral elements and 7 prism layers on the walls. The viscous effects were captured using the k-ε turbulence model with automatic wall functions; walls were assumed to be hydraulically smooth. The rotating impeller domains were connected to stationary diffuser domains using mixing plane interfaces.

As is common practice in the design process, a single impeller and a single diffuser passage were modelled, whereby the diffuser domain is connected to the volute also using a mixing plane interface. The more computationally expensive full-annulus calculations, where all the impeller and diffuser passages are included in the model (and a General Grid Interface [8] with non-conformal meshes is used between the diffuser and the volute) have been found generally to give similar results. This is demonstrated in Figure 2, where the predicted overall total-to-total pressure ratio and efficiency of the stage are compared for a turbocharger compressor using the two calculation approaches. Close agreement can be observed between two sets of results. The mesh topologies and calculation setups are shown for both cases in Figure 3.

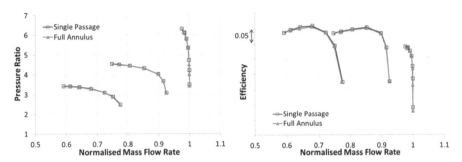

Figure 2. Predicted performance of a turbocharger compressor using single passage and full-annulus calculations.

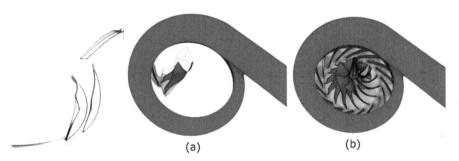

Figure 3. Mesh topology and calculation setup for (a) single passage and (b) full-annulus calculations.

3 SIZING OF THE VOLUTE

The volute geometry consists of a spiral duct with increasing flow area around the circumference, followed by a diffuser cone downstream of the throat. The first part may be considered as a single 'bladed' component with a geometrical throat equal to the volute throat area. Therefore, the behaviour of the volute at a particular flow condition depends on the variation of the stream-tube thickness from the volute inlet to its throat. At the design operating point, the volute is typically designed to achieve uniform static pressure around the circumference (no acceleration or deceleration of the flow), with any rise in static pressure occurring in the downstream diffuser cone. This is also normally close to the point of minimum total pressure loss across the volute and occurs only at one operating point at each rotational speed. At lower mass flow rates, with higher inlet flow angles, the throat is too large, resulting in deceleration of the flow along the path, meaning that the volute acts like a diffuser. Conversely, at mass flow rates higher than the design value, the throat is too small and the volute behavior is similar to a nozzle.

The performance variation of a well behaved volute is therefore predominantly a function of inlet flow angle. The total pressure loss and static pressure recovery coefficients in volutes can be respectively defined as:

$$Y_p = \frac{P_{04} - P_{06}}{P_{04} - P_4} \quad C_p = \frac{P_6 - P_4}{P_{04} - P_4} \tag{1}$$

where, suffices 4 and 6 denote volute inlet and outlet planes. The variation of Y_p and C_p with mass flow rate and inlet flow angle for a typical volute, downstream of a vaneless diffuser, is shown in Figure 4. The results are predicted using CFD calculations at three different operating speeds across an automotive turbocharger compressor map. It can be observed that the performance parameters from different speed lines fall onto unique curves with respect to inlet flow angle, with the minimum loss occurring at the design flow angle.

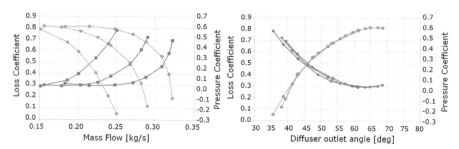

Figure 4. Variation of volute performance parameters with mass flow and inlet flow angle.

The design flow angle can be linked to the required volute throat area at the selected design point by considering the conservation of mass and angular momentum from volute inlet to its throat. Assuming a frictionless, incompressible flow with no circumferential pressure gradient, the conservation of mass and angular momentum from volute inlet to the throat ('th') can be written as:

$$V_{t4}r_4 = V_{th}r_{th}$$
$$V_{th} = V_{t4}r_4/r_{th} \tag{2}$$

$$V_{th}A_{th} = 2\pi r_4 b_4 V_{m4}$$
$$V_{th} = 2\pi r_4 b_4 V_{m4}/A_{th} \tag{3}$$

Combining equations 2 and 3 gives:

$$A_{th}/r_{th} = 2\pi b_4/\tan\alpha_4 \tag{4}$$

The ratio A/r (which has dimensions of length) relates the cross-sectional area at the volute throat (or for simplicity at the top-centre, 360° plane) to the radius from the centroid of that area to the machine axis, and is often used to denote the size of a given volute. This is determined only by the width of the diffuser channel and the flow angle at diffuser outlet at the chosen matching condition.

The variation of the total-to-total isentropic efficiency calculated at the outlet of the diffuser for the example of the compressor stage used in Figure 4 is shown against volute A/r in Figure 5. The locus of the peak efficiency at the diffuser outlet is denoted by the dashed line. A smaller A/r value matches the volute peak performance to the peak performance of the upstream stage at higher speeds. However, if high stage efficiency is required at lower operating speeds, a larger A/r might be needed, accepting the suboptimal performance at the higher speeds.

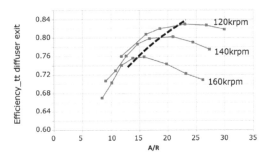

Figure 5. Total-total isentropic efficiency at the diffuser outlet versus A/r.

The choice of the design inlet angle of the volute does not only move the minima of the loss bucket, it also changes the shape of the loss and recovery curves, as illustrated in Figure 6. The larger volutes tend to have a higher minimum loss but a flatter loss and pressure recovery characteristics. Smaller volutes, on the other hand, tend to a have lower minimum loss but sharper (less flat) characteristics, negatively impacting the performance at off design condition.

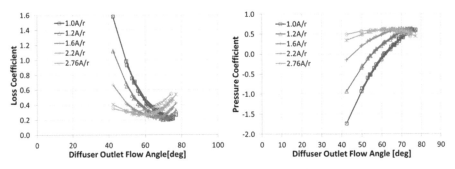

Figure 6. Impact of A/r on volute performance.

The effect of matching point choice is demonstrated in Figure 7, comparing two different sized volutes with the same impeller and diffuser. The smaller volute results in higher efficiency at high speeds and lower efficiency at part-speed relative to the results for the larger scroll. The choice of a suitable volute A/r, or in other words establishing the matching of the volute to the rest of the stage, depends on the overall requirement from the compressor map. It is commonly known that in practice a purely efficiency-based optimization may not be possible as the impact of the volute performance on the stage performance and operating range near the choke and the surge conditions should also be considered.

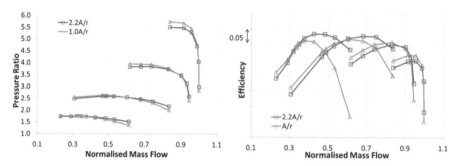

Figure 7. Impact of volute sizing on compressor performance.

4   VOLUTE PERFORMANCE MODELLING

The present objective is to estimate the volute performance using reduced order models instead of CFD simulations in order to reduce the calculation overhead. For a typical volute mesh used by the authors, including the volute in the CFD models increased the calculation time by 100-200% at the design point and up to 250% at near surge condition. It should be emphasized that the current models are applied during the intermediate design steps and the performance of the final design needs to be checked with full stage CFD calculations.

In its simplest form, for a volute operating downstream of a vaneless diffuser, if the parabolic functions of the loss and pressure recovery coefficients versus inlet flow angle (see Figure 4) are known (for example from previous calculations) they can be used to estimate the overall stage performance with reasonable accuracy. However, such information may not be available in advance. Furthermore, for a vaned diffuser, a_4 is sensibly constant as it is set by the diffuser geometry. In these cases, physics-based models are required to estimate volute performance. Models currently available in the literature are reviewed here and, where discrepancies in the loss and recovery characteristics are observed, corrections are proposed with the aim to reduce the errors across the range of volutes studied.

Meridional and tangential head losses: Two main sources of loss in a volute are the meridional and tangential head losses. Together, they form the 'bucket' or para-bolic shape of the loss coefficient with respect to the inlet flow angle that was seen in Figure 4. The meridional loss component is associated with the fact that only the tan-gential flow is collected by the volute and the incoming meridional head is lost through dissipation, mainly energizing the vortical flow in the volute. Its level can be esti-mated as:

$$\Delta P_{0_{mer}} = C_{mer} \frac{\rho V_m^2}{2} \tag{5}$$

Not all of the meridional head is lost; a part is spent to stabilise the flow [1] and this is accounted for in Equation 4 by a correction factor, C_{mer}. A value of 0.65-0.7 is found to give satisfactory results across the range of volutes considered here.

The tangential head loss accounts for diffusion losses between volute inlet and the throat as a sudden expansion loss. Therefore, it is by definition zero when the main throughflow (tangential velocity) accelerates between inlet and throat. For a decelerating throughflow, its value can be estimated as:

$$\Delta P_{0_{tan}} = C_{tan} \frac{\rho(V_{t4} - V_5)^2}{2} \text{ for } \cdot V_5 < V_{t4}$$

$$\Delta P_{0_{tan}} = 0 \text{ for } \cdot V_5 V_{t4} \tag{6}$$

The variation in meridional, tangential and combined loss coefficients, calculated using equations 4 and 5, are shown against volute inlet flow angle for a centrifugal compressor operating with a vaneless diffuser in Figure 8. For comparison, the volute performance predicted by CFD calculations is also shown in the figures. The shape of the combined loss coefficient agrees well with the CFD results. The absolute levels are, however, under-predicted as extra loss terms associated with exit cone and friction have not been included in the 1D model.

At high mass flow rates (low flow angles) the meridional loss is dominant and explains most of the total loss. Whereas at low mass flow rates it is the increasing tangential loss that causes the total loss to rise.

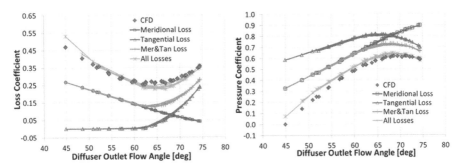

Figure 8. Meridional and tangential loss and pressure recovery coefficients for a volute operating downstream of a vaneless diffuser.

The variation of the performance parameters with inlet mass flow at three different operating speeds are compared with the corresponding CFD results for a volute operating downstream of a vaned diffuser in Figure 9. In general, when vaned diffusers are present, the meridional head loss appears to be dominant over most of the operating range. Although the value of meridional and tangential total pressure loss increase at high mass flow rates, the loss coefficients show opposite trends due to the variations in the dynamic head at diffuser vane outlet. In general, the CFD predicted trends are captured well by the 1D model.

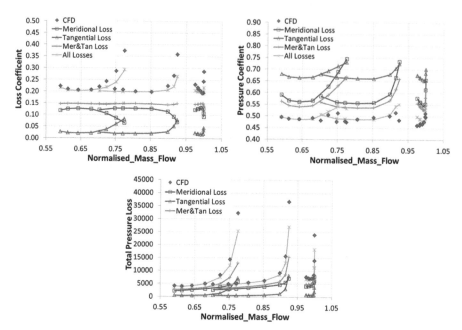

Figure 9. Meridional and tangential loss coefficient, pressure recovery coefficient and total pressure loss for a volute operating downstream of a vaned diffuser.

Friction losses: The friction losses are associated with surface friction in the volute assuming a hydraulically smooth surface. These losses are calculated by considering the volute as an equivalent pipe with a hydraulic diameter d_H, length L and an average throughflow (tangential) velocity $\overline{V_t}$ calculated as follows.

$$d_H = \sqrt{4A/\pi}$$
$$L = \pi(r_2 + R)/2 \tag{7}$$
$$\overline{V_t} = (V_{t4} + V_5)/2$$

Sutherland's law of viscosity and the law of Blasius for turbulent pipe flow are used to calculate the fluid dynamic viscosity and the friction coefficient respectively:

$$\mu = \frac{C_1 T_4^{\frac{3}{2}}}{T_4 + C_2}, C_1 = -1.458 \times 10^{-6}, C_2 = 110.4$$
$$Re = \frac{\rho_4 \overline{V_t} d_H}{\mu} \cdot C_f = \frac{0.3164}{Re^{0.25}} \tag{8}$$

The corresponding total pressure loss coefficient is calculated as:

$$\Delta P_{0fric} = \frac{1}{2} C_f \left(\frac{L}{d_H}\right) \rho_4 \overline{V_t}^2 \tag{9}$$

The predicted volute loss and static pressure recovery with and without friction losses are compared to corresponding CFD results in Figure 10 for a stage with a vaneless diffuser.

296

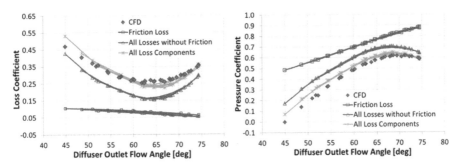

Figure 10. Impact of friction losses on the volute performance.

The variation of the frictional total pressure loss coefficient with flow angle is also depicted in the figure, showing a linear increase of the loss coefficient as the flow angle is reduced at high flow rates. The influence of the friction losses on the volute performance is relatively strong (about 35% of the overall loss) as this particular compressor operates at low Reynold's number.

Exit cone loss: The exit cone loss is treated as a sudden expansion loss from throat to the cone exit and is calculated as:

$$\Delta P_{0_{exit}} = \frac{1}{2}\rho_4 (V_5 - V_6)^2 \tag{10}$$

where V_6 is the velocity at the cone exit and is estimated from the known cone area ratio and the mass flow. A blockage factor of $Bl_{exit} = 0.06$ at the cone exit is found to improve agreement with the CFD results.

$$A_6 = A.AR_{cone}$$
$$V_6 = \frac{}{\rho_4 (A_6 (1 - Bl_{exit}))} \tag{11}$$

It can be observed in Figure 11 for a vaneless stage that the contribution of the exit cone loss to the overall volute loss increases as the flow rate increases.

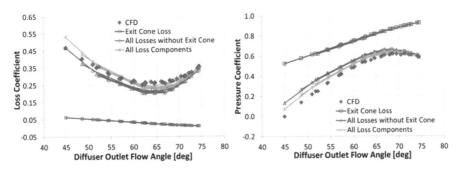

Figure 11. Impact of exit cone losses on the volute performance.

Trailing edge mixing loss: When vaned diffusers are present, an extra loss term associated with the mixing of the wakes downstream of the vane trailing edges needs to be included. In the case of CFD calculations with a mixing plane between diffuser and volute, these losses are automatically captured in the entropy generation across the mixing plane. The mixing loss is often modeled as sudden expansion loss (see for example [9] and [10]). Following a similar philosophy, in the current work the wake mixing loss has been related to the predicted blockage downstream of the diffuser vanes:

$$Bl_{diff} = 1 - \bar{V}_{m_{area}}/\bar{V}_{m_{mass}}$$
$$Y_{mix} = C_{mix}Bl_{diff}^2, \cdot C_{mix} = 2.5 \tag{12}$$

$$\Delta P_{0\,mix} = Y_{mix}DPD \tag{13}$$

The predicted mixing loss and its impact on the overall volute performance as a function of inlet mass flow are compared with CFD results in Figure 12. The mixing loss has generally a parabolic shape with a steep rise in loss levels towards choke. Without this term the sudden drop in the pressure recovery that occurs near choke, as captured by CFD calculations, could not be reproduced. In this case, neglecting the mixing losses under-predicts the overall loss by about 10% near the design point and 50% near choked condition. Including the mixing losses, removed the error at the design point and reduced the maximum error to 20% at the choked condition compared with the CFD results.

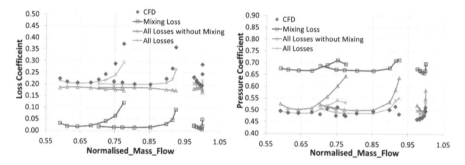

Figure 12. Impact of diffuser mixing loss on the volute performance.

Incidence losses: Although the existing models reasonably capture the overall characteristics of the volute loss and pressure recovery coefficients, the shape of the curves towards low flow angles is not generally reproduced correctly. This was particularly true for extreme cases of vaneless diffuser stages with volutes matched at high pressure ratios with a design flow angle of 70° or higher. The comparison of the CFD predicted losses at volute throat and volute outlet with 1D predictions (shown in Figure 13) suggests that this is due to a major increase in loss in the exit cone as choke approached.

The sudden increase in the loss levels generally accompanied a strong increase in blockage $(1 - \bar{V}_{m_{area}}/\bar{V}_{m_{mass}})$ with mass flow rate at the inlet of the volute. The blockage in this case is mainly related to the circumferential variation of the velocity and pressure field around the tongue and varies with the incidence of the flow on the tongue. Investigation of the range of compressors available suggested that the incidence loss was based not only on the difference between the inlet flow angle and the design flow angle (representative of the tongue incidence) but also on the design flow angle itself. The incidence loss was found to become significant when design flow angle was larger than 50° and its value varied as parabolic function of the incidence angle $(\alpha_i = \alpha - \alpha_d)$. The model formulation applied to represent the observations is given as:

$$\alpha_d > 50$$
$$\alpha_i = \alpha - \alpha_d$$
$$Y_{inci_{50}} = 0$$
$$Y_{inci_{70}} = 0.0015 * (\alpha_i + 10)^2 \ where \ \alpha_i < -10 \qquad (14)$$
$$= 0 \ where \ \alpha_i - 10$$
$$Y_{inci} = \frac{Y_{inci_{70}} - Y_{inci_{50}}}{70 - 50}(\alpha_d - 50) + Y_{inci_{50}}$$
$$\Delta PO_{inci} = Y_{inci}(P_{04} - P_4) \qquad (15)$$

The impact of the incidence loss on the performance of a volute operating downstream of a vaneless diffuser with a design flow angle of 70.5° is shown in Figure 13. The strong effect of the incidence loss on the shape and level of the total loss profile is evident in the figure. In this case, including the incidence losses increased the overall loss at the highest flow rate by a factor of two, reducing the overall error in predictions from 45% to about 11%.

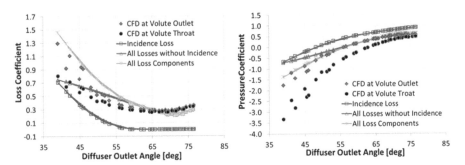

Figure 13. Impact of diffuser mixing loss on the performance of a volute operating downstream of a vaneless diffuser.

In the case of a well-designed volute operating with a fixed flow angle downstream of a vaned diffuser, the incidence loss is zero. However, the losses can increase if the volute is mismatched to the diffuser vane, then effectively operating at an incidence. Figure 14 shows the corresponding losses for a well-matched volute as well as a volute operating with 25° mismatched to its upstream vaned diffuser. The predicted loss level and its impact on the overall loss can be clearly seen for the mismatched volute.

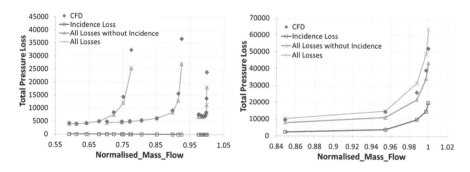

Figure 14. Left: Volute matched to diffuser outlet flow angle. Right: Volute mismatched by 25°.

5   CONCLUSIONS

The application of 1D performance models for external circular volutes have been investigated and the impact of the volute sizing as well as the relevant loss terms on volute performance have been discussed. The following points can be made:

- The common loss terms including meridional, tangential, friction and exit cone loss predicted the overall shape of the characteristics but marked differences remained between 1D and CFD predictions in a number of cases.
- When vaned diffusers were present, a mixing loss associated with the vane trailing edge wakes was needed to predict the losses correctly.
- Addition of an incidence loss term has been found to be necessary especially in highly loaded stages with vaneless diffusers or when the volute was mismatched to its upstream vaned diffusers.
- The full set of models can be used effectively for cases with both vaned and vaneless diffusers.
- The models presented in this paper have been tuned for a family of external volutes with circular cross sections. While the physical models used here are felt to be general, the contribution of various loss terms may be effected by changes in the cross-sectional shape, extent of overhang, design of the tongue etc.

ACKOWLEDGEMENTS

The authors acknowledge PCA Engineers Ltd for supporting the work and for permitting the publications of the results.

NOMENCLATURE

A	**area (m^2)**	r	**radius (m)**
b	diffuser outlet width (m)	Re	Reynolds number
Bl	blockage	T	static temperature (K)
c_f	friction coefficient	V	velocity (m/s)
C_p	pressure recovery coefficient	Y_p	total pressure loss coefficient
d_H	hydraulic diameter (m)	α	flow angle (degrees)
L	characteristic length (m)	α_i	incidence angle (degrees)
P	static pressure (Pa)	μ	dynamic viscosity (Pa-s)
P_0	total pressure (Pa)		
Subscripts			
4	diffuser outlet	f, fric	Friction
5, th	volute throat	Inci	Incidence Loss
6	stage outlet	mass	mass averaged
area	area averaged	m, mer	meridional
d	design condition	mix	mixing Loss
exit	exit cone	t, tan	tangential

REFERENCES

[1] Japikse, D., 1996, "Centrifugal Compressors Design and Performance", ISBN 0-933283-03-2.

[2] Weber, C. R., and Koronowski, M. E., 1986, "Meanline Performance Prediction of Volutes in Centrifugal Compressors." Proceedings of the ASME 1986 International Gas Turbine Conference and Exhibit. Volume 1: Turbomachinery. Dusseldorf, West Germany. June 8–12, 1986. V001T01A091. ASME.

[3] Van den Braembussche, R. A., Ayder, E., Hagelstein, D., Rautenberg, M., and Keiper, R., 1999, "Improved Model for the Design and Analysis of Centrifugal Compressor Volutes." ASME. *J. Turbomach.* July 1999; 121(3): 619–625.

[4] Van den Braembussche, R. A. (2006) Flow and Loss Mechanisms in Volutes of Centrifugal Pumps. In Design and Analysis of High Speed Pumps (pp. 12-1 – 12-26). Educational Notes RTO-EN-AVT-143, Paper 12. Neuilly-sur-Seine, France: RTO.

[5] Xiaoyang Gong & Rui Chen, 2014, "Total Pressure Loss Mechanism of Centrifugal Compressors", Mechanical Engineering Research; Vol. 4, No. 2; ISSN 1927-0607 E-ISSN 1927-0615.

[6] Khoshkalam, N., Mojaddam, M. and Pullen, K. R. ORCID: 0000-0001-8501-9226 (2019). Characterization of the Performance of a Turbocharger Centrifugal Compressor by Component Loss Contributions. Energies, 12 (14),2711. doi: 10.3390/en12142711

[7] Sanz Solaesa, S., 2016, "Analytical Prediction of Turbocharger Compressor Performance: A Comparison of Loss Models With Numerical Data", Master of Science Thesis MMK 2016:176 MFM 169, KTH Industrial Engineering and Management.

[8] ANSYS manual, Version 19.2, ANSYS Inc., 2019.

[9] Aungier, R. H., 2000, "Centrifugal Compressors: A Statagy for Aerodynamic Design and Analysis", ISBN 0-7918-0093-8.

[10] Cumpsty, N. A., 1989, "Compressor Aerodynamics", ISBN 0-470-21334-5.

SC-VNT™ a route toward high efficiency for gasoline engines

N. Bontemps[1], D. François[1], J.C. Sierra[1], P. Davies[1], L. Lazzarini Monaco[2], A. Fuerhapter[2]

[1]Garrett Advancing Motion, Thaon-Les-Vosges, France
[2]AVL List GmbH, Graz, Austria

ABSTRACT

CO2 targets and anticipation of Eu7 Legislation is accelerating the widespread adoption of VNT™ turbochargers for Gasoline engines. VW set the industry best BSFC of 222 g/kWh when it released its 1,5L EA211 Miller engine with a Garrett GT12V in the Vienna Motor Symposium in 2017 [1,2]. Since this time engine ratings and VNT product ranges have been expanded and it is now expected that we will see new releases from several OEMs in multiple applications from 2020 onwards.

As Powertrain's become increasingly electrified the idea of an engine dedicated for hybridisation (DHE) is emerging and concepts are being studied of how to reach significantly lower BSFCs, in a range of loads and speeds that such an engine would operate at in a future hybridized vehicle.

Single stage compression associated to Miller cycle will eventually reach a limit. One possible solution to take the next step in BSFC improvement towards 200 g/kWh is to split the compression between two compressors and control compressor outlet temperatures and compressor work by cooling the charge between the two stages.

This paper shows how such a machine can be realized with a single shaft with a single VNT turbine and will discuss results from a collaboration project between Audi, AVL and Garrett.

1   INTRODUCTION

After more than one century of continuous improvements of the internal combustion engines (ICE) in terms of power, costs, thermal efficiency and finally fuel consumption, concern over global warming is now leading to a worldwide reinforcement of CO_2 emissions for all means of transport using combustible fuels especially for passenger cars.

From 2021, phased-in from 2020, the Europe fleet-wide average carbon dioxide emission target for new cars will be 95 g CO_2/km (Figure 1).

Future emission standards include more stringent requirements for HC, CO, NO_x and particulates and this is accelerating the pace of innovation process at engine developers and car manufacturing OEMs.

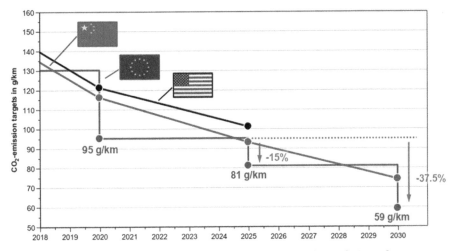

Figure 1. CO$_2$ legislation for Europe, China and North America.

Furthermore, monitoring of CO emissions into the Real Driving Emissions (RDE) and the possible introduction of a specific Conformity Factor (CF) is questioning the use of fuel enrichment for high temperature component protection again and will result in running Gasoline engines at lambda 1 operation across the entire engine map.

Various technological solutions are available to satisfy these expected norms. One major challenge for engine and component manufacturers however is the detailed understanding not only the behaviour of the individual components but also the potential interactions and synergies when combining different technologies in a single engine application.

In such a context, it is feasible that an extremely heavily Millerized gasoline engine could be a good candidate to pair with heavily hybridised powertrains (HEV, PHEV) to make vehicles more efficient.

2   GASOLINE VNT MASS ADOPTION AND LIMITS

2.1   Mass production

In this context, gasoline VNT technology was first presented in 2016 [1] by VW & 2017 [2] by Garrett. The technology was launched into mass production in 2018. When used in conjunction with an engine using the Miller cycle, the VNT turbocharger is now a well-known technical breakthrough to enable the engine to run in stochiometric conditions across the whole engine map since the exhaust gas temperature can be significantly reduced (see Figure 2).

The main contribution to this Brake Specific Fuel Consumption (BSFC) and CO$_2$ improvement is the result of the level of Millerization. As already shown in other papers, high degrees of Millerization lead to an increased engine geometrical compression ratio, a reduced effective compression ratio, and so the expansion ratio, a better knock resistance and, finally, an earlier combustion phasing. All this steer to an increase in Brake Thermal Efficiency (BTE).

303

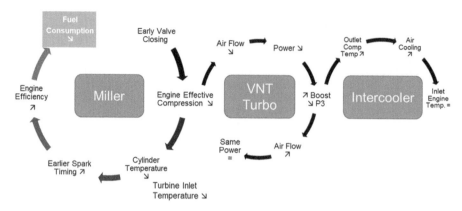

Figure 2. Miller cycle + VNT principle.

As previously stated, the variable turbine geometry market dedicated to gasoline engines is set to grow very quickly. Figure 3 is an estimation of the Market evolution within the next years and the sharing of gasoline turbochargers between waste-gated and variable geometry turbine units.

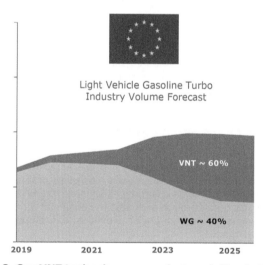

Figure 3. Gas VNT turbocharger market evolution in Europe.

Garrett is in series production with a first generation and committed to developing a second generation of gasoline VNT. However, based on simulations and first results measured on development engines, it is now proved that the performance potential of a typical Miller layout associated to a single compression stage gasoline VNT turbine is currently limited in the range of 90 kW/l with a minimum BSFC in the area of 225 g/kWh. At such a performance the turbocharger is running close to its maximal limits in terms of speed and compressor outlet temperature.

2.2 Millerization limits with a single stage compressor and VNT

It is commonly known that Millerization requires higher boost to compensate the lower engine volumetric efficiency induced by the early or late intake valve closing. Although Miller cycle adoption creates a virtuous cycle, the investigations to push for a higher Miller ratio with more powerful engines, or to reach better thermal efficiencies, become a significant challenge for both compressor and turbine stages.

On the compressor side, higher Miller ratio is obviously leading to:

- Increase the shaft speed because of higher boost demand. The maximum shaft speed may be a limiting factor and the speed margin, or the turbocharger speed control strategy, is a key parameter to deal with.
- Operating points moving from high efficiency island to the top right corner of the well-known compressor map. Immediate consequences are: lower compressor efficiencies with points running at higher compression ratio and higher compressor outlet temperature. As for Diesel applications, this is a limiting factor and is likely to drive the engine maximum Miller ratio.

Increasing engine millerization has some consequences on turbine component also:

- It is expected that the higher boost/air claim requires more work demand from compressor stage. This causes a closing of the turbine guide vanes which can lead to a bad turbine efficiency if the vanes are too much closed
- Consequently, the low turbine efficiency will induce a higher exhaust backpressure increasing the pumping work and affecting the fuel consumption negatively
- The higher engine backpressure and subsequent higher residual gas concentration within the combustion chamber will enhance the knock tendency. Consequently, the delayed spark timing will induce a higher exhaust gas temperature which may compromise the lambda 1 conditions
- The retarded combustion phasing is driving a poor combustion efficiency which will affect the fuel consumption also
- The resulting higher air demand requires higher boost pressure even more heading to close the VNT nozzle, and so forth...

All these points are creating a negative spiral to be avoided at all cost: based on a focused turbocharger optimization with GT-Power, it is shown that an innovative boosting system being capable of generating high boost pressure without the disadvantage of higher exhaust backpressure may be a solution. Without considering a conventional two-stage turbocharging solution, a two-stage compression system with a single turbine wheel appears to be a good candidate.

2.3 Split compression

As explained in the first part of this paper, increasing Millerization of gasoline engines drives the need for high compressor compression ratio soaring up to 5.0 or more. For instance, for Diesel applications, current maximal single stage compression ratio is close to 4.5:1.

For very high boost pressure requirement, on Diesel engines, it is common practise to employ a two-stage system composed with two compressor wheels and two turbine wheels in two separated turbochargers arranged in series. The different operating modes of the turbochargers are controlled with bypass valves.

However, such systems have large turbine housing thermal inertia which lead to excessively long cat light-off, which is significant disadvantage for future gasoline applications.

The requirement for strong Millerized engines still requires a turbocharger that can deliver large air mass flow range at high compression ratio with a good performance. It is then possible to mount the two-stage serial compressors on the same shaft using a single VNT turbine wheel.

3 SPLIT COMPRESSION (SC) CONCEPT WITH INTER-STAGE COOLER

The combination of a two-stage compressor with a single stage turbine is not a new idea but the system has never shown a significant benefit, mainly because of prohibitive rotating inertia when this concept is associated to a previous 'old' gasoline engine, where fuel enrichment was still allowed.

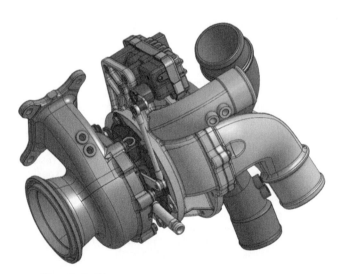

Figure 4. SC-VNT unit with inter-stage cooler.

Thanks to the uncapped potential of Millerized gasoline engines, at same power and same operating conditions, the thermal efficiency benefit requires lower air mass flow. Since such engines have a lower specific air demand we can use smaller wheels and turbocharger inertia can be controlled to an acceptable level.

3.1 Split compression interest

The two compressors staged in series concept lies in a separation of compression ratio between the first and the second stages.

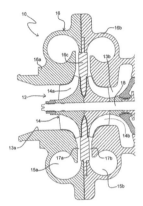

Figure 5. Split Compressor concept.(*from European Patent EP 3 249 234 B1*).

The air is first pressurized within the first conventional stage and then is routed to an inter-stage cooler. The cooled air exiting this exchanger is then driven to the inlet of the second-stage compressor where the air is further compressed before being cooled again by the main charge air cooler before supplying the engine inlet manifold.

This paper will demonstrate that inclusion of an inter-stage cooler is of significant beneficial.

Split compression allows us to achieve compression ratios up to 6:1 without over-speeding the turbo, keeping a wide composite compressor map and ensuring a good overall compression efficiency (see Figure 6, 7 and 8).

As shown in Figure 4, the first and second impellers are arranged in a back-to-back design to reduce the packaging impact. This design has led the engineers to optimize the feeding of the second impeller to improve the air repartition and limit the pressure drops to a minimum.

Since the overall boost pressure is the result of two compressions, a two-stage series compressor can achieve the desired pressure ratios at a lower shaft speed than a single-stage compressor. The final speed is mainly depending on the matching selection and the inter-stage cooler features and efficiency.

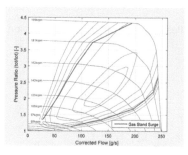

Figure 6. LP compressor wheel.

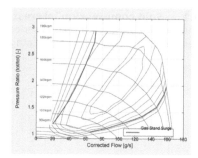

Figure 7. HP compressor wheel.

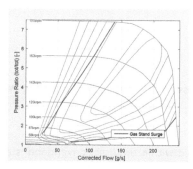

Figure 8. Composite SC map with interstage cooler.

3.2   Inter-stage cooler interest

There are three main consequences of applying a high split compression ratio with interstage cooler on intake air charge:

- When air is compressed, it increases in temperature. Compressor perform-ance typically approximates to isentropic compression with an efficiency around 70% in this high flow and high compression ratio area.
- The higher the air temperature increase over the course of the compression, the worse the compression efficiency will be. The maximal acceptable impel-ler outlet temperature is a physical limit that may be a limiting factor to the degree of Millerization that we can exploit, as already mentioned in § 2.2
- By cooling the compressed air between each stage, the compression pro-cess becomes more isothermal which minimizes the compression work (Figures 9 & 10). The work done by the compressor is less than if it was a single-stage. This reduces the demanded turbine power and improves engine backpressure. However, the pressure drop through the interstage cooler has to be taken into account and reduces the benefit on compres-sion work.
- The second stage impeller size can be smaller and the overall rotating inertia will be reduced compared to a system without an interstage cooler.

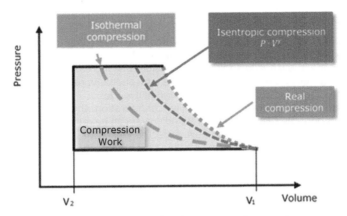

Figure 9. Single wheel compression work.

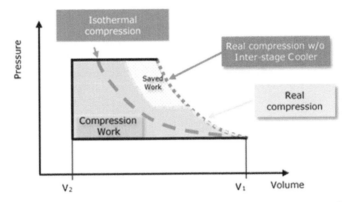

Figure 10. Interest of interstage cooling on a two-stage compression work.

On figure 10, the yellow/green shaded area represents the savings achieved with the two-stage and intercooler compression versus a single stage on the top Figure 9.

Splitting the compression work into two-stages with intercooling between stages can be the most energy efficient arrangement. In addition to the thermodynamic benefit, intercooled compression systems result in lower discharge temperatures, which reduce the need for special compressor materials and the size of the second stage compressor wheel.

3.3 SC-VNT with interstage cooler in conjunction with a Miller gas engine

The lower compression temperatures result in higher boosting efficiency and a lower turbine work demand and allow us to reach extremely high boost pressures while maintaining or even decreasing engine backpressure. The consequently lower residual gas concentration and lower in-cylinder charge temperature are a lever to optimize the combustion phasing, to increase the engine compression ratio and then the overall engine efficiency while respecting the stochiometric boundary conditions.

Different approaches are possible:

- Compared with a given single stage boosting Miller engine, the SC-VNT turbocharger can be used to raise the Miller ratio of this engine. Consequently, the induced engine compression ratio increase will improve the overall efficiency even more in all operating conditions resulting in fuel economy and CO_2 reduction over the WLTC and RDE cycles
- With unchanged layout of the Miller configuration, the boosting pressure and thus both the BMEP as well as Rated Power can be raised with a SC-VNT under lambda 1 conditions. It is now expected that with a well-matched SC-VNT turbocharger and an adapted Miller ratio, the performance potential should reach around 120 KW/L with an excellent minimum BSFC around 208 g/kWh (see Figure 11) measured on a 2.0L gasoline Demo engine on dyno and without either exhaust gas recirculation (EGR) or Water Injection (WI) which are both systems increasing the engine costs.

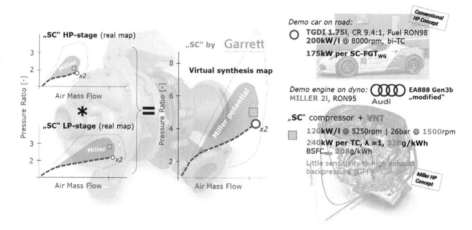

Figure 11. Preliminary results with SC concept on two gas engines.

4 RESULT

Given the already mentioned potential and benefits of the SC-VNT in § 2.2 and § 3.3, a Miller Cycle engine (Audi EA888 Gen3), which achieves 70 kW/L specific power at stoichiometric operation with a conventional WG-turbocharging system, was used as the baseline for the investigations.

The initial assessment of this engine, with geometric compression ratio of 12:1 (approximately) has demonstrated the potential increase of specific power from 70 kW/L to approximately 110-115 kW/L at stoichiometric operation and the reduction of the minimum BSFC from roughly 215 g/kWh to 208 g/kWh, which is, in itself, already impressive (Figure 12).

Considering future trends, which lead to some degree of electrification, such high performance will not be necessary for all passenger car segments. Therefore, a further step towards improving the overall efficiency was given and the geometric compression ratio of the engine was increased to roughly 14:1. The results show that, even though the maximum specific power was reduced (90-100 kW/L), there was a significant gain in efficiency from 41 up to 42,4% BTE, as it can be observed in Figure 12 and Figure 13.

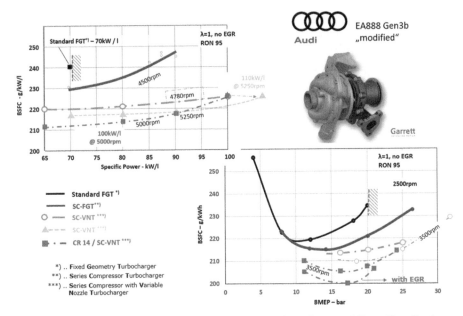

Figure 12. Improvement of TGDI performance by advanced Boosting System (SC-VNT).

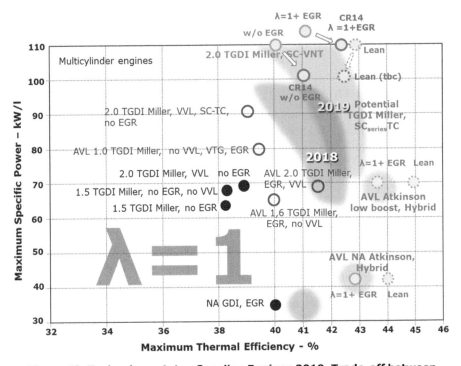

Figure 13. Technology status Gasoline Engines 2019. Trade-off between maximum efficiency and maximum specific power.

311

Considering the almost mandatory future synergy with a full Hybrid, an aggressive Miller concept, combined with such advanced boosting system, shows excellent potential of increase efficiency without a significant reduction in specific power.

Once this step has been made, an engine optimized for best efficiency (e.g. compression ratio > 15:1, stroke/bore ratio >1.2, reduced friction, etc...) that can reach 43-44% thermal efficiency and also a specific power higher than 80 kW/L no longer seems unrealistic.

5   CONCLUSION

Nowadays, all the predictions tend to show that more than 50% of Gasoline engines sold into the European market will be boosted by VNT turbochargers by 2025. Applying Millerization and VNT will set the Benchmark BSFC at 225 g/kWh @ 70 -90 kW/L.

Adopting high engine compression ratio, ultra Millerization and Split-Compression in the form of a SC-VNT turbocharger and Interstage Cooling allows us to reach BSFC levels approaching 200 g/kWh @ 100-120 kW/L with Lambda 1 conditions, respecting the maximum turbine inlet temperature by 980C and without using EGR or Water Injection.

More importantly this project has demonstrated BSFC below 220 g/kWh across an extremely broad operating range of speeds and loads. Further potential exists to go lower than 200 g/kWh and lower again in temperature with the additional application of EGR.

Such a concept could be ideal for a Load Point Shifting Hybrid Vehicle equipped with a Dedicated Hybrid Engine (DHE) where it is expected that the engine might be OFF for large portions of the WLTC cycle and the engine will be called upon to be highly efficient when it is ON in the extra urban and Highway of the homologation cycle and during Real World driving.

ACKNOWLEDGEMENTS

The Authors and Co-Authors would like to acknowledge the kind support of AUDI AG Ingolstadt during the execution of this project and for permission to publish the data in this study.

REFERENCES

[1] F. Eichler, W. Demmelbauer-Ebner, J. Theobald, B. Stiebels, H. Hoffmeyer, M. Kreft, Volkswagen AG: The New EA211 TSIevo from Volkswagen. 37.Internationales Wiener Motorensymposium 2016, Vienna

[2] G. Fraidl, M. Brunner, P. Kapus, W. Schöffmann, H. Sorger, G. Teuschl, M. Weißbäck „ICE 4.x – Implementation as Modular Electrification". 40.Internationales Wiener Motorensymposium 2019, Vienna

[3] K. Prevedel, P. Goetschl, O. Dumböck, AVL List GmbH, Austria: Lambda 1, Extended RDE, EU7, Hybrid: Electrified Boosting as Enabler? 6. International Engine Congress 2019, Baden-Baden

[4] K. Prevedel, AVL List GmbH, Austria: Lambda 1: Goal or boundary condition with regard to EURO 7? 24. Supercharging Conference 2019, Dresden

Study on loss characteristic of penetrating exhaust manifold for low-speed two-stroke marine diesel engine

Yingyuan Wang, Ruiqi Zhang, Mingyang Yang, Kangyao Deng*

Shanghai Jiao Tong University, China

Bo Liu, Yuehua Qian

China Shipbuilding Power Engineering Institute Co. Ltd, China

ABSTRACT

The penetrating exhaust system is commonly used in low-speed two-stroke marine diesel engine which has great advantage of fuel economy. However, few study has been carried out on either performance prediction or optimization design on this type of exhaust system. Penetration depth and diffusion angle are the most important configurations of this exhaust system on the loss characteristics, which mean the length of the branch pipe penetrating in the main pipe and the cone angle exhibited by the diffusion of the end of the branch pipe respectively. This paper investigates loss characteristics and flow mechanism of the exhaust manifold of low-speed two-stroke marine diesel engine via experimental and numerical method. Results show the minimum loss is achieved when the penetration depth and diffusion angle are 0.5 and 6° respectively. Furthermore, it is found that the trend of flow loss as 'decrease-increase' is the trade-off among the mixing loss due to the shearing of injecting flow and main flow, the momentum loss due to impinging on the bottom of main pipe, and the separation loss in the branch pipe. The investigation of the paper can provide guidance of the configuration optimization and loss modeling of the exhaust manifold of the low-speed two-stroke marine engine.

1   INTRODUCTION

Low-speed two-stroke diesel engine is widely used for marine industry because of its great advantage on fuel economy [1]. In a low-speed two-stroke diesel engine, the configuration of engine exhaust system affects airflow in the exhaust pipe and the conversion of exhaust energy, thereby affecting the performance of the boosting system. Compared with pulse boosting systems, constant pressure boosting systems has their advantages of high turbine efficiency and are widely adopted by mainstream low-speed diesel engine manufacturers in their products, such as Win-GD RTA series diesel engines and MAN B&W ME series diesel engines [2-4].

The penetrating exhaust system is the main feature of constant pressure boosting system in low-speed two-stroke diesel engine. The feature of the penetrating exhaust manifold is that the branch pipe penetrates into the exhaust main pipe. Moreover, the diameter of exhaust branch increases gradually in the main pipe. Therefore, the flow is decelerated and the pulsed kinetic energy in exhaust gas can be recovered to some extent. Transforming the pulsed kinetic energy into pressure potential energy also makes the exhaust gas flows into the nozzle ring stably throughout the cycle, avoiding the periodic fluctuation of flow direction of the airflow entering the turbine blade, and hence higher turbine efficiency. Moreover, penetrating exhaust system decelerating the pulsed airflow, together with the large volume exhaust main pipe stabilizes the airflow entering the turbine, so it is potential to match a smaller turbine, meanwhile its full cycle turbine efficiency is higher. However, few research has been performed on the performance of the penetrating exhaust manifold published in literatures, regardless of the performance prediction methodology.

For the special geometry of the penetrating exhaust branch pipe, the relative penetration depth r and the diffusion angle φ are adopted to describe the characteristics. The relative penetration depth is a dimensionless parameter:

$$r = \frac{L}{D} \qquad\qquad (1)$$

The diffusion angle is the cone angle exhibited by the diffusion of the end of the penetration exhaust branch pipe, which is defined in Figure 1.

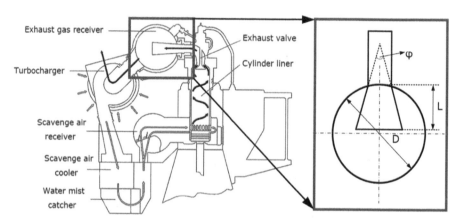

Figure 1. Definition of the relative penetration depth and the diffusion angle [15].

In order to predict the performance of the penetrating exhaust manifold, the model of T-junction together with an extra pipe are usually employed. Comprehensive studies have been carried out on the performance and the prediction method of T-junction of exhaust manifolds [5-7]. Moreover, the influence of configuration on the loss characteristic of the T-junction are experimentally studied, including lateral branch-common branch area ratio, chamfer of interface and lateral branch shrinkage rate, Reynolds number, and branch angles [8-11]. In order to study the influence of compressibility of the flow on the loss characteristics, Mach number of the flow is also experimentally studied. Besides experimental study, numerical method is also used for the study on the loss characteristics of T-junction for the establishment of the performance prediction model [12, 13]. Although extensive studies have been carried out on the performance of T-junction, and it is then applied for the prediction of penetrating exhaust manifold, the configuration that the penetration of branch in the main pipe cannot be considered. The extra pipe may be possible to capture the loss characteristics of the penetrating configuration, but it is highly dependent on the experience and also closely dependent on the engine operational conditions. It has been already confirmed that it seems not possible for the simplified method to capture the characteristic of the penetrating exhaust manifold over wide operational conditions and the credibility of the engine performance prediction has to be sacrificed [14].

Therefore, it is valuable to obtain the performance of the manifold in order to establish the reliable performance model of such type of exhaust manifold.

The object of this study is to investigate the performance of penetrating exhaust system and the mechanism of internal flow loss. This paper is organized in two sections. Firstly, the experimental method is applied for the investigation on the influence of geometrical parameters on the loss characteristics of the penetrating exhaust manifold. Secondly, the numerical method is used to understand the flow mechanism of the influence. This study enlightens the establishment of the loss prediction model of the penetrating exhaust manifold in future.

2   EXPERIMENTAL FACILITY AND ARRANGEMENT

2.1   Penetrating exhaust system test bench

In order to deeply study the flow characteristics of the marine low-speed two-stroke diesel penetrating exhaust system under steady and unsteady conditions, a penetrating exhaust flow test bench was built. The test bench is composed of a motor, a compressor, a penetrating exhaust main pipe, a penetrating exhaust pipe branch, a pulse generator and a measurement control system, wherein the motor drives the compressor as a gas source. Figure 2 is a schematic diagram of the experimental bench.

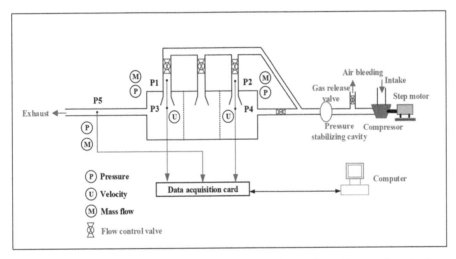

Figure 2. Schematic diagram of the penetration exhausting system test bench.

The penetration exhaust system test bench is based on the 6EX340EF marine two-stroke diesel engine, which has six exhaust branches. The test bench consists of three exhaust branches to satisfy the research requirement of the adjacent exhaust branches. Because the engine is bulky, the test bench is a model of the original engine according to the similarity theory. The total pressure is widely applied for the flow loss evaluation for the static flow component. When the flow velocity is low, the total pressure can be conveniently evaluated by the static pressure plus the dynamic pressure which is calculated by the velocity. However, as the velocity is high and the compressibility can't be ignored, the total pressure has to be evaluated accurately by the static pressure and Mach number together. Moreover, it was found that the penetrating exhaust flow of a marine two-stroke diesel engine belongs to the compressible high-speed airflow[8-11], in order to study the influence of the compressibility of the flow on the loss in the manifold, Mach number is chosen as the flow parameter in the study for the comparison among different cases. After being molded, the exhaust main pipe has a diameter of 204mm and a length of 612mm, which adopts modular design, and each module is inserted into a penetrating exhaust branch pipe. The test bench consists of three modules as shown in Figure 3. The left and right modules are made of aluminum alloy material, and the hot wire anemometers and static pressure sensors are arranged on the front side. The middle module is made of polymethyl methacrylate (PMMA) material by 3D printing, and its light transmittance is over 95%, satisfying the requirement of optically measurement.

Figure 3. Penetrating exhaust system test bench measuring section.

The measurement and control system consists of sensors, a data acquisition card and a computer. The mass flow rate of the branch pipe and the main pipe are assumed to measure respectively by the pitot tube. In addition, the outlet velocity of the branch pipe is to measure by the Dantec constant temperature hot-wire anemometer 55P11. Furthermore, the hotwire is calibrated by the Dantec Hotwire calibration unit before and after every case. The averaged calibration coefficients are used for the velocity measurement. Moreover, the pressure scanning valve TSI9116 is used to measure the static pressure at different points in the main pipe. A rotating speed sensor is also used for the purpose of controlling the compressor and further controlling the Mach number at the outlet of the branch pipe.

2.2 Experimental procedure

The airflow circulation process of the penetrating exhaust test bench is: the front end of the test bench is connected with a centrifugal compressor. The air enters the inlet of the compressor, and after being compressed by the compressor, flows along the pipeline to the rear end. Then it flows through the total gas valve and enters the main pipe and the branch pipe respectively. Mach number of the airflow in the branch pipe is controlled by adjusting the compressor rotating speed. Moreover, the change of relative penetrating depth and diffusion angle is achieved by changing the branch pipe modules. Table 1 is the experimental scheme of total pressure loss measurement.

In low-speed engines, the exhaust gas flows from the outlet of the exhaust main pipe and flows to the turbine through the front branch pipe of the turbine. In order to simulate the exhaust gas back pressure due to the turbine, in the experimental bench, a shrink orifice was designed at the outlet of the main pipe and a pipe with smaller diameter was connected downstream. At the end of the pipe, a shrink conical pipe was added for the purpose of reducing flow area. Pitot tube and static pressure scanning valve measuring points are set at the downstream pipeline to measure exhaust back pressure. In further experimental studies, a valve will be installed at the downstream pipe to control the back pressure.

Table 1. Experimental scheme of total pressure loss measurement.

Outlet Mach number	Relative penetrating depth	Diffusion angle				
0.1	0	0	-	-	-	-
	0.2	0	3	6	10	15
	0.5	0	-	-	-	-
	0.73	0	-	-	-	-
0.15	0	0	-	-	-	-
	0.2	0	3	6	10	15
	0.5	0	-	-	-	-
	0.73	0	-	-	-	-
0.2	0	0	-	-	-	-
	0.2	0	3	6	10	15
	0.5	0	-	-	-	-
	0.73	0				
0.25	0	0	-	-	-	-
	0.2	0	3	6	10	15
	0.5	0	-	-	-	-
	0.73	0	-	-	-	-

3   DISCUSSIONS ON EXPERIMENTAL RESULTS

In order to study the influence of geometrical parameters on the performance of the exhaust manifold, a pressure loss coefficient of a multi-branch exhaust manifold is defined as following:

$$K = \sum_N \left(\frac{1}{m_{tot}} \frac{m_i}{*} \frac{P_{0\_in} - P_{0\_out}}{P_{0\_out} - P_{out}} \right)$$

Where N is the number of branch pipes in exhaust manifolds; m_i is the exhaust flow of the branch i; m_{tot} is the exhaust flow of the pipe before turbine, that is, the sum of the flow rates of the exhaust branches; P_{0_in} is the total inlet pressure of the branch i; P_{0_out} is the outlet total pressure of the pipe before turbine; P_{out} is the outlet static pressure of the pipe before turbine. P_{0_in}-P_{0_out} represents the total pressure loss in each branch, in addition, P_{0_out}- P_{out} is the dynamic pressure of the pipe before turbine which is selected as a reference pressure. Because there are several branch pipes with different mass flowrate, and the pressure loss of each branch pipe is not the same as well, K is introduced to describe the total pressure loss. It is a mass weighted average pressure loss in essence.

For a penetrating exhaust manifold, there are several key geometrical parameters, which are the penetrating depth, the diffusion angle and the ratio of the diameters between the branch pipe and the main pipe. Especially, the first two parameters are considered to be the most important ones for the influence. Therefore, the penetrating depth and the diffusion angle are going to be studied in following sections.

3.1   Influence of relative penetrating depth
For the purpose of exploring how the relative penetrating depth effects the total pressure loss coefficient of the penetrating exhaust system, steady flow experiment is carried out with the same configuration while only the relative penetrating depth increases from 0 to 0.73. In addition, averaged Mach number in the branch pipe is adjusted from 0.1 to 0.25 to analyze its effect on the pressure loss.

The total pressure loss coefficient versus the relative penetrating depth at different Mach number is shown in Figure 4. It can be seen from the figure that the loss coefficient decreases firstly with the increase of the relative penetration depth, but begins to increases as the depth increases to large value. Consequently, there is a minimum pressure loss corresponding to the optimized penetrating depth, which appears near the depth as about 0.4. Moreover, the trend of 'decrease-increase' is more evident at high Mach number condition. It implies that the penetrating configuration become much more important for high Mach number condition. Furthermore, it can also be observed from the figure that the pressure loss increases consistently with the Mach number and hence mass flow rate in the branch pipe. Specifically, when the relative penetration depth is changed from 0 to 0.2 for Mach number as 0.25, the total pressure loss coefficient reduces by 14%. On the other hand, the change in the total pressure loss coefficient is 9.6% when Mach number is 0.1. Detailed variations are list in Table 2. According to the discussions, it can be concluded that the penetration exhaust system has a significant effect on reducing the total pressure loss of the exhaust flow of the constant pressure boosting system. More importantly, there is an optimized penetrating depth for the exhaust manifold, and the optimized value is approximately independent of the averaged Mach number in the branch pipes.

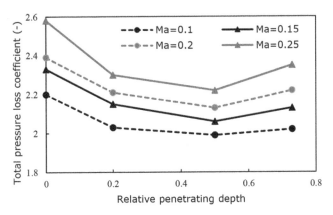

Figure 4. Effect of relative penetrating depth on total pressure loss coefficient.

Table 2. Benefit of penetrating exhaust system comparing to the non-penetrating exhaust system.

Ma	K(non-penetrating)	ΔK(penetrating)	Percentage
0.1	2.2	0.21	9.6%
0.15	2.33	0.27	11.6%
0.2	2.39	0.26	10.9%
0.25	2.58	0.36	14.0%

Considering the flow loss is directly related with the mixing process of flow coming out of the branch pipe with the flow in the main pipe, it is enlightening to study the flow velocity coming from the branch. Figure 5 and Figure 6 show the flow velocity at the exit of the penetrating exhaust branch pipe and at the bottom of the main pipe, respectively for different relative penetration depths. It can be seen from the Figure 5 that as the relative penetration depth increases, the airflow velocity at the outlet of the branch pipe nearly remains the same. It is expectable because the diameters of the penetrating branch pipe are the same for all the four configurations. However, Figure 6 manifests that the flow velocity near the inner wall surface of the exhaust main pipe increases consistently with the relative penetration depth. Furthermore, the larger the Mach number in the branch pipe is, the higher the velocity is. Particularly, the velocity increase at a larger rate with the penetrating depth for larger Mach number case. Considering the physical process of the flow coming out of the branch pipe, the flow is injected into the main pipe and moves towards the bottom of the inner surface of the main pipe. The injected flow with high velocity is shearing with the main flow on the interface and hence producing mixing loss. More importantly, the flow velocity is still high when it impinges with the bottom of main pipe. Consequently, it stagnates at the wall surface and flow momentum in the direction of the injection is almost lost. The process produces large amount of flow loss in the manifold. According the analysis of the process, it can be implied that the velocity of injection together with the length of shearing layer and the velocity of impinging at bottom are the most important factors determining the flow loss in the exhaust manifold. Specifically, when the penetrating depth is small, the shearing layer between the injecting flow and the main flow is relatively long although the injecting velocity is almost the same. As a result, high flow loss is generated due to the long shearing layer. On the other hand, the impinging velocity is high when the penetrating depth is large, thus the high flow loss is produced. Consequently, the trend of 'decrease-increase' is resulted, as demonstrated in Figure 4.

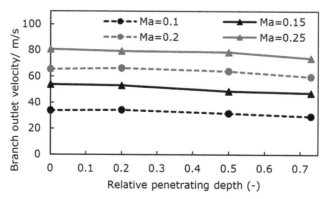

Figure 5. Effect of relative penetrating depth on branch outlet velocity.

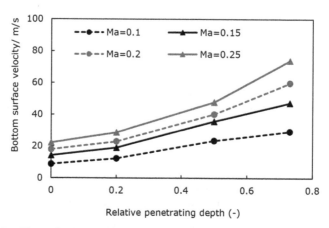

Figure 6. Effect of relative penetrating depth on bottom surface velocity.

3.2   Influence of diffusion angle

Besides the relative penetrating depth, diffusion angle is another critical parameter that effect the total pressure loss coefficient. Steady flow experiment is carried out with the same penetrating depth (0.2) and the diffusion angle changes from 0 degree to 15 degree. Again, 4 different Mach numbers from 0.1 to 0.25 in the branch pipe are investigated.

Figure 7 shows the variation of pressure loss coefficient versus the diffusion angle at four Mach numbers. It can be seen from the figure that the total pressure loss coefficient decreases rapidly with the increase of the diffusion angle, and then increases slowly for all the cases of Mach number. Specifically, the critical angle corresponding to the minimum flow loss is 6°. Different from the influence of penetrating depth shown in Figure 4, the magnitude of the variation is approximately the same at four Mach number. It implies that the variation of the flow loss is independent from the Mach number in the branch pipe, although the loss is considerably higher for higher Mach number.

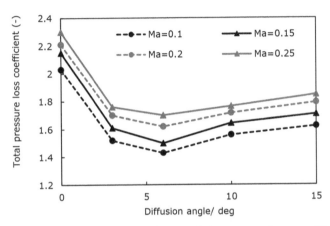

Figure 7. Effect of diffusion angle on total pressure loss coefficient.

In order to understand the influence of diffusion angle on the flow loss, Figure 8 and Figure 9 also shows the injecting velocity from penetrating exhaust branch pipe and the impinging velocity at the bottom of inner surface of main pipe. It can be seen that both flow velocities firstly decrease rapidly with the increase of the diffusion angle, and then remain relatively stable as the angle further increases. This trend is consistent with the trend of changes in total pressure loss. This is because the configuration of the diffusion decelerates the flow velocity and hence the injecting velocity is reduced significantly. According to the analysis previously, the strength of shearing between the injecting flow and the main flow will be alleviated. Furthermore, the impinging velocity will be reduced corresponding as well. Therefore, the flow loss is dramatically reduced, as shown in Figure 7. As the angle increases beyond 6 degrees, however, the flow normally encounters separation in the branch pipe. Consequently, the blockage of the pipe increases significantly and hence the injecting velocity and hence the impinging velocity don't reduce any longer with the diffusion angle, as shown in Figure 8 and Figure 9. In addition, the flow separation in the branch pipe may introduce the extra flow loss. Therefore, the trend of flow loss as 'decrease-increase' is produced, as demonstrated in Figure 7.

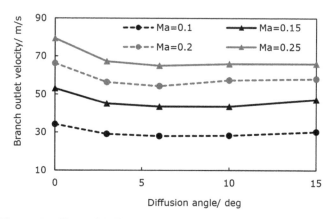

Figure 8. Effect of diffusion angle on branch outlet velocity.

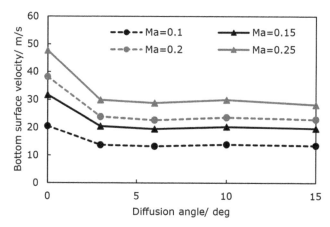

Figure 9. Effect of diffusion angle on bottom surface velocity.

According to the discussions on the influence of penetrating depth and the diffusion angle, it can be concluded that the trend of flow loss as 'decrease-increase' is the trade-off among the mixing loss due to the shearing of injecting flow and main flow, the momentum loss due to impinging on the bottom of main pipe, and the separation loss in the branch pipe. Consequently, there is the optimum penetrating depth and diffusion angle for the configuration of penetrating exhaust manifold. Moreover, the potential for the reduction of flow loss is higher at high Mach number condition.

The influence of the injection velocity, the length of shearing layer and the impinging velocity on the pressure loss is shown in Figure 10. It is clearly manifested that the flow loss increases consistently with three factors, that is branch outlet velocity v_1, bottom surface velocity v_2, and shearing layer length D. The square of v_2 represents the kinetic energy loss due to impinging while the product of D and the square of v_1 represents the shearing work because the shearing stress is proportional to the square of v_1. The algebraic sum of these two parts is the loss of the exhaust manifold which is described by total pressure loss coefficient K. This linear relationship is reflected in the figure that each data point almost falls near a same plane. Therefore, it is a sound proof that the discussion on the loss mechanism is reasonable.

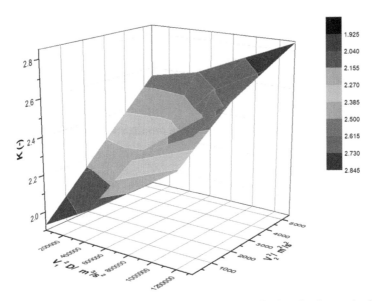

Figure 10. Influence of the injection velocity and the impinging velocity on the pressure loss.

In order to validate the discussion on the loss mechanism, unsteady 3D CFD method is applied via large eddy simulation method. Figure 11 shows the velocity contours in the penetrating exhaust manifold. The diffusion angle is 0 degree and penetrating depth is 0.5. As can be observed from the figure, when the flow is injected from the branch pipe, strong shearing happens between the injecting flow and the main flow. Moreover, the shearing layer becomes wave-like due to the K-H instability. This shearing layer corresponds to high entropy generation. More-over, the strength of the hearing layer is directly linked with the velocity coming out of the penetrating pipe. At the meantime, it can be seen clearly from the vel-ocity distribution that a stagnation area is formed. High entropy generation is expected to happen due to the impingement near this location. This is consistent with the loss mechanism obtained by analyzing the experimental data. Further-more, due to the impinging of the injecting flow on the bottom surface of main pipe, different scales of vortexes are generated near the bottom and then are mixed with the flow in the main manifold, which again produces high entropy gen-eration. Therefore, the flow field proves that the velocity at the exit of the pene-trating pipe and the impingement velocity are the key factors influencing the flow loss in the manifold system.

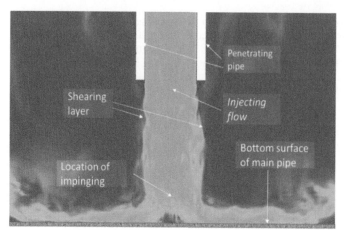

Figure 11. Velocity distribution in the penetrating exhaust manifold.

4   CONCLUSIONS

The penetrating exhaust system is commonly used in low-speed two-stroke marine diesel engine which has great advantage of fuel economy. This paper studies pressure loss coefficient for penetrating exhaust system via experimental research. Four conclusions are obtained from the study, as following:

(1) The variation of total pressure loss coefficient shows the trend of 'decrease-increase' for both relative penetration depth and diffusion angle. The optimum values for the relative penetrating depth and diffusion angle is 0.5 and 6°, respectively.
(2) The 'decrease-increase' trend is resulted from the trade-off among the mixing loss due to the shearing of injecting flow and main flow, the momentum loss due to impinging on the bottom of main pipe, and the separation loss in the branch pipe. Large-eddy-simulation method on the penetrating exhaust manifold is applied for the detailed flow field analysis and the loss mechanism is well validation.
(3) In subsequent studies, a pulse generator will be installed in each exhaust branch pipe to simulate the phase difference of exhaust gas from different cylinders in an actual low-speed engine. The influence of the multi-branch exhaust pipe on the performance and flow field structure of the exhaust manifold will be studied, and its impact on the pulsed exhaust energy recovery and the turbine operation will be analyzed.

ACKNOWLEDGEMENT

This work was supported by China Ship Power Research Institute Co. Ltd. and Key Laboratory for Power Machinery & Engineering of State Education Ministry, Shanghai Jiao Tong University.

REFERENCES

[1] Woodyard D. Pounder's marine diesel engines and gas turbines[M].

[2] Liu Zheng, Wang Jianxin, Automotive engine fundamentals[M], Tsinghua University Press, 2011,113–114.

[3] Liu Fenghua. Research on Performance Matching Optimization of Marine Two-stroke Low Speed Diesel Engine [D]. 2014.

[4] Lamas M I, Rodríguez Vidal C G. Computational Fluid Dynamics Analysis of the Scavenging Process in the MAN B&W 7S50MC Two-Stroke Marine Diesel Engine[J]. Journal of Ship Research, 2012, 56(3):154–161.

[5] Ito, H. and Imai, K. Energy losses at 90°pipe junctions[J]. J. Hydraul. Engng, ASCE, 1973, 99(9): 1353–1368.

[6] Bing, J.F. and Blair, G. P., A Improved Branched Pipe Model for Multi-cylinder Automotive Engine Calculation[J]. Proc. Insten. Mech. Engrs, Part D, Journal of Automobile Engineering, 1985, 199: 65–77.

[7] Ethier C R, Prakash S, Steinman D A, et al. Steady flow separation patterns in a 45 degree junction [J]. J. Fluid Mech., 2000, 411: 1–38.

[8] N. I Abou-Haidar et al, Pressure losses in combining subsonic flows through branched ducts[J]. Trans. ASME, Turbomachinery. 1992, 114 (1): 264–270.

[9] Perez-Garcia, Sanmiguel-Rojas, A Viedma. New coefficient to characterize energy losses in compressible flow at T-junctions[J]. Applied Mathematical Modeling, 2010, 34: 4289–4305.

[10] Perez-Garcia, Sanmiguel-Rojas, A Viedma. New experimental correlations to characterize compressible flow losses at 90-degree T-junctions[J]. Experimental Thermal and Fluid Science 2009, 33: 261–266.

[11] Perez-Garcia, Sanmiguel-Rojas, Hernandez-Grau, et al. Numerical and experimental investigations on internal compressible flow at T-type junctions[J]. Experimental Thermal and Fluid Science, 2006, 3: 61–74.

[12] C.T. Shaw, D.J. Lee, S.H. Richardson, et al. Modeling the effect of plenum-runner interfacegeometry on the flow through an inlet system[J]. SAE Technical paper series 2000, 2000–01–0569.

[13] Y. Zhao, D.E. Winterbone. A study of multi-dimensional gas flow in engine manifolds[J]. Proc.Inst. Mech. Eng, 1994, 218: D04892.

[14] Duan Shaoyuan, Deng Kangyao, Yang Mingyang. Performance and Loss-mechanism Analysis of Plug-in Exhaust Pipe for Low Speed Marine diesel engine [J].Diesel Engine, 2018, v.40;No.279(03):7–11.

[15] Duan Shaoyuan, Study on the performance and loss mechanism of the low speed Marine diesel engine injection exhaust pipe [D]. Shanghai Jiao Tong University, 2015.

Heat transfer modelling in vehicular turbochargers for engine simulation

Y. Watanabe[1], N. Ikeya[1], D. Filsinger[2], S. Okamoto[1]

[1]IHI Corporation, Vehicular Turbocharger Business Unit, Japan
[2]IHI Charging System International GmbH, Engineering division, Germany

ABSTRACT

Heat transfer within turbochargers affects the measured turbocharger efficiency. Differences of actual and measured efficiency cause reduced accuracy in the matching process. A so-called thermal network representing the thermal behavior of the turbocharger consisting out of several thermal nodes is created based on geometric design data. The thermal network considers the thermal connectivity between the components and, therefore represents the thermal behavior of the turbocharger. It is validated by experiments utilizing data from various thermocouples applied to a turbocharger. Tests were done for different hot gas temperature ranges. A heat transfer model, named HTTR, combined with heat transfer effect corrected turbine and compressor maps integrated into the thermal network allows improved accuracy in engine simulation in all engine operating conditions even in cold start and transient states. This report shows the validation of the HTTR by comparing experimental results from hot gas test benches and calculation results utilizing GT-POWER. The accuracy of the temperature distribution in the turbocharger and the efficiency affected by heat transfer is verified by varying the turbine inlet temperature. Under steady state conditions the HTTR was confirmed.

1   INTRODUCTION

Measurements of turbocharger efficiency on hot gas test benches are effected by heat transfer. This effect is pronounced for low load and low turbocharger speed operation. Figure 1 shows the heat transfer effect in a turbocharger in a H-S chart. In the compressor stage, heat transfer occurs between the housing and the working fluid (air) as well as to the ambient. (a) represents the heat flow rate transferred at the compressor outlet. (b) represents the actual compressor work to the working fluid. (c) represents the heat flow rate at the compressor inlet. In the turbine stage, heat transfer occurs between the housing and the working gas and again to the ambient. (d) represents the heat flow rate at the turbine inlet. (e) represents the actual extracted turbine work from the working gas. (f) represents the heat flow rate transferred at the turbine outlet. While the actual compressor/turbine work is illustrated with the green line, $(X_{1act}$ to $X_{2act})/(X_{3act}$ to $X_{4act})$ in Figure 1, the test result, which is obtained from temperature measurements, are expressed with the black line, $(X_{1test}$ to $X_{2test})/(X_{3test}$ to $X_{4test})$. The heat exchange with ambient is not explicitly shown in Figure 1, but also contributes to a discrepancy between measured and actual efficiency. The difference between the green and the black lines indicate that heat transfer leads to apparently lower compressor efficiency and higher turbine efficiency. This trend becomes most apparent with low flow rates, but relatively high heat flow when the turbocharger is operated at low speeds and high turbine inlet temperatures (Otobe [1]).

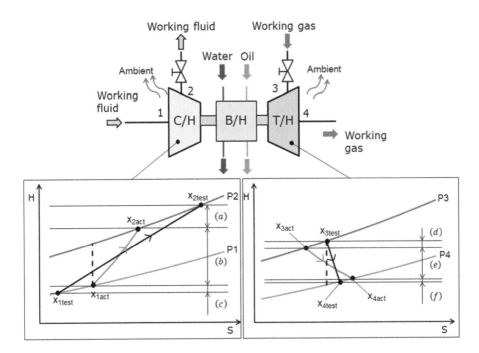

Figure 1. Schematic of turbocharger and H-S diagram with and without heat transfer.

The Study from Casey [2] defines the heat transfer from the turbine to the compressor as constant. The simple and practicable heat transfer correction method from Lüddecke [3,4] is derived by considering temperature variation of the cooling water and the lubrication oil. These methods are simple and easy to apply for producing heat corrected turbocharger component performance maps. However, engine simulation need to account for changes in turbine inlet temperature whereas turbine maps and also simple heat flow corrected maps are usually generated for a constant turbine inlet temperature. Both, the heat transfer and also the gas temperature are not constant during engine operation which leads to varying heat transfer. For a more precise 0D/1D engine simulation, the thermal network approach utilizing detailed temperature measurements from hot gas test stand operation is therefore adopted.

2   METHODOLOGIES

2.1   Modeling approaches
The commercial software ANSYS workbench is used for the heat transfer analysis employing the Finite Element Method. For modeling the temperature distribution of the turbocharger system it is divided into several nodes considering the actual temperature level from CAE, compare Figure 2.

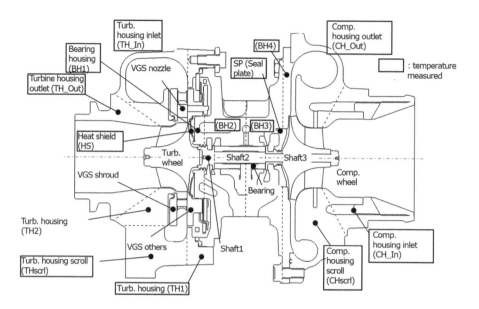

Figure 2. Overview of temperature nodes.

The outlet of turbine housing for example is diagonally divided in two zones respecting the specific temperature field in the scroll. Overall, 23 nodes have been defined in the model. Additionally, the media such as air, gas, coolant water, lubricant oil and ambient air are considered. Moreover details such as the open volume of the nozzle link mechanism of the variable turbine and the air gap on the back side of the heat shield are treated. In these "stagnation" areas the temperature is estimated by averaging the temperature values of the connected components. In total 32 temperature regions and 88 thermal connections are defined.

Solver

The HTTR is set up as simultaneous equations of temperatures. The 4th order Runge-Kutta equation is used to solve the 1st order simultaneous ordinary differential equation (1). Heat conductance is defined depending on the state of the connected objects.

$$mc\frac{dT}{dt} = rT + Q \qquad (1)$$

Solid to solid

Each node named in Figure 2 is represented by a constant temperature value. The heat transfer between each node depends on the temperature differences between them. The thermal contact resistance is defined in equation (2). To modify the thermal contact considering roughness and hardness of the surface, a coefficient is introduced

$$r = \frac{1}{\frac{L_i}{2\lambda_i A_{ij}} + Rc_{ij} + \frac{L_j}{2\lambda_j A_{ij}}} \qquad (2)$$

Free convection

Free convection and forced convection are considered. The thermal contact resistance of heat convection and heat transfer rate is defined in Equation (3) and (4). Heat

328

transferred by natural convection is represented by Equation (5) using the Rayleigh Number which accounts for the state of flow influenced by buoyancy. Equation (5) is describing natural convection between turbine housing, compressor housing and bearing housing with the ambient, respectively.

$$r = A \cdot h \tag{3}$$

$$h = Nu \cdot \frac{\lambda}{D} \tag{4}$$

$$Nu = 0.54 Ra^{1/4} \text{ at } 10^4 < Ra < 10^7 \tag{5}$$

$$R_a = Gr \cdot Pr \tag{6}$$

Forced convection

The formulation of forced convection in circular pipes with fluid flow, such as oil and coolant water, is based on the Sleicher-Rouse Equation (7) for turbulent flow. Applying the boundary layer approximation for laminar flow, the Mori-Nakayama Equation (8) accounts for heat transfer in curved pipes, which is used for the flow in the compressor and turbine scroll.

$$Nu = 5 + 0.015 Re^a Pr^b \tag{7}$$

$$\left(10^4 < Re < 10^6, 0.1 < Pr < 10^4\right)$$

$$a = 0.88 - \frac{0.24}{4 + Pr}, b = \frac{1}{3} + 0.5e^{-0.6Pr}$$

$$\begin{cases} Nu = \frac{Pr^{0.3}}{24} Re^{0.8} \left(\frac{d}{D}\right)^{0.1} \left[1 + \frac{0.098}{\{Re(d/D)^2\}^{0.2}}\right] : 0.6 < Pr \leq 1, Re(d/D)^2 > 0.1 \text{ (for gas)} \\ Nu = \frac{Pr^{0.4}}{41} Re^{\frac{5}{6}} \left(\frac{d}{D}\right)^{\frac{1}{12}} \left[1 + \frac{0.061}{\{Re(d/D)^{2.5}\}^{1/6}}\right] : Pr > 1, Re(d/D)^{2.5} > 0.4 \text{ (for liquid)} \end{cases} \tag{8}$$

Radiation

Radiation, which is most prominent at high temperatures, is also considered. It is mainly relevant for the turbine housing. Modeling is done according to Equation (9).

$$h = \varepsilon\sigma\left(T_w^2 + T_a^2\right)\left(T_w + T_a\right) \tag{9}$$

2.2 Experimental setup

Multiple thermocouples were implemented in 13 positions (boxes in Figure 2) in the turbocharger corresponding to nodes in the thermal network. The position (depth) of thermocouple was intended to be in the center of the lumped mass of the node of the thermal network. To prevent heat losses by radiation from the inlet/outlet pipes connected to the test facility, insulation is applied. Figure 3 shows the set-up of the tested turbocharger and the connecting pipes with insulation applied. The test conditions are shown in Table 1. The turbocharger is equipped with a variable turbine. A description of such technology and its features can be found in Starke et al 2016 [5]. For several VGS nozzle opening conditions full compressor map measurement was performed. Figure 4 shows the compressor map with the various numbered operating points. With increasing numbers the operating conditions approach surge starting from choke conditions. Higher rotational speed operation is labelled with higher numbers. In Figure 4 (b), the turbo speed is regulated that the same Mach number as in (a) is achieved.

Three representative points (#3, #17, #38) discussed in section 3 are emphasized in red.

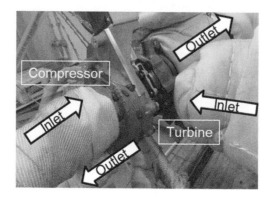

Figure 3. Test stand and insulation.

Table 1

Parameter	Value
Turbine inlet temperature °C	600,400
Oil inlet temperature °C	90
Coolant water temperature °C	90

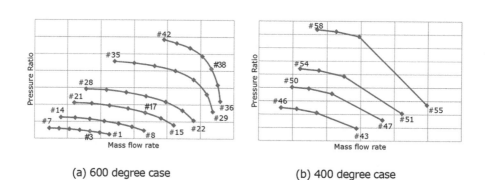

(a) 600 degree case   (b) 400 degree case

Figure 4. Operation condition on compressor map.

2.3   Calculation setup

The HTTR is implemented into the commercial software GT-POWER using the so called user code object. The user code is called from the main program for each iteration when the state is updated. The boundary conditions are as well shared for every time step. The volumetric flow rate, the temperature of the water and the lubricant flowing to the turbocharger are needed to describe the boundary conditions. In this study, these boundaries are kept constant for calculating steady state conditions. Typically, compressor and turbine performance are described by maps representing flange to flange conditions. In the HTTR, the maps are referring to conditions just before the wheels. This means that the performance is calculated based on adiabatic work (b) and (e) as illustrated in Figure1. The GT-POWER model simulates a hot gas turbocharger test controlling compressor inlet mass flow and turbo shaft speed by using valve objects that allow varying diameter. The pipe object in front of the compressor/turbine object represents the respective housings. The measured pressures and temperatures are compared with calculation results at the pipe objects before the compressor/ turbine objects. Calculation conditions are adjusted to the experimental conditions as described in 2.2. Temperature, pressure and ambient conditions at compressor/turbine housing inlet are direct input for each operating point. The boundary conditions do not have to be constant, but can be varying with time, allowing transient simulation. In this study, verification of the HTTR model is done by utilizing steady state hot gas bench measurements for several conditions at a temperature level of 600 degree.

Mechanical loss

During hot gas test, it is difficult to measure the mechanical loss of the turbocharger directly because the oil temperature is also affected by not only the mechanical loss between shaft and oil but also by heat transfer from the turbine. The mechanical losses are measured in dedicated tests that are controlled in a way that no heat transfer occurs. Based on such measurements such loss is treated separated from the heat transfer and parameterized as a function of the turboshaft speed. The effect of thrust is considered. Exemplary modeling is described by Höpke et al. [6].

3   RESULTS AND DISCUSSIONS

Figure 5 shows the effect of heat flow on component efficiency. Just as also described by Otobe et al 2010 (10) the turbine and compressor efficiency is strongly affected. In Figure 6(a) and (b) the opening position of the variable geometry nozzle is varied. (a) is full open, and (b) is a middle open position with the related effect on mass flow parameter. The results in the charts are labeled according to the numbering of the operating conditions illustrated in Figure 4. Both Figure 6(a) and (b) show that as turbocharger speed increases, the temperature around the turbine housing also increases. While the temperature in the turbine housing inlet region increases, the temperature at the housing outlet region decreases. This is in accordance with increasing turbocharger speed and related higher enthalpy drop by work extraction at the turbine. The deviation of temperature in the heat shield, HS, is relatively high, more than 10% but less than 20% under all conditions. This component is clamped between turbine housingand bearing housing at its outer diameter. Therefore, the HS experiences a large temperature gradient in radial direction. The node model representing the HS will only consider one uniform temperature value. Moreover, due to the strong gradient in this component the sensitivity of the measurement result regarding the measurement position is high, which explains the deviation in this node. All other nodes around compressor/turbine scroll show a good accuracy ±3%. The characteristic of the temperature distribution does not change with varying VGS opening position.

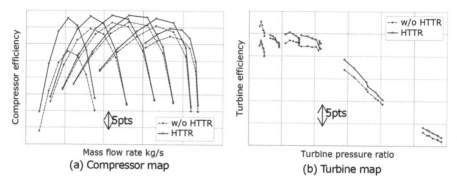

(a) Compressor map

(b) Turbine map

Figure 5. Difference of turbocharger efficiency with and without HTTR.

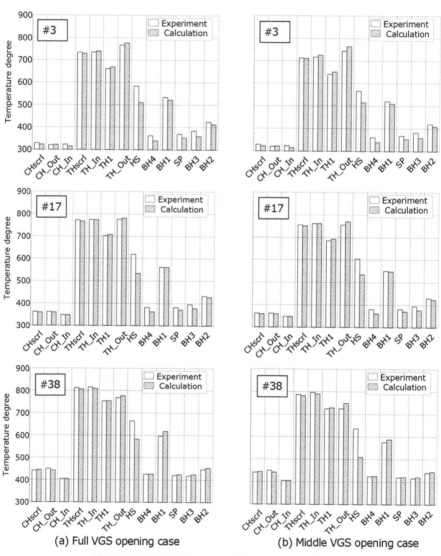

(a) Full VGS opening case

(b) Middle VGS opening case

Figure 6. Comparing results of temperature.

Figure 7 shows the heat transfer between the solid and the media. Positive sign means that there is heat flow from solid to fluid, the opposite heat flow direction is expressed by a negative sign (fluid to solid). Largest heat flows are observed between the compressor/turbine inlet/outlet and the gas. The compressor work calculated with measured temperatures results is larger than the actual compressor work respecting that the inlet is heated and outlet is cooled, (#21, #25 - #41). For relatively low speed conditions, both the compressor inlet and outlet are heated from the wall, (#1-#15). As expected the general direction of the heat flow is from the hot turbine to the colder compressor. For rather low speeds the heat transfer at the turbine is about 30 times higher than the heat transfer at the compressor. The difference in heat is flowing to the water, the oil and the ambient. The larger the heat transfer, the higher the influence to the apparent compressor/turbine efficiency is. Therefore, the effects are exaggerated for relatively lower turbo speed and related lower compressor outlet temperature. The ratio of the total heat flow is shown in Figure 8(b). dQAir is getting smaller as the operation number increases, it is smaller than zero after #15. This means that with increasing compressor work the heat flow to the compressor outlet becomes smaller. In contrast, dQGas_ex is getting larger as the operation number increase. It becomes larger than after #27. This means that with larger turbine work and linked to this enthalpy drop the gas at the outlet is heated by the heat flow from turbine housing. This can be recognized since the performance at relatively low turbo speed conditions measured on the hot gas test bench is underestimating the compressor efficiency and overestimating the turbine efficiency. dQAir for #1 to #7 in Figure 7 (b) which correspond to the lowest turbo speed is 3 to 7% of the total heat flow. In this test, the heat flow at the lowest speed leads to a temperature increase of about 3 degree of the compressed air. This evaluation was performed by comparing η_{tot} between experiment and calculation. The total pressure value was used for calculating compressor/turbine efficiency.

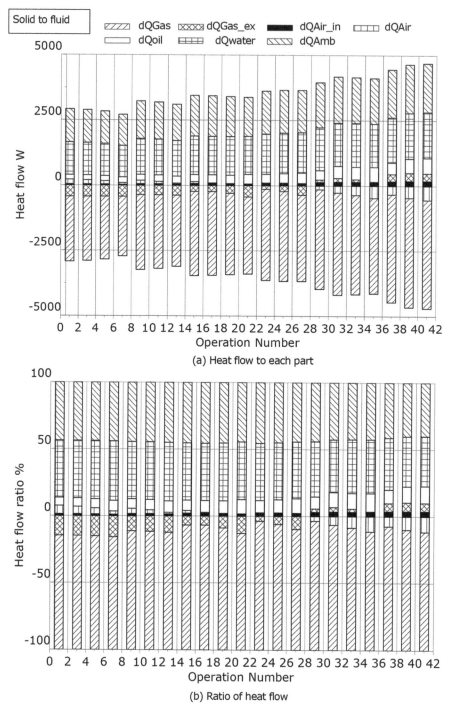

(a) Heat flow to each part

(b) Ratio of heat flow

Figure 7. Distribution of heat transfer.

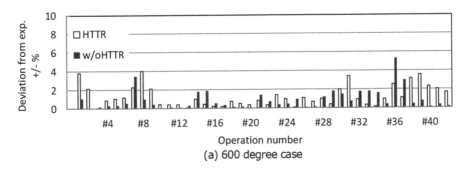

(a) 600 degree case

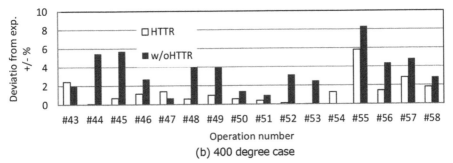

(b) 400 degree case

Figure 8. Full VGS opening.

Figure 8 to 10 show the results of comparing the deviation from the total turbocharger efficiency η_{tot} from experiment. (a) gives the results for a turbine inlet temperature of 600 degree while (b) gives the results for 400 degree. For the 600 degree case, the deviation with and without the HTTR shows the same accuracy with some deviation in tendency. In general it can be observed that close to choke and surge the deviations are much higher. The change in compressor efficiency to change in air mass flow at choke (e.g.#36) or surge (e.g.#7) point to the next measuring point is 3 to 10 times larger than others. Moreover, the measurement error is relatively high because of more unstable operating conditions. The maximum deviation with the HTTR is less than around ±4% in the 600 degree turbine inlet temperature case, and less than ±2% in the 400 degree case, except #55. In contrast, the maximum deviation without HTTR is ±8% for 600 and 400 degree. Because of the described heat transfer the calculated turbine efficiency varies if the turbine housing inlet temperature or the wheel temperature is used. To illustrate the effect of the turbine inlet temperature change, Figure 11 left shows the enthalpy drop in the stator respectively between the turbine housing inlet and the wheel inlet. For the 600 degree C case the difference in enthalpy is approximately two times higher than for the 400 degree C case. In case the compressor work and the turbo speed are unchanged, the adiabatic turbine work changes with turbine efficiency. Equation 10 describes the difference between adiabatic turbine work calculated by the housing or the wheel inlet temperature. ΔL_{tad} is given in dependency from T3 and pressure ratio of the turbine. In Figure 11 right the normalized enthalpy drop versus turbine inlet temperature is shown for constant pressure ratios. It is confirmed that the effect of heat flow on turbine efficiency is pronounced for lower pressure ratios. For pressure ratios higher than two it is observed that the influence of heat transfer increases only incrementally.

$$\Delta L_{tad} = (3.46 \times 10^{(-5)} \ln \pi_t - 6.74 \times 10(-5))T_3 + 3.02 \ln \pi_t + 1.06 \tag{10}$$

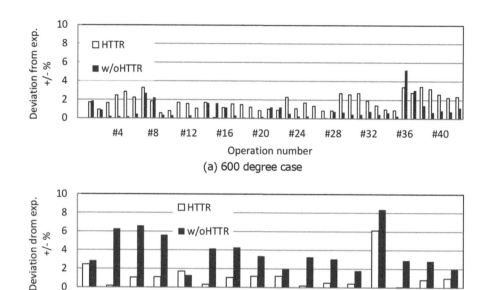

Figure 9. Middle VGS opening.

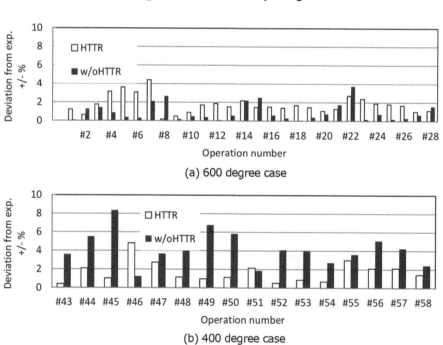

Figure 10. Small VGS opening.

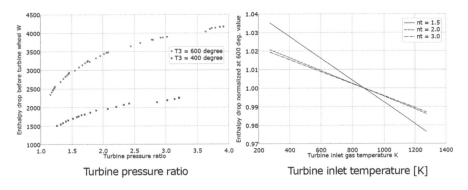

| Turbine pressure ratio | Turbine inlet temperature [K] |

Figure 11. Left: Enthalpy drop in the stator of the turbine; right: normalized enthalpy drop versus temperature.

4   CONCLUSIONS

The heat flow in turbochargers is investigated with the herewith introduced HTTR model. The HTTR model represents the turbocharger and its components as a thermal network. In this paper it is evaluated utilizing temperature measurements from experiments on a hot gas test bench. In summary, the deviation between predicted and measured temperatures at the compressor and the turbine scroll is less than ±3%. An exception is the heat shield behind the turbine due to its specific positioning in the assembly. The turbocharger geometry is discretized with thermal nodes that represent regions with different temperatures. The calculated temperatures were compared with measured values from hot gas test stand measurements. The HTTR could show its benefit by assessing the total efficiency of the turbocharger. The HTTR is capable of calculating the enthalpy change and the related heat flow to and from the housings and therefore determines the turbocharger efficiency better by employing the temperature before and after the wheel, which are not known from experiment. It can therefore compensate this heat flow effect that is dependent from turbine inlet temperature and pressure ratio.

For future work, HTTR with engine simulation will be reported. Both steady state and transient state simulation needs to validate how HTTR contribute matching accuracy. The HTTR accuracy improvement helps developing intercooler and cata-lyst by good compressor and turbine outlet temperature simulation. To improve HTTR accuracy, the effect of lubricant oil and coolant water will be validated with the HTTR. Finally, the HTTR is expected to help prediction the temperature of the catalyst during cold start.

NOMENCLATURE

A_{ij} = Contact area

c = Heat capacity

d, D = Representative length

Gr = Grashof number

h = Heat transfer coefficient

H = Enthalpy

L_i, L_j = Representative length

L_{tad} = Adiabatic turbine work

m = Mass

Nu = Nusselt number

Q = Heat transfer

r = Heat conductance

Ra = Rayleigh number

R_{cij} = Thermal contact resistance

Re = Reynolds number

S = Entropy

T = Temperature

t = Time

ε = Emmissivity

η = Efficiency

η_{tot} = Total turbocharger efficiency

λ_i, λ_j = Thermal conductivity

π_t = Turbine pressure ratio

σ = Stefan-Boltzmann constant

Δ = Difference

ABBREVIATIONS

0D/1D = Zero/One dimensional

B/H = Bearing housing

BH1 = Bearing housing nearby heat shield

BH2 = Bearing housing nearby VGS

BH3 = Bearing housing nearby shield plate

BH4 = Bearing housing nearby

C/H = Compressor housing

CAE = Computer aided engineering

CHscrl = Compressor scroll

CH_Out = Compressor housing outlet

CH_In = Compressor housing inlet

dQGas = Heat flow between component and gas before turbine wheel

dQGas_ex = Heat flow between component and gas after turbine wheel

dQAir = Heat flow between component and gas after compressor wheel

dQAir_in = Heat flow between component and gas before compressor wheel

dQoil = Heat flow to lubricant oil

dQwater = Heat flow to coolant water

dQAmb = Heat flow from turbocharger surface to ambient

HS = Heat shield

HTTR = Heat transfer model

P1 = Compressor inlet pressure

P2 = Compressor outlet pressure

P3 = Turbine inlet pressure

P4 = Turbine outlet pressure

SP = Seal plate

T1 = Compressor inlet temperature

T2 = Compressor outlet temperature

T3 = Turbine inlet temperature

T4 = Turbine outlet temperature

T/H = Turbine housing

THscrl = Turbine scroll

TH_Out = Turbine housing outlet

TH_In = Turbine housing inlet

TH1 = Turbine housing nearby bearing housing

VGS = Variable geometry system

REFERENCES

[1] Otobe, T., Grigoriadis, P., Sens, M., Berndt, R. (2010), „Method of performance measurement for low turbocharger speeds", 9th International Conference on Turbochargers and Turbocharging of the IMechE 2010.
[2] Sirakov, B., Casey, M. V. (2013) Evaluation of Heat Transfer Effects on Turbocharger Performance. Journal of Turbomachinery, Vol.135, ASME.
[3] Lüddecke, B., Filsinger, D., Ehrhard, J., Bargende, M., Heat Transfer correction and torque measurement for wide range performance measurement of exhaust gas turbocharger turbines, Aufladetechnische Konferenz, Dresden, 2012.
[4] Lüddecke, B., Filsinger, D., Bargende, M., On wide mapping of a mixed flow turbine with regard to compressor heat flows during turbocharger testing. Proceedings of 10th International Conference on Turbochargers and Turbocharging of the IMechE, Woodhead Publishing, ISBN 978-0-85709-209-0, 2012.

[5] Starke, A., Leonard, T., Hehn, A., Model, M., Hoppe, L., Kotzbacher, T., Weiß, M., Segawa, K., Bamba, T. & Iosifidis, G., The Next Generation of Variable Geometry Turbochargers from IHI, Aufladetechnische Konferenz, Dresden, 2018.
[6] Hoepke, B., Uhlmann, T., Pischinger, S., Lueddecke, B., Filsinger, D. Analysis of Thrust Bearing Impact on Friction Losses in Automotive Turbochargers; Journal of Engineering for Gas Turbines and Power, Vol 137, August, 2015.

Trends in passenger car powertrains and their impact on turbocharger developments

N.C. Baines[1], E.M. Krivitzky[2], J. Bai[3], X. Zhang[3]

[1]Consultant, UK
[2]Thayer School of Engineering, Dartmouth College, USA
[3]Xeca Turbo Technologies (Beijing) Co. Ltd., China

ABSTRACT

Drive trains for passenger cars are changing rapidly as the market moves away from diesel towards hybrid gasoline/electric, battery electric and hydrogen fuel cells. This will have implications for turbocharging that are only starting to become evident. Hybrid cars and electric cars with range extender ICEs will still offer opportunities for turbocharging. The use of an electric machine can reduce the low speed torque requirements of the ICE and the turbocharger can be re-optimised to emphasise efficiency over range and inertia. The benefits will vary with individual circumstances, but in the most favourable conditions it might be as much as 5-7 percentage points for the compressor and turbine efficiencies.

There is good evidence that battery electric and fuel cell vehicles are less favourable to CO_2 emissions than hybrid vehicles in the short and medium term, but the final decision will be based on geopolitical and market factors rather than technical evidence. Such a switch will create opportunities for air cycle machines and electrically-driven compressors based on turbocharger technology, but in market terms these are unlikely to compensate for the loss of the passenger car turbocharger market.

1   HISTORICAL OVERVIEW

Twenty five years ago most cars in all markets of the world were powered by gasoline engines. Since that time, two trends became overwhelmingly important: the switch to diesel engines in the EU (but not elsewhere), and the growth of passenger cars in China.

The adoption of diesel engines (Figure 1) has been extensively discussed, see for example (1), but in broad terms it was the result of concerns about CO_2 emissions, the potential for diesel to be more fuel efficient, and the availability of diesel fuel following a switch from oil to gas as a heating fuel. It was driven by favourable regulatory and tax environments that held the cost of diesel fuel below that of gasoline. For the most part, these factors did not operate outside the EU. An initial surge in diesel use in Japan rapidly declined, and in the USA there was no significant growth at all. Critically, the impacts of other diesel pollutants including NOx and PM were ignored or played down at this time.

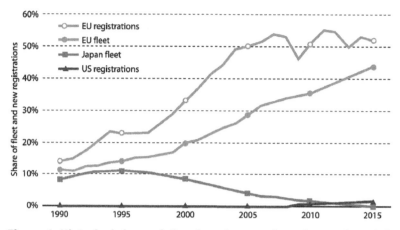

Figure 1. Historical share of diesel engine cars in major markets (1).

A naturally aspirated diesel engine is larger and heavier than an equivalent gasoline engine, but turbocharging allows the gap to be closed. When the cumulative effect of NOx and particulate emissions became a cause for concern, regulators responded with a series of diminishing limits (2). The fundamental solutions to emissions creation are to reduce the combustion temperature, control the temperature distribution to reduce "hot spots", and avoid over-fuelling that gives rise to incomplete combustion. All measures to achieve these goals also have the effect of reducing the specific power output of the engine. Turbocharging is therefore a key technology. It does not itself reduce emissions, but it allows the use of emissions-reducing technologies (EGR, variable valve timing, etc.) while maintaining an acceptable engine performance.

Unfortunately, a fixed geometry turbocharger is not well-matched to an internal combustion engine (3), and such an engine suffers from low torque at low speed, resulting in poor driving characteristics. The historical response was to increase the fuelling rate at low speed, but the scope for this has been limited by emissions regulations and concerns for fuel economy which make the engine uncompetitive. The problem was addressed more effectively by the use of a turbocharger wastegate (still a very commonly used device) or variable nozzle turbine. More complex turbocharging systems (4) were also targeted at the premium car segment which has attracted a greater share of diesel engines.

In the past ten or so years, there has been increasing evidence of the effects of emissions such as NOx, PM and ozone on human health (5), and the potential harm of quantities well below the regulatory limits. Evidence also emerged that emissions were not always measured consistently or accurately, that emissions testing often did not reproduce real driving conditions sufficiently well to give realistic results, and that manufacturers sought active measures that would demonstrate that tested engines were considerably cleaner than they were in real driving.

The cumulative effect of these was to halt and reverse the growth of diesel engines in passenger cars (Figure 2). Fewer diesel engine cars are now being sold, and some regulators have discussed banning the use of such cars in some cities where emissions are a particular concern, or of banning the sale of new diesel cars completely.

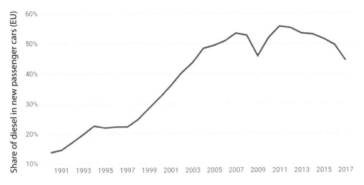

Figure 2. Share of diesel engines in new cars in the EU (6).

Turning now to the total number of cars, Figure 3 presents a historical breakdown between several of the largest car-owning countries. In countries with a large and stable car-owning population, the number of new cars bought has remained fairly steady since 2005, but with a significant weakening in the USA since 2015. In China there has been a very marked growth between 2008 and 2017, and a fall in sales since that time.

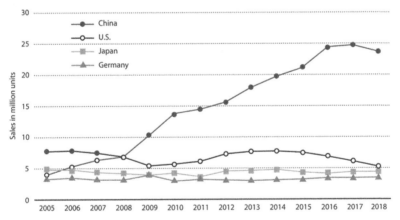

Figure 3. Historical breakdown of car sales in important market sectors between 2005 and 2018. Sales in million units (7).

This figure shows new car sales and does not accurately reflect cars on the road. Average car life has increased in the time period considered, which means that the growth in the number of cars on the road is likely to be greater in Germany, Japan and the USA which had greater car ownership at the start of the time period, but the magnitude of growth in China is still overwhelming. Of these, the vast majority (89%) have gasoline engines (8).

The adoption of turbocharging in gasoline engines has lagged well behind that of diesel engines. The technical reason for this is the likelihood of damaging pre-ignition and knock of gasoline fuel that puts a limit on boost pressure (9). Until

recently there has not been the same incentive to turbocharge. A naturally aspirated gasoline engine is smaller and lighter than the equivalent diesel engine of
the same power and so has inherently better specific power and torque. Gasoline
fuel in the cylinder burns relatively slowly, and some residual exhaust will be
trapped in the cylinder, creating an internal EGR that limits emissions. Attention
to the combustion process and valve timing can increase this to the extent that it
has a significant effect on emissions.

But with tighter emissions requirements applied to gasoline as well as to diesel
engines, the incentive to turbocharge has now arisen, and with controlled fuel injection, the combustion limitations have been, if not eliminated, at least eased to the
point where charge boosting is possible. Boost pressures are lower than in diesel
engines and there is no incentive to move to complex turbocharging systems. In
almost all cases, a wastegate is sufficient.

The result is that turbocharging is now being very widely employed in small engines
that constitute the majority of the market. Where regulations apply fleet-wide, manufacturers can continue to provide a minority of premium cars without turbochargers,
but further penetration of turbocharging into all sizes of cars with gasoline engines is
confidently expected.

2   THE PRESENT SITUATION

The current situation in all car markets is one of change. Conventional gasoline,
and in the EU diesel, engines form by far the largest fractions. There is
a continued downwards pressure on CO_2 emissions which seems set to continue
in the foreseeable future in all markets (Figure 4). Because CO_2 is the result of
oxidising carbon in the fuel, the only solutions are to reduce the amount of hydrocarbon burnt by making hydrocarbon fuel-powered vehicles more efficient, or sub-
stituting other fuels.

Various forms of electrification are currently receiving much attention. Hybrid gasoline-electric vehicles still occupy less than 5% of sales in markets other than
Japan. Battery electric vehicles (BEVs) are being actively promoted as a route to
zero-CO_2 emissions. At the present time BEVs are sold in small numbers and the
charging infrastructure is still being developed, and the final market share is quite
uncertain.

Fuel cell vehicles (FCVs) substitute zero carbon fuel (hydrogen) or lower carbon fuel
(e.g. methanol) to achieve CO_2 reductions. As with BEVs, the number of cars sold is
small, fuel cell technology is still progressing, and the infrastructure for dispensing
alternative fuels is at a very early state in its development.

The adoption of alternative powertrains and fuels calls into question the prospects for
turbocharging passenger car engines, which is the focus of this paper. It is therefore
appropriate to consider each type from this perspective.

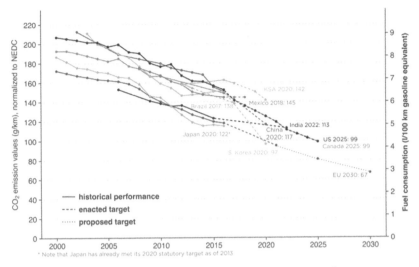

Figure 4. Trends in CO$_2$ emissions and regulatory targets for passenger cars (10).

3 POWERTRAIN OPTIONS

3.1 Conventional gasoline and diesel engines

It is most likely that conventional gasoline engines will continue to be available for passenger cars and will be the most common powerplant for years to come (11). Existing cars will continue to be used until the end of their economic life. For alternative fuel cars, technological developments and the widespread infrastructure for fuel delivery are still to be achieved. Taxation regimes and regulations may have to be implemented or adjusted to close the price gap with conventional cars for the customer. Consumer resistance to change generally means a time delay of roughly 10 years in the widespread use of new technologies.

New direct injection gasoline engines are significantly more efficient than their predecessors, with a consequent benefit to CO$_2$ emissions. Turbocharging allows the use of high, controlled levels of EGR and variable valve timing (e.g. Miller and Atkinson cycles) that can reduce NOx, CO and UHC emissions. A recent survey of new recent production engines (12) has shown that there are many technology options that potentially improve the efficiency of conventional gasoline engine powertrains. Some of these are now widely adopted but no engine incorporates all of the potential improvements. Table 1 is based on (12) with added comments on the relevance of turbocharging to the measures proposed.

Some of the measures proposed are purely engine-dependent. Others depend on the use of a turbocharger to be effective, such as cooled EGR and Miller cycle. Many of them have implications for the turbocharger or its matching, for example requiring changes in boost pressure. In general they have the same or even greater demands on turbocharger range and inertia as current technology engines. The effects of many options on the match between the compressor and turbine may require a switch from wastegate to variable geometry.

Table 1. Current and future gasoline car engine technologies, based on (12), and their implications for the turbocharger.

Engine technology	Implementation	Turbo implications
Variable valve timing	Widespread	
Integrated exhaust manifold	Widespread	
High geometric CR	Partial	Match
Friction reduction	Partial	
High stroke/bore ratio	Partial	
Boosting	Partial	Yes
Cooled EGR	Partial	Boost pressure/ match
Variable valve lift	Partial	
Miller cycle	Partial	Increased boost
VG turbocharger	Uncommon	Yes
Cylinder deactivation	Not developed	Match
Variable CR	Not developed	Boost pressure/ match
Gasoline compression ignition/lean burn	Not developed	?

Diesel engines in cars are in decline, even though developments in combustion have significantly reduced CO_2 and other emissions. It has also been argued (13) that emissions testing in the EU has been seriously defective and that diesel engine vehicles can be as clean as gasoline for NOx and at the same time lower in CO_2. Nevertheless, there is strong consumer and political resistance that will prevent a resurgence of diesel. Diesel engines are universally turbocharged, in the larger sizes with multiple and multistage turbochargers, and the net effect of the switch to gasoline will be a reduction in the total market for passenger car turbochargers despite the growth in gasoline.

3.2 Micro hybrid
In a micro hybrid engine the conventional starter and lead-acid battery are upgraded so that the electrical machine can provide assistance with starting and accelerating from rest. In stop-go commuter traffic this can provide 2-10% improvement in fuel economy (14) and an improved driving experience. It is a very simple development requiring minimal change in the ICE, and has few implications for the turbocharger. It is based on the existing car 12 or 48 volt electrical system, and can be expected to be widely implemented.

3.3 Mild hybrid
This is a step up from micro hybrid where the electric machine boosts the ICE during acceleration as well as starting, and regenerative braking is used to recover electrical energy. A separate EV-only mode is not available. Fuel economy improvements of

10-20% are estimated for city driving (15). Using an electric machine to assist in accelerating the vehicle addresses an area of weakness in turbocharging: the speed of response of the turbocharger. Current automotive turbochargers are compromised in size and inertia of the rotating system in order to reduce the "turbo lag". The use of an electrical machine alleviates this problem to some extent, and can benefit the turbocharger efficiency and the fuel economy of the vehicle.

3.4 Full hybrid and plug-in hybrid

A full hybrid car has an electric machine that is large enough to provide propulsion in an EV mode with the ICE turned off. The electrical system is high voltage and uses Li-ion or NiMH batteries of about 1-10 kWh capacity. Charging is by regeneration and ICE in a simple HEV, but is augmented by stationary external charging in a PHEV. Even in the latter case, the EV-only range is less than about 40 km and EV operation for any distance is limited to about 80 km/h. Fuel economy improvements of 20-50% in city driving can be expected (15).

The ICE is required for journeys beyond the EV range, but also for acceleration at all road speeds. When the electric machine is used together with the ICE, it acts as a power boost device. In addition to the turbo lag previously mentioned, the turbocharger does not provide sufficient boost at low engine speed, but in an HEV system this limitation is overcome by the electric machine.

In a conventional automotive gasoline or diesel engine, the turbocharger is small to maximise the boost at low engine speed, small to minimise turbo lag, boost limited by wastegate or variable geometry, and wide range in order not to compromise the high speed performance of the engine. This is a complex and contradictory set of requirements. In an HEV, low speed torque is augmented by the electric machine. The turbocharger can be larger and have a smaller range without compromising the engine performance. By removing some operational compromises, the turbocharger efficiency can be improved, with further benefit to the fuel economy and CO_2 emission of the vehicle.

The option of an electrically-driven supercharger in addition to (but not in place of) the electrical machine is also possible for this type of powertrain, and this would allow further optimisation of the turbocharger for maximum efficiency following a reduced range requirement. This option has not been much developed for production, but based on existing mechanical supercharger experience, it appears that an electrical supercharger will require a power rating of roughly 5 kW minimum for a small car engine of 1.5 litre or less displacement. Current HEVs in this segment have a battery capacity of 1-1.5 kWh, from which it is apparent that an electric supercharger could be used only sparingly during the drive cycle, for example during starting. For a larger PHEV with a 10 kWh battery, an electric supercharger could play a larger role during the drive cycle. Even so, in view of the extra cost and complexity this remains a less likely option.

The electric turbocharger, in which the turbocharger is mechanically linked to an electric machine that can provide additional boost at low engine speed or generate electricity at high engine speed, has also created some interest (16), (17). The electric machine removes some of the operational compromises of the turbocharger, allowing its performance to be improved in the same way and for the same reasons as using an electrical machine coupled to the vehicle transmission. However, the electric turbocharger does require a major redesign of the turbocharger to incorporate an electrical machine. Because it requires battery storage, it integrates better with a hybrid powertrain than a conventional ICE where additional energy storage and an electrical power control system are necessary.

3.5 Battery electric with range extender

Battery electric vehicles are subject to limits on driving range due to the battery capacity. For cars that are regularly required to make long journeys where recharging facilities are not available or recharging is not practical, a range extender may be considered. The range extender is a small ICE carried in the vehicle that is coupled only to a generator used to charge the battery, not directly to the powertrain.

The range extender engine is required to run theoretically at one operating point only, and practically at ranges of speed and power much smaller than a conventional car engine. Turbocharging provides the same advantages for such an engine as it does for any other gasoline engine, but the turbocharger can be highly optimised for design point operation without inertia or range requirements.

An electric supercharger has been applied to a range extender engine to increase the power output (18). If the car is regularly charged externally with electricity and the range extender is only occasionally used, this may be justified, but otherwise the supercharger will be powered indirectly by fuel burn.

3.6 Battery electric

The battery electric powertrain is now widely expected to take a considerable share of the future market, and its potential for zero-emission operation is given as the reason for this. However, the enthusiasm for BEVs must be tempered by several considerations.

Limits on energy storage and battery availability are serious concerns. The energy density of an Li-ion battery is a very small fraction of that of gasoline fuel, so that a large battery is required. A car of 400 km range requires a battery of about 60 kWh, weighing approximately 500 kg and costing about USD12000 at current prices (Figure 5). Some projections have suggested that the price might be halved by 2030, but if there is also a large growth in the number of BEVs, the demand for batteries will not easily be met by the availability of battery materials such as lithium and cobalt and limitations in the supply chain. These may keep the price high and this magnitude of cost reduction may not materialise.

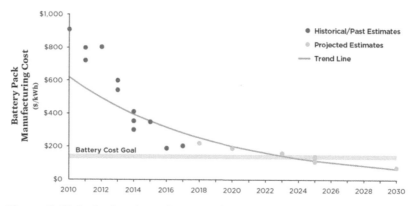

Figure 5. Historical costs and cost projections for Li-ion batteries (19).

Lithium itself is not a rare element, but it is highly reactive and the costs of extraction and refinement are considerable. Mining for lithium also has an environmental impact (20). The CO_2 emissions of a BEV, considered in isolation, are very attractive, but the total lifecycle emissions, including the manufacture of the battery and the safe disposal at the end of its life, require closer examination. The total lifecycle emissions also depend on the source of electricity. Only if the electricity is from 100% renewable sources is the total CO_2 emission very low. Electricity generation from fossil fuels make BEVs less attractive from this point of view.

A recent study (21) has focused on limitations in the supply of batteries and how they can be optimised for minimum CO_2 emission. The same battery capacity might be used in a small number of BEVs or divided among a much larger number of HEVs that individually use much smaller batteries. It is also argued that BEVs are only zero carbon from "tank to wheel", but that the total lifecycle emission of CO_2 must include electricity generated from non-renewable sources to manufacture the battery, charge and recharge it over its useful life, and finally dispose of it. The study concludes that comparing a fleet of HEVs with the same size fleet made up of a small number of BEVs with the same total battery capacity as the HEV fleet, with the balance made up of conventional gasoline engine cars, the HEV option is much more effective in reducing CO_2 emissions.

BEVs have, at first sight, no place for turbochargers. However, optimum battery life and performance requires a careful control over the battery temperature, for which air cycle machines based on turbocharger technology have a potential role that has not yet been much investigated or exploited.

3.7   Hydrogen fuel cell

FCVs are also promoted as the basis for zero emission vehicles. A fuel cell uses hydrogen as the fuel, which combines with atmospheric oxygen to produce electricity that is used by an electric machine to power the vehicle. Water vapour is the only exhaust product. For vehicle use the fuel cell is combined with battery storage to smooth out transients (sudden power demand transients may reduce the life of the fuel cell) and to enable regenerative braking to be used (22).

A car with 500 km range requires 3-5 kg H_2 (23) with a volume of 150-300 litres if stored as a high pressure gas or 75-100 litres as a cryogenic liquid. Cryogenic liquid systems offer considerable space reduction, but the delivery system is much less convenient. Most vehicle prototypes have been based on high pressure gas storage.

Currently most hydrogen is produced by steam reforming of methanol, and the total lifecycle CO_2 emissions of an FCV depend on the source of energy used for this purpose and the final disposal of the unwanted carbon byproducts. Methanol is a fuel that has received attention particularly in China which is rich in coal (from which methanol can be produced) but poor in oil and gas (24, 25). The molar hydrogen/carbon ratio of methanol is approximately twice that of gasoline, and it represents a significant step towards zero carbon transport.

Hydrogen is also available as a byproduct of chemical processes such as PDH (propane dehydrogenation) (26) and from the electrolysis of water. If electrolysis is achieved using renewably-generated electricity, it represents a zero carbon fuel. Hydrogen energy storage is being actively pursued as a means to cope with the variability of renewable energy generation (27).

The fuel cell is pressurised and an air compressor is required for the purpose, which is electrically driven. The compressor is based on turbocharger technology. The exhaust can in principle be expanded in a rotordynamic turbine, as is done in a conventional

ICE turbocharger, but the temperature of the fuel cell exhaust is much lower than that of an ICE so there is insufficient energy available to drive the compressor without the assistance of an electric machine (3). Water droplets condensing in the exhaust cause erosion and represent a considerable challenge to the design and operation of an expander. Figure 6 shows the compression and expansion systems, together with the hydrogen fuel, water management and thermal management subsystems that are necessary for a fuel cell vehicle.

Fuel cells have already achieved mass production and are widely used in many applications. The application of FCV for passenger cars are more likely for the SUV segment, but it is also possible as a powertrain for trucks and sedans. Given the developments in fuel delivery infrastructure that will be necessary, it is more likely for taxis than private cars initially, and will take several years before it becomes common.

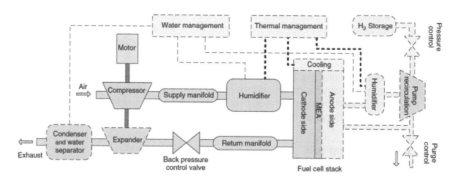

Figure 6. Schematic of a fuel cell system in a fuel cell vehicle (23).

4   COMPARISON OF CO$_2$ EMISSIONS FROM DIFFERENT POWERTRAINS

The concerns for CO$_2$ emissions has lead to studies and projections for various power-trains. To be credible, such analysis must include emissions not just from the vehicle during operation ("tank to wheel") but also the CO$_2$ emitted from the fuel extraction, refining and distribution processes ("well to tank"). For BEVs, this means electricity generation and distribution. The figures reproduced from (26) in Table 2 are based on "well to wheels". It is important to understand that these figures are projected from of current trends and technologies and are based on many assumptions and are subject to large uncertainties. As the authors note (26): "Vehicle electrification with the several hybrid options (HEV, PHEV, and FCV) offer 40%–50% reductions if electricity and hydrogen production still involve significant (though lower than today) GHG emissions. Only the much greener electricity [and] hydrogen ... pathway options offer more significant reductions". In other words, the final comparison depends on geopolitical decisions on national and global scales.

Table 2. Well to wheels GHG emissions estimated for the average new US car in 2030. Data from (28).

Powertrain/fuel	CO$_2$/km ratio (based on a gasoline, naturally aspirated, SI engine)
Gasoline, naturally aspirated SI engine	1.00
Turbocharged gasoline SI engine	0.90
HEV	0.62
PHEV[1]	0.48-0.36
BEV[2]	0.41-0.22
FCV[3]	0.70-0.35

1 Dependent on the relative use of gasoline and electric drives
2 Dependent on CO$_2$ intensity of electricity generation
3 Lower number based on clean H$_2$ (with carbon capture and sequestration)

That said, there are several important conclusions to be drawn from this study. The scope for development of conventional gasoline engines is considerable, but the total lifecycle CO$_2$ emissions are still higher than the alternatives. The HEV offers considerable reductions and the PHEV further reductions, though largely dependent on how the vehicle is used. Both the FCV and BEV are zero carbon tank to wheels, and the figures here reflect the CO$_2$ emitted in electricity or hydrogen generation. This accounting demonstrates that they are both considerable CO$_2$ emitters.

5   IMPLICATIONS FOR TURBOCHARGING

The various powertrains that are being considered for future cars bring with them different implications for turbocharging. Some factors may be considered constant. Where an internal combustion engine is employed, the benefits of turbocharging for downsizing the engine and vehicle and enabling the use of methods to reduce NOx and UHC emissions are so well established as to require no further discussion.

The needs for high turbocharger efficiency, life and durability and low manufacturing cost will remain. Here the interrelated nature of turbocharger design requirements is important. Turbocharger efficiency, for example, is a trade-off with other factors such as operating range, size and inertia, and the final design is always a compromise between conflicting requirements (3). When considering different powertrains it is important to be aware of the changing requirements of the turbocharger because this may modify the design compromises in ways that can be favourable to efficiency.

The turbocharger requirements for various powertrains have been discussed above. In Table 3 they are summarised. Experience of turbocharging conventional car engines has shown the importance of maximising the stable operating range of the compressor and minimising the inertia of the compressor and turbine. The former improves the low speed torque of the engine while maintaining efficiency at high speed, high load conditions. The latter improves the transient response desirable for city driving and overtaking.

Table 3. Summary of powertrain options and their implications for developments in turbocharger technology.

Powertrain	Turbo range	Turbo inertia	Turbo derived technology
Turbocharged SIE	+ +	+ +	
Micro hybrid	+ +	+	
Mild hybrid	+ +	0	
Full hybrid	0	−	E-supercharger [?]
Plug-in hybrid	0	− −	E-supercharger [?]
BEV + range extender	− −	− −	E-supercharger [?]
BEV	N/A	N/A	Battery conditioning
FCV	N/A	N/A	Gas compressor Exhaust expander [?]

Ratings are on a scale from + + (very important) to − − (not at all important).

Both of these factors have implications for the efficiency of the turbocharger which, in current products, is routinely sacrificed in pursuit of wide range and low inertia. Anything that reduces the need for these can provide a benefit in turbocharger efficiency, providing the manufacturers are wise to this. Range and inertia will continue to be highly desirable in conventional engines, but increasing hybridisation means that more use can be made of the electric motor for low speed operation and during transients. The extent to which this can be done depends on the power rating of the electric motor and the capacity of the battery.

BEVs and FCVs do not require turbochargers, and the widespread deployment of these technologies would considerably reduce the size of the turbocharger market. Even here, there are opportunities for products derived from turbocharger technology, notably for gas compression and possibly expansion in a FCV, and for conditioning the battery in a BEV. The electric supercharger is also an option for many ICE-based powertrains, but it is only a serious contender in those with a large battery capacity.

5.1 Compressor developments

The design of passenger vehicle turbocharger compressors has been heavily influenced by two operability requirements: a wide stable operating range and fast transient response (32). The design-level consequences of imposing these two requirements have limited the compressor performance ceiling. Here the potential compressor performance gains are considered when these operability requirements are removed.

The transient response requirement is manifest as a low inertia requirement, realized via small diameter compressor and turbines wheels, which necessitate a higher design-point rotational speed to maintain the work transfer rate. Alleviating the transient response constraint allows the rotational speed to be selected primarily to optimize the aerodynamic performance rather than balancing efficiency potential with inertia. Most modern, on-road turbocharger compressors are designed with rotational speeds 30-60% above the optimal design speed. Figure 7 shows the design-point efficiency trade-off with specific speed (and hence rotational speed for a given boost pressure and flow rate). Changing from current typical specific speed design values of 1.2-1.3 down to a more optimal value has the potential to yield 2-5 points of efficiency.

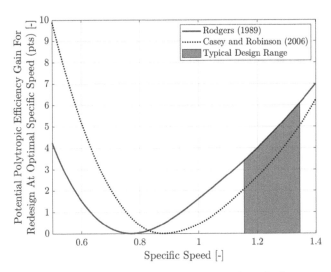

Figure 7. Potential efficiency improvements through redesign at optimal specific speed. Adapted from (29)-(31).

Wide, stable operating range is achieved through careful management of secondary flow and loss development, directly influencing design decisions at the architectural foundation. Earlier work (32) discussed the critical design compromises necessary to accommodate wide range with directly identifiable performance impacts pertaining to compressor inlet (eye) sizing, diffuser configuration, and volute sizing.

For optimal performance given rotational speed and flow requirements, the inlet size is determined to minimize the inlet relative velocity (33), (34). Stages capable of wide operating range must provide sufficient annulus area necessary to meet high-flow capacity targets beyond the peak-efficiency point, often leading to inlet sizes greater than optimal, particularly for stages with high inlet relative Mach numbers. Figure 8 illustrates this effect, where the optimal design point (red square) for the original design speed (blue curve) is insufficient to meet the flow capacity. The inlet size must be increased by 25% to meet the high-flow requirements (yellow star). If the specific speed and hence rotational speed is reduced, which is allowable by de-weighting the transient response requirement as discussed above, the inlet relative Mach number for the optimal design point is markedly reduced (dashed curve).

There are two performance benefits: first, with the reduced inlet relative Mach number, higher impeller diffusion is enabled (35), (36), leading in part to the performance improvements observed in Figure 8, and second, reduced inlet Mach numbers push operation away from conditions requiring careful shock management (37), (38), further enabling higher loading and thus performance improvements. As discussed in (32), the blading design is often compromised such that the optimal inlet loading is not achieved at the peak efficiency point. This intentional mismatch results from the need to extend stable operation to lower flow rates without shifting choke. It is difficult to quantify the efficiency penalty due to the mismatch, but the authors' experience suggests that it varies from 0 to 1 point of efficiency with sensitivity diminishing at reduced inlet relative Mach numbers.

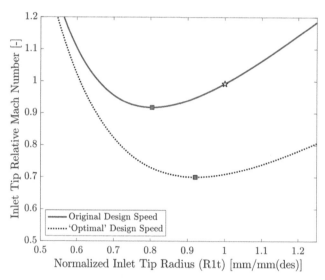

Figure 8. Inlet relative Mach number and inlet tip radius for typical and optimal turbocharger design speeds.

Vaneless diffusers are used in wide-range centrifugal compressors for their broad loss characteristics, while vaned diffusers exhibit strong gradients in loss and recovery when flow separation occurs, thus limiting their viable range. In applications where the widest range is not required, low and moderate solidity airfoil diffusers are used. The higher diffusion rate reduces the mean velocity and can increase stage performance by 2-4 points, (39)-(41). Similarly, it is common in turbochargers to match the stage with an oversized volute to avoid excessive losses at high volume flow rates where a properly sized volute would be too small. A volute that is too small acts as a nozzle, accelerating the flow around the circumference, increasing stage losses (33), (43). By properly matching a volute at the design point, efficiency improvements up to an efficiency point or more have been realized.

5.2 Turbine developments

In a conventional turbocharger for a passenger car engine, the turbine design is most compromised by the requirement of low inertia. This applies even more strongly to the turbine than the compressor because the density of the turbine rotor material is about three times higher than that of the compressor impeller. The magnitude of the compromise can be estimated from the collected test data shown in Figure 9. In this graph, the flow coefficient is defined as the meridional velocity of the gas at exit of the rotor divided by the blade speed. The gas velocity depends on passage area, and therefore on the tip radius, because the hub radius is set to a minimum that is fixed by the blade thickness at the hub.

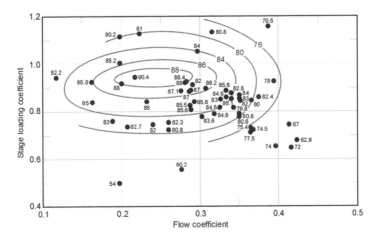

Figure 9. Collection of test data for radial inflow turbines, plotted as stage loading versus flow coefficient. Numbers show measured total to static efficiency (44).

Flow coefficient thus depends on the square of the tip radius, and the inertia varies approximately as the 4th-5th power of radius, depending on the design of the rotor. The design flow coefficient for small automotive turbocharger turbines is typically in the range of 0.35-0.4. Halving this value improves the efficiency by approximately 12 percentage points but with a roughly four-fold increase in inertia.

However, this performance gain cannot be achieved in reality in turbocharging because the requirements for turbine life and installation size will not change. The turbine is highly stressed and works in a hot gas environment, and any increase in blade height will increase the blade hub stresses, approximately as the square of the blade height. This will require thicker blades and very likely fewer blades to avoid crowding at the hub, and an increase in blade loading. Limitations on overall size to fit the engine compartment will also restrict the ability to increase the rotor diameter. The exact trade-off in each case depends on the specific details of the application, but Figure 10 shows the results of a numerical optimisation of a passenger car turbocharger turbine made to investigate this point. This study provided a set of optimised designs of different inertias and design point efficiencies. The results show that even a modest increase in inertia can provide a very worthwhile increase in efficiency.

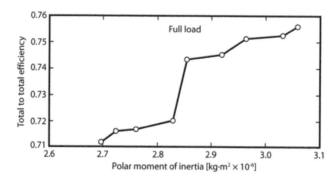

Figure 10. Results of a numerical optimisation to investigate the relationship between rotor inertia and efficiency for a passenger car turbocharger turbine operating at its best efficiency point (45).

5.3   Turbocharger optimisation

The benefits of changes to the turbocharger design for a hybrid vehicle can only be determined by optimising the full powertrain and its management, and results given here are only indicative of what might be achieved, based on the information presented here and the authors' experience in compressor and turbine design.

For the compressor, it seems reasonable that the reduction in specific speed that follows from allowable increases in inertia may lead to 1-3 percentage points increase in efficiency. A reduced demand on range may enable as much as 4 points increase, divided between the impeller inlet (1), diffuser (2), and volute (1). For the turbine, the trade-off is conceptually a simpler one between inertia and efficiency, and 5 percentage points appears to be a realistic goal.

These targets are most likely to be achieved in a range extender engine that is only required for charging and runs at constant conditions. For hybrid engines, the benefit will depend on the size of the battery, the power rating of the electric motor, and the operational charge/discharge strategy of the powertrain controller.

6   CONCLUSIONS

The switch from conventional ICEs to alternative powertrains and fuels in passenger cars will lead to changes in the technology and market for turbochargers. The inertia of the market means that powertrain changes do not happen quickly, and it is probable that ICEs will be the basis of the majority of powertrains for years to come. The switch from diesel to gasoline is already taking place and is expected to continue. The proportion of gasoline engines that are turbocharged will increase. The market for cars globally, but most particularly in China, is decreasing. The present emphasis on wide range and low inertia for conventional ICE turbochargers will continue and the scope for turbocharger technology development will be small.

An increase in hybrid gasoline-electric engines continues to provide market opportunities for turbochargers, but a change in the turbocharger design strategy, emphasising efficiency over size and operating range, is possible and desirable. Such changes are based on known technology and do not require large-scale changes to the turbocharger configuration. The scope for this depends on the rating of the electric machine and the on-board battery capacity. In the most favourable conditions, it might be as much as 5-7 percentage points for the compressor and turbine.

A switch to BEVs and FCVs will certainly reduce the turbocharger market. CO_2 emissions result from the fuel preparation and the outcome will depend on decisions removed from the vehicle itself. Consumer resistance to these powertrains based on first cost, range, and refuelling options may be expected. Such a switch will create opportunities for air cycle machines and electrically-driven compressors based on turbocharger technology, but these are very unlikely to fully compensate for the loss of the passenger car turbocharger market.

GLOSARRY

BEV	Battery electric vehicle
CR	Compression ratio
EGR	Exhaust gas recirculation
EV	Electric vehicle
FCV	Fuel cell vehicle
GHG	Greenhouse gases
HEV	Hybrid electric vehicle
ICE	Internal combustion engine
NOx	Oxides of nitrogen
PHEV	Plug-in hybrid electric vehicle
PM	Particulate matter
SIE	Spark ignition engine
UHC	Unburnt hydrocarbons

REFERENCES

[1] Cames M, Helmers E 2013 Critical evaluation of the European diesel car boom □ global comparison, environmental effects and various national strategies. In *Environmental Sciences Europe. Bridging Science and Regulation at the Regional and European Level* **25**:15. https://doi.org/10.1186/2190-4715-25-15.
[2] ACEA 2019 Euro Standards. https://www.acea.be/industry-topics/tag/category/euro-standards. Accessed 24/4/2019.
[3] Baines N C 2005 *Fundamentals of Turbocharging*, Concepts NREC.
[4] Baines N C 2018 A comparison of boosting technologies in the automotive light duty sector. Turbocharging Seminar, Shanghai, China, 16-17 October.
[5] World Health Organization. Health and Sustainable Development, Air Pollution. https://www.who.int/sustainable-development/transport/health-risks/air-pollution/en/. Accessed 24/4/2019.
[6] ACEA 2019 Share of diesel in new passenger cars. https://www.acea.be/statistics/tag/category/share-of-diesel-in-new-passenger-cars. Accessed 24/4/2019.
[7] Statistica 2019 Sales of passenger cars in selected countries worldwide from 2005 to 2017. https://www.statista.com/statistics/257660/passenger-car-sales-in-selected-countries/. Accessed 31/ 10/2019.

[8] Ministry of Ecology and Environment of the People's Republic of China 2019, China Vehicle Environmental Management Annual Report. http://www.mee.gov.cn/gkml/sthjbgw/qt/201806/t20180601_442293.htm. Available at http://dqhj.mee.gov.cn/jdchjgl/zhgldt/201806/P020180604354753261746.pdf. Accessed 4/4/2019.

[9] Heywood J B 2018 *Internal Combustion Engine Fundamentals*. 2nd ed. McGraw Hill.

[10] ICCT 2019. Chart library: Passenger vehicle fuel economy. http://www.theicct.org/chart-library-passenger-vehicle-fuel-economy. Accessed 20/6/2019.

[11] Els P 2019 Turbocharging - fuel efficiency and the path to EV's. https://www.automotive-iq.com/powertrain/articles/turbocharging-fuel-efficiency-and-path-evs. Accessed 24/4/2019.

[12] Stuhldreher M, Kargul J, Barba D, McDonald J, Bohac S, Dekraker P, Moskalik A 2018 Benchmarking a 2016 Honda Civic 1.5-liter L15B7 turbocharged engine and evaluating the future efficiency potential of turbocharged engines. SAE paper 2018-01-0319.

[13] Molden N 2018 Q&A, Automotive IQ Real Driving Emissions conference. https://www.automotive-iq.com/powertrain/articles/qa-nick-molden-founder-and-ceo-of-emissions-analytics-united-kingdom. Accessed 16/4/2019.

[14] Sepe R B, Morrison C M, Miller J M et al. 2001 High efficiency operation of a hybrid electric vehicle starter/generator over road profile. In *IEEE Industry Application Society Annual Meeting*, Chicago: 921–925.

[15] Chan C C 2014 Overview of electric, hybrid and fuel cell vehicles. *Encyclopedia of Automotive Engineering*, John Wiley & Sons, Ltd. DOI: 10.1002/9781118354179.auto061.

[16] Winward E, Rutledge J, Carter J, Costall A, Stobart R, Zhao, D, Yang Z 2014 Performance testing of an electrically assisted turbocharger on a heavy duty diesel engine. Inst Mech Engrs 12th International Conference on Turbochargers and Turbocharging, London.

[17] Cooper A, Bassett M, Hall J, Harrington A et al. 2019 HyPACE - Hybrid Petrol Advance Combustion Engine - Advanced boosting system for extended stoichiometric operation and improved dynamic response. SAE Technical Paper 2019-01-0325. DOI:10.4271/2019-01-0325.

[18] Bassett M, Hall J, Warth M, Mahr, B 2014 Development of a range extender engine family. FISITA World Congress, F2014-TMH-016, Maastrict, The Netherlands, 2-6 June.

[19] Union of Concerned Scientists 2017 Accelerating US Leadership in Electric Vehicles. www.ucsusa.org/EV-incentives. Accessed 24/4/2019.

[20] Lithiummine.com. Lithium mining and environmental impact, 2018. http://www.lithiummine.com/lithium-mining-and-environmental-impact.   Accessed 24/4/2019.

[21] Emissions Analytics 2019. Hybrids are 14 times better than battery electric vehicles at reducing real-world carbon dioxide emissions. https://www.emissionsanalytics.com/news/hybrids-are-better. Accessed 19/6/2019.

[22] Zhao H, Burke A F 2009 Optimization of fuel cell system operating conditions for fuel cell vehicles. *Journal of Power Sources* **186**(2), 408–416.

[23] Zhao H, Burke A F 2014 Fuel cell powered vehicles. *Encyclopedia of Automotive Engineering*, John Wiley & Sons, Ltd. DOI: 10.1002/9781118354179.auto066.

[24] Ministry of Industry and Information Technology of the People's Republic of China 2019. Guidance from the eight departments on the application of methanol vehicles in some regions. Document no. 61. http://www.miit. gov.cn/n1146295/n1652858/n1652930/n3757016/c6684042/content.html.   Accessed 12/6/2019.

[25] Anon 2019. Methanol hydrogen production + fuel cell technology has become the new favorite of the automotive industry. http://www.cbea.com/yldc/201904/587801.html. Accessed 12/6/2019.

[26] Wan V, Asaro M 2015 Propane dehydrogenation process technologies. IHS Chemical PEP Report 267A. https://ihsmarkit.com/pdf/RP267A-toc_237249110917062932.pdf. Accessed 23/2/2020.

[27] Zhang F, Zhao P, Niu M, Maddy J 2016 The survey of key technologies in hydrogen energy storage. *Int Journ Hydrogen Energy* **41**(33): 14535–14552.

[28] Heywood J, MacKensie D (eds) 2015. On the road toward 2050. Potential for substantial reductions in light-duty vehicle energy use and greenhouse gas emissions. MIT Energy Initiative Report. http://mitei.mit.edu/publications/. Accessed 14/2/2020.

[29] Rodgers C 1991 The efficiencies of single stage centrifugal compressors for aircraft applications. ASME paper 91-GT-77.

[30] Casey M V, Robinson C J 2006 A guide to turbocharger compressor characteristics. In Bargende M (ed) *Dieselmotorentechnik*, 10th Symposium, 30-31 March, Ostfildern, TAE Esslingen.

[31] Casey M V, Robinson C J, Zwyssig C 2010 The Cordier line for mixed flow compressors, ASME paper GT2010-22549.

[32] Krivitzky E M 2012 The quest for wider-range, higher-efficiency automotive turbocharger compressors. Concepts NREC, Automotive Engineering International.

[33] Dixon S L, Hall C A 2010 *Fluid Mechanics and Thermodynamics of Turbomachinery*. 6th ed. Butterworth-Heinemann.

[34] Japikse D 1996 *Centrifugal Compressor Design and Performance*. Concepts ETI, Inc., Wilder VT.

[35] Dean R 1977 On the unresolved fluid dynamics of the centrifugal compressor. *Advanced Centrifugal Compressors*, ASME 77–149815, ASME Publications.

[36] Rusch D, Casey M V 2013 The design space boundaries for high flow capacity centrifugal compressors. *Trans ASME Journ Turbomach* **135**(1): 031035–1.

[37] Hazby H, Casey M V, Numakura R, Tamaki H 2015 A transonic mixed flow compressor for an extreme duty. *Trans ASME Journ Turbomach* **137**(5): 051010.

[38] Krivitzky E M, Larosiliere L M 2012 Aero design challenges in wide-operability turbocharger centrifugal compressors. SAE Technical Paper 2012-01-0710. https://doi.org/10.4271/2012-01-0710.

[39] Tamaki H 2019 A study on matching between centrifugal compressor impeller and low solidity diffuser and its extension to vaneless diffuser. *Trans ASME Journ Eng Gas Turbines Power* **141**(4): 041026.

[40] Japikse D 1982 Advanced diffusion levels in turbocharger compressors and component matching. *Proc 1st Int Conf on Turbocharging and Turbochargers*, London.

[41] Ziegler K U, Gallus H E, Niehuis R 2003 A study on impeller-diffuser interaction – Part I: Influence on the performance. *Trans ASME Journ Turbomach* **125**(1): 173–182.

[42] Hu L, Yang C, Krivitzky E, Larosiliere L et al. 2009 Numerical study of ultra low solidity airfoil diffuser in an automotive turbocharger compressor. SAE Technical Paper 2009-01-1470. https://doi.org/10.4271/2009-01-1470.

[43] Dai Y, Engeda A, Cave M, Di Liberti J-L 2010 A flow field study of the interaction between a centrifugal compressor impeller and two different volutes. *Proc Inst Mech Engrs, Part C: Journ Mech Eng Sci* **224**(2): 345–356.

[44] Chen H, Baines N C 1994 The aerodynamic loading of radial and mixed flow turbines. *Int Journ Mech Sci* 36: 63–79.

[45] Baines N, Dubitsky O 2016 A novel non-radial turbocharger turbine created using numerical optimization. Turbocharging Seminar, September 22-23, Beijing, China.

14th International Conference on Turbochargers and Turbocharging
Institution of Mechanical Engineers, ISBN: 978-0-367-67645-2

Boosting the JCB Fastrac "World's fastest tractor"

A. Skittery

JCB Power Systems, UK

R. Cornwell, R. King

Performance Development, Ricardo Automotive & Industrial, UK

ABSTRACT

The JCB Fastrac, which first rolled off the production line 28 years ago, is firmly established as a premium British fast tractor, pushing agricultural productivity and availability to ever-higher levels by enabling rapid "between-jobs" road travel speed. The JCB 672 engine builds on JCB's successful and record-breaking DieselMax family of rugged, robust diesel engines. With the addition of advanced boosting technology, the pairing of Fastrac and 672 engine takes the definition of "fast tractor" to new levels. The applied technologies – which retain relevance to the real-world challenges of emissions control and transient response – comprise high pressure ratio single stage turbocharging, electric supercharging, water injection and air pulse transient torque enhancement. The paper will describe the virtual and physical development steps undertaken by JCB and Ricardo in taking an off-highway diesel engine from market-competitive low-30s kW/l to over 100kW/l, in pursuit of record-breaking performance, as well as integration onto the vehicle, vehicle performance and driveability.

1   BACKGROUND TO TRACTOR SPEED RECORDS

The title of "fastest tractor" was previously problematic to definitively bestow due to the different ways a "record" speed can be judged. Some would argue that it is purely the peak instantaneous speed attained that defines "fastest". However, in the land speed record community both principal adjudicating authorities, the FIA and Guinness World Records, currently require a vehicle to pass through timing gates placed a measured kilometre apart and the average speed between the two gates is recorded. Furthermore, the measurement must be taken twice, once in each direction, and with both runs completed within one hour. It is the average of these two consecutive runs which is documented as "the record".

In addition, the definition of "a tractor" is open to interpretation. The OED (1) defines "tractor" as "a powerful motor vehicle with large rear wheels, used chiefly on farms for hauling equipment and trailers".

Until June 2019, the following notable vehicles made claims on the title of "World's Fastest Tractor";

- AGCO's Valtra T234 tractor, using special winter tires from Nokian Heavy Tyres, set a Guinness World Records adjudicated record of 80.88mph (130.165km/h) on 19 February 2015 on the emergency airfield of Vuojärvi, Finland, driven by Juha Kankkunen. The total length of the packed snow strip was approximately 2.3km and a 50m speed trap was set at the middle of the strip (2).
- "Track-tor", developed for the BBC's Top Gear television programme in March 2018 achieved a Guinness World Records adjudicated record speed of 87.271mph (140.45km/h) under the category of "fastest tractor (modified)"

at Bruntingthorpe Proving Ground, a former RAF aerodrome in Leicestershire, UK. Driven by Top Gear's "The Stig", and designed by Matt LeBlanc, the vehicle was powered by a 500hp 5.7l Chevy V8 gasoline engine. It achieved 91.919mph (147.929km/h) in one direction and 82.623 mph (132.969km/h) on the return run, with an average taken of the two runs (3).

- "The Pioneer", an Allis Chalmers D19 tractor was driven by Kathy Schalitz at an East Coast Timing Association (ECTA) adjudicated event, "The Arkansas 1-Mile Challenge", at Blytheville International Airport, Arkansas, USA in June 2018. Powered by a 262 cubic-inch Buda gasoline engine, heavily modified its owner, Dave Archer, and using a push-vehicle to start, it recorded a peak speed of 108.5mph over a one-mile track under the ECTA category of D/GCT (4), (5).

Figure 1. OED definition of a tractor.

2   TARGET SETTING

JCB has a strong background in land speed record breaking, with its FIA sanctioned world record set on 23 August 2006 with the JCB DieselMax land speed record (LSR) car (6) which achieved 350.097 mph (563.418 kph) on the Bonneville Salt Flats, Utah, USA.

The company has produced the Fastrac tractor since 1991, which enables rapid "between-jobs" travel with an on-road top speed of 43.5mph (70km/h). That JCB would aspire to have this product formally recognised as the "World's Fastest Tractor" was inevitable (7) and hence the "World's Fastest Tractor" (WFT) programme was conceived in the summer of 2018 with the intention to use JCB's DieselMax 672 engine (see Table 1).

Figure 2. The JCB DieselMax land speed record car, which still holds the FIA world record diesel-powered cars.

Table 1. Specifications of JCB DieselMax 672 engine.

Production JCB DieselMax 672 Engine	
Maximum rating	212kW (284hp)
At speed	2000 rpm
Peak torque	1150 Nm (848 lb-ft)
Peak torque speed	1400rpm
Maximum no load governed speed	2250rpm
Nominal idling	850rpm
Thermodynamic cycle	Diesel 4 Stroke
Air intake	TCA
Arrangement	6 cylinders, 7.2□
Nominal bore	106mm
Stroke	135mm
Maximum rating	212kW (284hp)

JCB's desire was to comfortably exceed the existing Guinness World Records record of 87mph, as well as exceeding any other claimed peak speeds, therefore a target speed of at least 150mph was set, whilst preserving a vehicle shape and from such that it was still demonstrably a tractor. Unlike outright wheel-driven LSR racing, a push start could be considered unacceptable for a vehicle category supposedly all about tractive effort. Adhering to this philosophy places particular challenges on launch, pull-away and transient torque performance. Finally, the record attempt would adhere to the latest regulations specified by Guinness World Records for the category of "fastest tractor (modified)":

- Average measured speed through 1km timing gates, average of two consecutive runs, one in each direction, recorded within 1 hour
- Timing gates to be located at the centre of the track
- Vehicle must have an FIA approved roll cage
- Vehicle must unmistakeably be a tractor

While the stock Fastrac has a mass of over 8 tonnes, the JCB vehicle team envisaged a final "race" weight of closer to 5 tonnes, with the reduction achieved via material and structural optimisation in the chassis. The aerodynamic performance of the base tractor was somewhat unknown at the start of the project. Ricardo performed some initial CFD analysis of the base tractor (see Figure 3) and JCB identified a realistic potential reduction of frontal area, which collectively allowed an estimate to be made of C_dA. As a further measure, JCB conducted some basic coastdown tests of a production Fastrac, albeit in a narrow speed window. These data sources, together with clutch characteristics, gear and axle ratios, were fed into a Ricardo IGNITE vehicle performance model, allowing power and torque targets to be set.

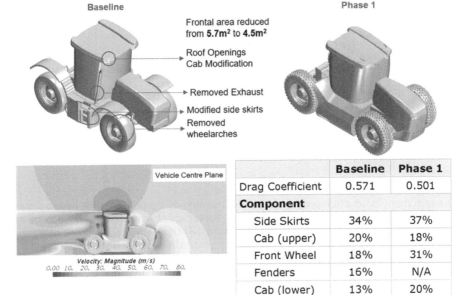

	Baseline	Phase 1
Drag Coefficient	0.571	0.501
Component		
Side Skirts	34%	37%
Cab (upper)	20%	18%
Front Wheel	18%	31%
Fenders	16%	N/A
Cab (lower)	13%	20%
Rear Wheels	4%	1%

Figure 3. Initial CFD analysis of the modified Fastrac, used to establish engine performance requirements.

To cater for the uncertainty in achievable mass and C$_d$A, plus the impact of varying track length, the vehicle model was run through a Design of Experiments (DoE) matrix to give contours of achievable average flying kilometre speeds, with solutions achieving over 150mph deemed worthy of further consideration.

As seen in Table 2, the engine target power needs to be between 950 and 1000hp to give acceptable performance with realistic vehicle characteristics. 1000hp was chosen as the ultimate target, with intermediate development steps of 800 & 900hp selected as gateways during the performance development programme. This phased approach to power upgrade aligned with vehicle development milestones. JCB envisaged two levels of "Fast Tractor": FT1 – with initial levels of chassis light-weighting and aerodynamic improvements, rapidly deployed in order to gain running experience as soon as possible, to be followed with a subsequent FT2, with final levels of light-weighting, aerodynamic and power development.

Table 2. DoE matrix/results from the IGNITE vehicle performance model.

| | Mass | C$_d$A | Achievable | | Moderate | | Radical |
			800hp	850hp	900hp	950hp	1000hp
Baseline	5244	2.245	142	146	149	150	150
Achievable	5000	2.1	146	150	152	154	156
Moderate	5000	2.0	147	152	154	156	157
	4500	2.1	148	153	155	157	159
Radical	3500	1.9	158	162	165	166	169

1.83mi bi-directional runway
0.633m wheel radius
0.025 rolling resistance

3   INITIAL FEASIBILITY

As with many production programmes, the first step in the performance development programme was to perform some initial spreadsheet calculations to determine required air flow and pressure ratio requirements. At this early stage generic compressor maps from Ricardo's library were used to indicate likely performance capability. For these early investigations, no consideration was given to turbine performance, the objective was simply to identify suitable compressor requirements. Initially rated power engine speed was set at 3200rpm a condition demonstrated by JCB during durability validation of the production engine.

Figure 4 shows that a very compelling and broad torque curve could be achieved with a series sequential (two-stage) turbocharger configuration, compared to single-stage (Figure 5). Given the duty cycle of a wheel-driven land speed is almost exclusively transient in nature, the ability of a smaller HP-stage turbocharger to initially spool up before handing over to a larger, rated power optimised LP machine, would be very attractive. However, examination of the tractor engine bay identified there would be huge challenges in packaging a two-stage system. Whilst the JCB

Dieselmax LSR streamliner used a two-stage system "wrapped" around each engine (Figure 6), Fastrac vehicle industrial design cues – which the team felt were mandatory to preserve in the WFT to ensure the finished vehicle was undeniably "a Fastrac" – meant that significant growth of engine bay dimensions was not possible (Figure 7). Early in the programme, therefore, the decision was taken to focus on single-stage turbocharging, and to concentrate on supplemental technologies to enhance low speed torque.

Once this primary decision had been made, the development programme moved on to identifying suitable boost system hardware using 1D simulation in Ricardo WAVE.

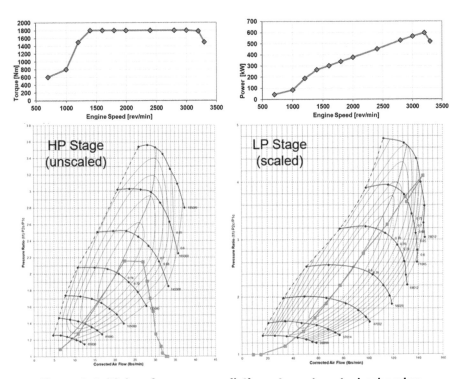

Figure 4. Initial performance prediction – two-stage turbocharging.

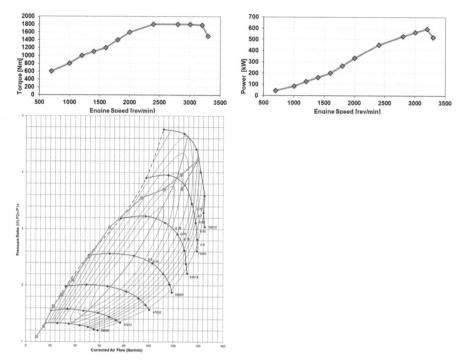

Figure 5. Initial performance prediction – single stage turbocharging.

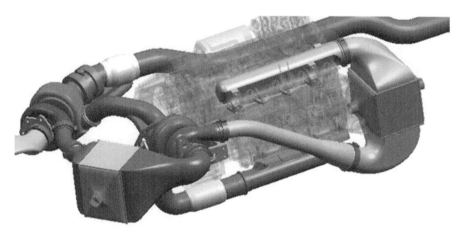

Figure 6. JCB Dieselmax LSR engine – two-stage turbocharging system packaged around each power unit.

Figure 7. Production Fastrac bonnet (L) was carried-over to the WFT (R), restricting the opportunity to create space for two-stage turbocharging.

4 BOOST SYSTEM SPECIFICATION

A WAVE model of the baseline 672 engine was updated with initially sized intake and exhaust system and an extended speed range. Combustion burn profile data derived from the Dieselmax LSR engine development programme, which utilized an ultra-low-compression ratio combustion system to limit Pmax, was also incorporated into the model.

The initial WAVE runs indicated the magnitude of the challenge, particularly when compared with values for the standard 672 engine. These initial results used a scaled production compressor map from a major Tier 1 supplier and can already be seen to be significantly challenged in terms of compressor map width (Figure 8). Whilst the engine for a land speed record application clearly does not have to be durable to the same degree as a production engine, there would still be significant concerns about running so close to both the surge line and the maximum speed condition; and these initial runs were still only at a preliminary target of 800hp.

As seen in Table 2, the engine target power needs to be between 950 and 1000hp to give acceptable performance (>150mph) with realistic vehicle characteristics. 1000hp was chosen as the ultimate target, with intermediate development steps of 800 & 900hp selected as gateways during the performance development programme. This phased approach to power upgrade aligned with vehicle development milestones.

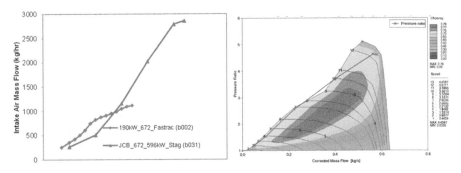

Figure 8. Initial WAVE results indicated almost 3x air mass-flow required and significant map width challenges.

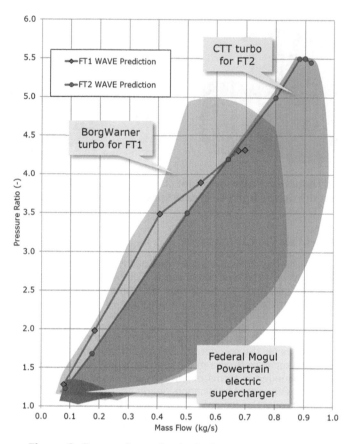

Figure 9. Comparison of selected compressor maps.

Figure 9 compares the maps for the 3 compressors used. In the lower-left corner is the electric supercharger (more details are given in section 5), which was used on both FT1 and FT2. The orange area and red line show the compressor map and predicted operating line for the BorgWarner turbocharge used on FT1, with which a peak power of 720hp was predicted. Finally, in Figure 9, the blue area and line show the compressor map and predicted operating line for the Cummins Turbo Technologies (CTT) turbocharger that was specified for FT2, with which a peak performance of up to 1000hp was predicted.

5   APPLICATION OF ELECTRIC SUPERCHARGER

Due to the low compression ratio (12:1) and the need to avoid white smoke emissions (due to incomplete combustion) at idle when the engine was cold, an electric supercharger was specified. A 24V COBRA water-cooled electric supercharger, produce by Federal Mogul Powertrain, was specified to both:

- Provide boost during cranking and idling to aid starting and prevent white smoke emissions
- Provide boost pressure at low to mid-loads at engine speeds below ~2000rpm, before the main turbocharge came on to boost

The electric supercharger was positioned upstream and in series with the turbocharger compressor (Figure 10). A throttle valve was used to close the main air inlet to prevent back-flow of air when the electric supercharger was operating. A differential pressure sensor detected when the differential pressure across the closed bypass valve was approaching zero, at which point the valve was automatically opened and the electric supercharger was deactivated.

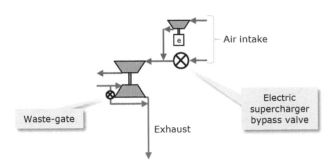

Figure 10. Arrangement of turbocharger and electric supercharger.

The turbocharger specified for the engine for FT1, which was capable of achieving a peak power of >720hp, worked well in conjunction with the electric supercharger and turbo lag was not seen as a significant issue during vehicle testing.

6   TURBO LAG MITIGATION ON FT2

The extremely large turbocharger for the FT2 (1000hp) specification engine causes the compressor maps of turbocharger and electric supercharger to overlap only to a negligible degree; increasing the risk of a significant "torque hole" under pullaway and gear change conditions. With a constrained track length and the extremely transient duty cycle of a speed record run, this could be a significant problem.

Two possible solutions were explored. Firstly, the steady state operating point on the compressor map can be shifted away from surge/low efficiency using an air side engine bypass or bleed valve. The operating principle here was that, opening the bypass valve at speeds between 1100 and 2400rpm, the airflow through the compres-sor exceeds the airflow consumed by the engine, allowing the wastegate to be closed without risking surge. The resulting potential increase in low speed (and hence also transient) torque is shown in Figure 11.

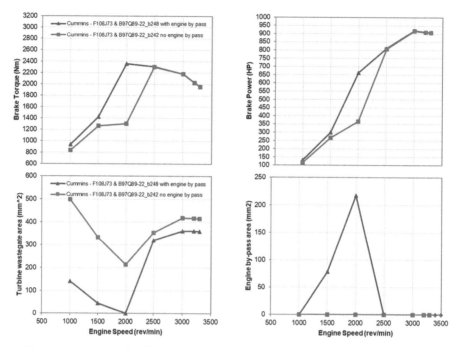

Figure 11. Improving low speed torque with engine bypass/bleed valve.

Within a few days of this idea being postulated however, Ricardo were beginning to run realistic transient manoeuvres on the engine testbed. This revealed that under transient behaviour the torque hole was significant even with the wastegate shut; the turbine was not receiving energy rapidly enough to come close to the compressor's surge line, therefore a compressor matching enhancement such as the bleed valve would not help. A more pragmatic solution addressing the lack of turbine energy was sought.

An alternative and interesting approach is the use of injected compressed air to acceler-ate the turbocharger. The Volvo "PowerPulse" system is now well established in the market on some diesel passenger car applications within the Volvo range and is described in full detail by Fleiss et al. (8). Volvo and other researchers identified that although injecting compressed air into the intake manifold has the benefit of adding air which can be consumed in the cylinders, it momentarily creates an adverse pressure ratio across the compressor which tends to cause turbo speed to falter. The Volvo system therefore injects stored compressed air upstream of the turbine (actually into the HP EGR circuit), and it was this strategy which was adopted for the 1000hp FT2 specification engine.

For FT2, externally charged air tanks were already in place for braking, due to the absence of any onboard air compressor on the vehicle for weight-saving purposes. A 300bar dive-air reservoir (scuba tank) was specified for the "air pulse" system, regulated downstream to 20bar. Air is injected in the exhaust manifold, upstream of the turbocharger turbine. On the engine testbed, experiments were conducted with air injection at various engine speeds and for various durations, considering of course the total stored air capacity to ensure the tank was not depleted too soon. Results are shown in Figure 12; the benefits in terms of lower engine speed to achieve 85% of peak torque can clearly be seen.

370

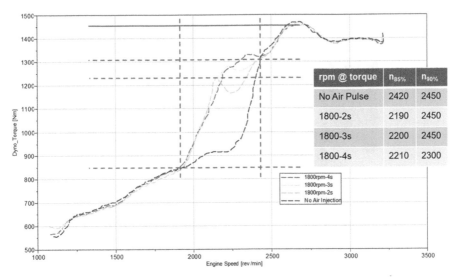

rpm @ torque	$n_{85\%}$	$n_{90\%}$
No Air Pulse	2420	2450
1800-2s	2190	2450
1800-3s	2200	2450
1800-4s	2210	2300

Figure 12. Torque response to varying air pulse durations.

Injecting air for 4 seconds at 1800rpm provides a sustained acceleration to the turbine, significantly improving the time-to-torque. A shorter duration of 2 seconds can be seen to be insufficient to sustain the turbo speed rise for the particularly transient tested, although may be suitable for lower gear transients.

Further fine tuning of the air pulse strategy took place during early vehicle testing at Elvington runway, resulting in 2-5 seconds pulse during pull away and after each gear change, depending on gear selected.

7   CHARGE AIR COOLING STRATEGY

On FT2 to maximise charge air density and minimize aerodynamic drag, the technique of ice charge air cooling (CAC) was carried over from the Dieselmax LSR car. Clearly with high boost pressure come high temperatures; compressor outlet temperatures of up to 265°C were seen on the testbed. The air-to-water charge air cooler was used on both FT1 and FT2 was developed by Denso, and its capable performance in terms of heat rejection, low pressure drop and low volume had already been demonstrated on FT1, rejecting its heat into a low temperature water to air radiator in the tractor's left side pod.

On FT2 the left side pod was used only for the oil coolers, the CAC is in circuit with a reservoir of ice and water at the front of the vehicle (Figure 13). In the Dieselmax LSR car the ice tank had to provide the entire vehicle's cooling requirements (both engine coolant, oil cooling and charge air cooling), and much time and effort was spent optimising the flows around the system such that the ice was just melted by the end of each run. For FT2 the ice tank's job was somewhat easier in that it only needs to cool the CAC circuit. The key constraint of the system primarily became how quickly the tank could be filled; with a willing "yellow army" of team members this was soon developed into a slick process (Figure 14)

371

Figure 13. Charge air cooling and ice tank installation.

Figure 14. Ice tank filling.

The low compression ratio of the JCBWFT engine (necessary to limit Pmax) increases the tendency to poor startability and misfire under cold conditions. As well as coolant pre-heating and electric intake air heating, the electric supercharger was run at startup and idle conditions, and the CAC water circulation pump is only energised above 20mph, to ensure the inlet air is not over-cooled.

8   HOT-SIDE ENGINE INTEGRATION

Simulation predicted exhaust manifold temperatures in excess of 910°C, well in excess of typical temperatures for medium/heavy duty or off-highway engines. The considerable weight of the turbocharger meant the conventional approach of mounting the turbo onto the exhaust manifold would not be suitable and so a lightweight, space-efficient turbo support bracket was designed and manufactured from laser-cut blanks. This allowed the design of the exhaust manifold to be focussed on pressure containment and expansion control.

Thermal expansion at such high temperatures would be a major problem – especially with a "long" engine like an I6. Two strategies were pursued. The prime path was to use expansion slip joints between the three manifold sections using "Fey" type sealings rings. The risk here was that available rings were only rated up to 750°C, so would be unproven in the JCBWFT application. The advantage of the slip joint solution is ease of assembly since the manifold can be assembled in stages. As a reserve solution, Inconel convoluted ("bellows") sections were sourced ready to be welded into a spare manifold module if required. The significant disadvantage of this approach would be difficulty in handling the complete manifold assembly with "floppy" (and easily damaged) bellows section in the workshop, or especially, if the manifold had to be changed trackside in a short time.

The actual manifold itself was fabricated by BTB out of a mixture of preformed pipe sections and custom 3D printed Inconel sections in the areas of highest temperature and greatest geometric complexity (Figure 15).

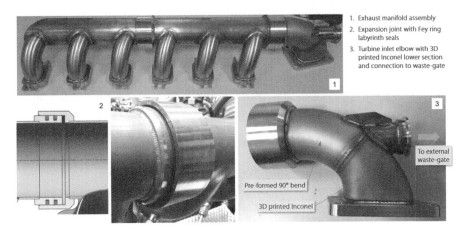

Figure 15. Exhaust manifold constructional details.

After initial and a careful alignment of the sections the slip joint manifold performed very well on the testbed, with no significant leakage at the joints seen and smooth expansion and contraction performance. Despite operating considerably above their rated temperature, the joint rings continued to work well during the initial record runs. Over time however the manifold did distort such that by the time of the commencement of record runs, the Fey rings were no longer running smoothly in their sleeves. This, coupled with a significant temperature excursion to over 1000°C due to insufficient cool-down time between runs, resulted in part of the manifold creating its own bellows section. It is testament to the quality of material and fabrication that even after this the manifold still did not leak or split (Figure 16)

Figure 16. Before and after view of exhaust manifold following sealing ring binding.

9   FINAL ENGINE AND VEHICLE PERFORMANCE

The engine performance with turbocharger plus electric supercharger and air pulse system was validated on Ricardo's testbed at Shoreham in summer 2019. Following installation and commissioning in the vehicle at JCB, the team decamped to Elvington airfield, North Yorkshire, home to a 3km (1.9 mile) runway ideal for record attempts. Loaded into the calibration laptop were torque curves - modulated for each gear to suit transmission and traction limitations – up to and including 920hp.

One aspect of speed record racing difficult to quantify is the so called "ram effect" of the vehicle forward speed in providing additional pressure at the air intake. In the case of the FT1 & FT2 the air intake on the vehicle is simply a large funnel with a "rock catcher" mesh in the nose of the bonnet, no traditional air filter was provided. The team were not able to formally quantify the ram effect due to lack of instrumentation (and lack of time/final vehicle nose geometry at the CFD stage) but it is believed that a ~1.5% boost in power at high speed can be attributed to this effect.

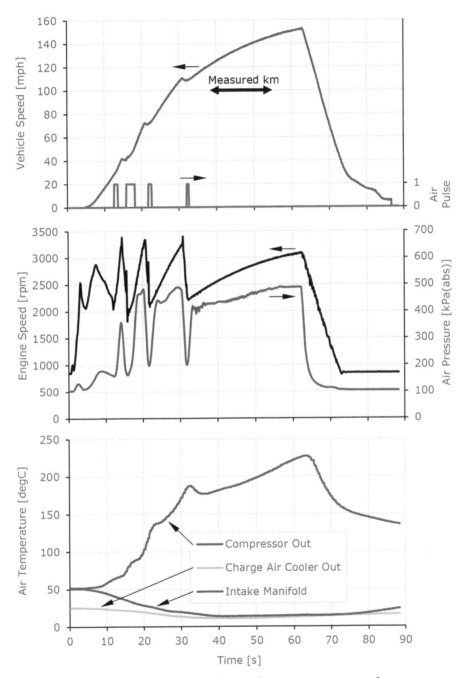

Figure 17. Vehicle and engine performance over record run.

Figure 17 shows the vehicle and engine performance over a record run. At the top is vehicle speed and air pulse operation. In the middle, engine speed and intake manifold pressure are shown. On the lower plot, compressor out, charge air cooler out and intake manifold air temperatures are shown. Note the initial offset between charge air cooler outlet and intake manifold temperatures at the start of the run is due to operation of the grid heaters to heat the intake air at low boost pressure.

As with the Dieselmax LSR, a progressive test run profile was followed, testing each calibration configuration (together with tyre and vehicle setup modifications), confirming safe and acceptable operation, before ratcheting up the power and torque. Fine tuning of the electric supercharger and air pulse systems was undertaken to minimize any occurrence of smoke puffs or "bogging down" on transients. Final power and torque curves from the vehicle (calculated and corrected from torque sensor in the driveline) show ultimate achieved performance was over 750kW (1000hp), 2370Nm, 104kW/l (Figure 18).

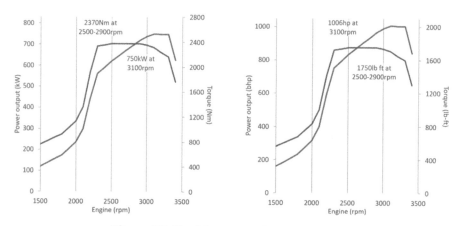

Figure 18. Final torque and power curves.

Compared to the standard engine rating of 1150Nm and 212kW (284hp) at a speed range of 1400-2000rpm, the development achievement – contributed in a major way by intelligent application of modern boosting technology – is immediately apparent. This outstanding turbocharging and engine performance led to a record-breaking run, average two-way km record speed of 135.191mph with a peak speed of 153.771mph.

10   SUMMARY

The WFT engine showcases JCB's 672 Dieselmax engine as a highly robust and capable power unit ready for the 21st century's demands for autonomous off-highway power. Despite advances in electrification, there still remains a place for highly capable IC engine machinery operating in remote off-grid locations with efficient emissions control and using reduced carbon fuels. The boosting technologies showcased therefore remain as relevant as ever.

The WFT programme proved a great collaboration between JCB and suppliers, showcasing engineering excellence, and inspiring the next generation of engineers. A significant proportion of the team were apprentices and young engineers, offered a career-defining, exciting engineering challenge.

ABBREVIATIONS

CAC	Charge Air Cooler
C_dA	Drag area; the drag coefficient (Cd) multiplied by vehicle frontal area (A)
COBRA	Controlled Boosting for Rapid response Applications, an electric supercharger produced by Federal Mogul Powertrain
DoE	Design of Experiments
HP	High Pressure
JCB	J.C. Bamford Excavators Ltd., universally known as JCB, a British multinational corporation, with headquarters in Rocester, Staffordshire, UK, manufacturing equipment for construction, agriculture, waste handling and demolition
LP	Low Pressure
LSR	Land Speed Record (as in "DieselMax LSR", the 350mph diesel-powered streamliner (6) that subsequently gave its name to the DieselMax family of JCB engines)
OED	Oxford English Dictionary
WFT	World's Fastest Tractor

ACKNOWLEDGEMENTS

Figure 19. JCB and Ricardo project team members celebrate with Guy Martin (centre) following his record-breaking run at Elvington, UK.

The authors would like to thank:

- Senior Management at JCB and Ricardo: Alan Tolley and Matt Beasley
- All the team at JCB and Ricardo

- Guy Martin, North One TV and Channel 4
- The project partners:
  - Delphi Technologies
  - BKT
  - Denso
  - Federal Mogul
  - ZF
  - GKN

- The turbocharger suppliers, BorgWarner and Cummins Turbo Technologies
- All the other suppliers who helped us make the World's Fastest Tractor

REFERENCES

[1] Oxford English Dictionary. Definition of tractor in English. [Online] 17 December 2019. https://www.lexico.com/en/definition/tractor.
[2] AgWeb.com (Wyatt Bechtel). World's Fastest Tractor Gets to the Field in a Flash. [Online] 28 April 2015. [Cited: 17 December 2019] https://www.agweb.com/article/worlds-fastest-tractor-gets-to-the-field-in-a-flash-NAA-wyatt-bechtel.
[3] Guinness World Records (Rachel Swatman). Top Gear smashes record for world's fastest tractor after driving at more than 140 km/h. Guinness World Records. [Online] 22 March 2018. [Cited: 17 December 2019] https://www.guinnessworldrecords.com/news/2018/3/top-gear-smashes-record-for-worlds-fastest-tractor-after-driving-at-more-than-14-519246.
[4] Farms.com (Diego Flammini). Farm family sets speed record with tractor. [Online] 9 November 2018. [Cited: 17 December 2019] https://www.farms.com/ag-industry-news/farm-family-sets-speed-record-with-tractor-387.aspx.
[5] WTAP.com (Megan Vanselow). Local family sets tractor speed record. [Online] 7 June 2018. [Cited: 17 December 2019] https://www.wtap.com/content/news/Local-family-sets-484795101.html.
[6] JCB. JCB DieselMax Ten Year Anniversary. [Online] 23 August 2016. [Cited 17 December 2019] https://www.jcbdieselmax.com/.
[7] Ricardo plc. Fastrac: A Q&A with Matt Beasley and Alan Tolley. Ricardo.com. [Online] 20 November 2019. [Cited: 17 December 2019] https://offhighway.ricardo.com/news-and-blogs/fastrac-a-q-a-with-matt-beasley-and-alan-tolley.
[8] Fleiss, M., Burenius, R., Almkvist, G., Björkholtz, J. The New Volvo 235hp Diesel Engine with Extreme Take-Off Performance. *24th Aachen Colloquium Automobile and Engine Technology*, 6 October 2015.

Additional papers

Development of a new, vaneless, variable volute turbine

S.D. Arnold

Engine Systems Innovation, Inc ., USA

ABSTRACT

A new type of variable geometry turbine has been designed, analyzed using CFD, prototyped and tested. The design is based on a spiral track which a flexible wall can use to traverse from an "open" position to a "closed" position. The "open" position creates a full 360-degree volute with a large throat section. As the flexible wall traverses to the "closed" position, the volute throat section reduces as does the remainder of the 360-degree volute. The flexible wall traverses in a 720-degree track. The moving wall changes the throat section and controls which 360 section the flow stream will use.

Computational Fluid Dynamics (CFD) was used to analyze the corrected flow and efficiency and to improve those parameters. Additionally, a Rotordynamics analysis was used to determine the bearing friction, which was then combined with the CFD results to yield the "turbine mechanical efficiency", which is commonly measured on turbocharger hot gas stands.

A protype unit was built utilizing 3D metal printing for the links that form the flexible wall. The unit was tested on a turbocharger hot gas stand which showed high efficiency and linear flow control. Another key advantage of the design is the elimination of the typical VGT "variable vanes". The main undesirable characteristics of variable vaned units are eliminated, including high cycle fatigue, vane and actuating arm wear, variation in the throat section at fully closed due to manufacturing variability, actuation force reversal, sticking, and most importantly, poor performance when nearly closed.

1   INTRODUCTION

The design of Variable Geometry Turbines (VGT) has a rich history of innovation as the engineering field was highly active many decades before any VGT became a commercial success. There have been a significant number of different methods and mechanisms invented and developed to dynamically change and control the flow capacity of a turbocharger turbine. Early patents staked out this incredible array of methods to implement variable flow capacity. Each turbocharger company sought to carve out their own Intellectual Property space to avoid being locked out of this potential game changer technology. Two basic methods have dominated the market; one being a cascade of movable vanes controlled by a ring (referred to herein as "multi-vane", and another, an axial sliding annular ring which changes the width of the turbine nozzle space (referred to as "axial vane" herein). The latter method has two sub-types, one with fixed vanes and another with no vanes. Each method has advantages and disadvantages, and a tremendous effort has been expended to overcome the attendant disadvantages. The key attributes are 1) flow range, 2) efficiency, 3) reliability, 4) control and 5) cost. A common disadvantage of existing VGT types is poor efficiency as the flow capacity is reduced. Another common disadvantage is the durability/reliability as there are many failure modes for each type.

A new type of VGT has been created specifically to improve those two areas. This paper covers the design, analysis, prototyping, and testing of the first Variable Volute Turbine (VVT).

2 DESIGN CONCEPT

The author developed many different VGTs through the years and acquired a deep appreciation for the difficulty of engineering a reliable and efficient device.[2][3] Vanes generated several failure modes.

1. high cycle fatigue of the turbine wheel,
2. sticking due to thermal distortion,
3. wear of the vanes and connections to the unison ring,
4. siren-type noise caused by the interaction of vanes and blades,
5. variation in flow capacity with vanes nearly closed,
6. reversal of aero torque resulting in actuation instability,
7. hysteresis in the mechanism resulting in actuation instability,
8. poor performance due to excessive side clearance,
9. poor performance as the vanes closed due to pressure-to-velocity conversion,
10. thrust bearing failure due to excessive static pressure buildup in the complete stage, or the opposite, and
11. thrust bearing failure due to loss of static pressure in the turbine wheel due to pressure-to-velocity conversion.

Since vanes are clearly the root cause of most VGT failures, it seemed paramount to eliminate them. Historically, turbocharger turbines were vaneless and used the throat section properties to establish the flow curve. The relevant throat section property is captured by its A/r; that is the Area (A) of the section divided by the radius (r) of the centroid of the section. This is derived from the Conservation of Angular Momentum Law

$$(L = Iw) \tag{1}$$

The turbine housing then reduces the section A/r through the 360° of the volute to zero. This results in the traditional scroll shape for the turbine housing.

One would infer from this information that the most efficient variable capacity turbine would use a vaneless housing and somehow reduce the volute and its attendant throat section to modify the turbine flow curve. Considering the odd, asymmetrical shape of the turbine housing, finding a way to reduce its geometric properties does not seem realistic. There have been several designs that use a pivoting blade to change the throat section, or force flow through one-side of a divided housing, but the results have not been exceptional. A true variable volute has never been created.

A new way of constructing a true variable volute turbine was conceived by considering a moveable wall forming the underside of the volute, then transitioning to the outer wall of the turbine housing near the location of a traditional vaneless volute tongue. This would allow the track to continue around the outside of the volute, which would be available to store the part of the flexible wall that was not in use closing the bottom of the inlet to extend the throat section. A base curve was generated using log spiral equations to define the track for the flexible wall as seen in Figure 1.

$$x(\emptyset) = a * e^{(b\emptyset)} * \sin(\emptyset) \tag{2}$$

$$y(\emptyset) = a * e^{(b\emptyset)} * \cos(\emptyset) \tag{3}$$

The constant "a" is equal to the base radius of the volute and "b" is a constant that was varied slightly to give the result desired through the complete 720°.

Figure 2 shows one half of the turbine housing with a track for the flexible wall. As the track goes counterclockwise from the traditional tongue position, it moves the throat area to a smaller A/r and simultaneously lengthens the tail of the volute to maintain a 360° volute. As the track goes clockwise from the traditional tongue position, it provides a storage space for the portions of the flexible wall that are not in use.

Once the flexible wall and track were conceived, a method of moving or actuating the device was needed. Numerous methods were investigated but all had major deficiencies. A gear to engage and move the flexible wall was considered, but since the wall has a continuous change in curvature, a gear drive seemed impossible initially...

When the wall moves, there is a continuous change in the curvature of the wall. However, when one considers the turbine housing, a given point on the turbine housing always has the same curvature. It was hypothesized that if gear teeth were designed into the flexible wall specifically for the curvature where the gear drive would be engaging, then it would work. At all other circumferential locations, the teeth could not mesh with another gear. But there is no requirement for the teeth to mesh other than at the specific location of the gear.

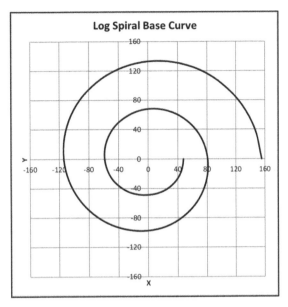

Figure 1. Log spiral base curve.

Figure 2. Volute housing showing the log spiral track.

There were two major design options to consider for the flexible wall. First was a continuous band formed to the curvature of the log spiral base curve. As the band would move to a smaller throat section, the band would need to bend and increase its curvature. As the wall moved to a larger throat section, the band would need to bend to reduce its curvature. Several versions of this design were modeled, including rollers spaced through the band, and various sections modeled to try to tailor the stiffness of the wall to not bend under aerodynamic loading, but to flex enough to be actuated. It was decided that this design was more difficult and would need more time and resources for development.

The second option was to do a link design analogous to a train where there were links supported on Rollers, and the links had interlocking geometry with links on either side. This was the chosen option for the first demonstrator unit. The design process required 4 different link designs; a Lead Link to guide flow off the link train, a Gear Link which includes the teeth to engage the drive gear, an Aero Link design identical to the Gear Link but with the teeth geometry removed, and an end gear.

2.1   Link Design—Aero Link
Key points for the Aero Link design were to:

1. Provide interlocking and conjugate action with the next link as well as the roller to allow motion from minimum curvature of track to maximum curvature of track,
2. To prevent as effectively as possible exhaust flow leakage through the link-to-link joint that includes the roller,
3. To minimize flow around the link assembly in the track,
4. To prevent accidental disconnection of the joints in the assembly through an interlock,
5. To be consistent with the MIM (metal injection molding) process and to minimize the amount of material in the link,
6. To minimize friction with the roller such that the roller rolls on the track rather than skidding.

Figure 3 shows two Aero Links interlocking with a roller. The roller contacts the track and raises the links slightly above the track so there is no sliding friction.

Note that the groove that accepts the roller is radially disposed on one end and is angled at the other end (two red lines). This creates a failsafe where links cannot be disassembled without moving the pair to an angle beyond the minimum angle when installed. The links must be assembled prior to assembly in the track and disassembled after removal from the track. The blue ellipse of Figure 3 also shows there are stops designed into the conjugate action which prevent the links from moving to an angle beyond the maximum angle when installed.

2.2   Link Design—Gear Link, Actuating Gear
Due to vast range of thermal conditions a VGT must contend with, the associated thermal growth, MIM precision, lack of a rotating center for the Gear Links, and the stack-up of many tolerances, the gear design needed to be relatively coarse. Note that the design needed to be kinematically functional from -40C to over 800C, and the housing and MIM links, gear and bushings would be made of two or more different metals.

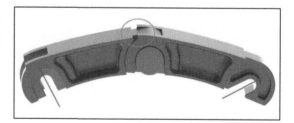

Figure 3. Two connected aero links.

To make a coarse gear, there needed to be significant radial extent to the Gear Link. This meant that the cutout in the Gear Link for the teeth had to coexist with the link's two end geometries, and associated roller. This required the gear and the link to be designed together because the spacing for the roller cutouts had to be equal to a multiple of the gear teeth spacing.

An aid to designing this was the use of the graphical gear design method, which was adapted to CAD. The graphical gear design method lacks high precision when used with pen and vellum, but the precision is restored when automating the process in the CAD environment.

Figure 4 is a section view through the gear and attendant gear links and rollers. One can see points A, B, C are part of the leftmost link, and that point D is part of the adjacent link. The two link's conjugate action boundary is defined by a radius about the roller center, noted as Point E. As the two links roll from a high value instantaneous center to a lower value instantaneous center, the distance between points C and D will grow and not form a useful gear tooth. That is acceptable as the only location the gear tooth must be correct is at the meshing location with the gear.

Figure 4. Cutaway gear mesh across multiple links.

The actuating gear, bushing, gear links, and rollers are shown in Figure 5, and the 4 types of links, "lead", "aero", "gear", and "end" are shown in Figure 6.

The circumferential location for the actuation gear is important as it can increase the number of links required. The optimum angle is 225° in the direction of the flow from the 100% open position. There are 14 gear links required and 7 aero links (no gear feature), one lead link and one end link. The lead link has a streamlined nose, and the end link is a modification of the gear link as there is no "conjugate action" link after it.

Figure 5. Gear meshing with gear links.

385

2.3 Volute and Housing Design

Since the basis of the design is "variable volute", one must define the maximum volute and the range that the variable volute will cover. We know the relationship between flow angle, tip width and throat section is:

$$\cot(\alpha) = \frac{A}{r \times 2 \times \pi \times b_5} \quad (4)$$

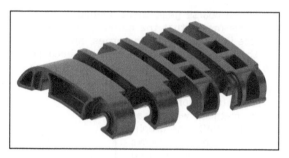

Figure 6. Lead, aero, gear and end links.

where α is the angle of the flow from radial, A is the section area, r is the radial distance to the area centroid, b5 is the tip width of the turbine. We also know the optimum angle for efficiency is 68°-70°.[4,5]

Our chosen turbine wheel had a rather large tip width of 16 mm, so the optimum A/r would be 40mm. As shown in Figure 7, this would make the maximum flow the most efficient. If the desire for the peak efficiency were to be at a more closed position, then the flow angle for the maximum volute should be more radial, such as 60°. If the desire were to be poor efficiency to prevent high boost at the maximum opening, then a flow angle even further radial, such as 50° should be chosen.

Another method to tailor the turbine efficiency is to change the tip width of the turbine. The previous method would be the best utilized during the design phase but changing the tip width would be the right choice when optimizing an existing design. Figure 8 shows the effect of trimming

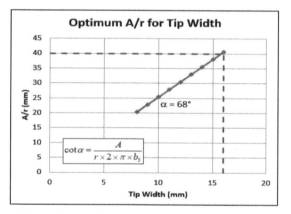

Figure 7. Relationship of flow angle, A/r, tip width.

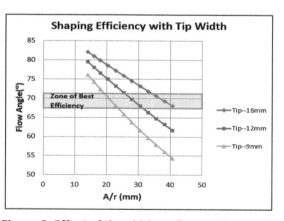

Figure 8. Effect of tip width on flow angle vs A/r.

the width of the housing and the wheel. Note that at a tip width of 16mm, the peak efficiency happens near the largest A/r. Width a tip width of 9, the peak efficiency will be around ¾ closed.

A conventional volute (Figure 9) was designed for a ⌀86mm turbine with a 16mm tip width and a 40mm A/r.

The volute A/r for the second 360° cannot be simply designed as it is a result of the log spiral and the tip width of the design. Figure 10 shows the resultant A/r for the complete 720°.

A method of assembling the unit together and make a robust seal was required. A bolted flange design was used with evenly spaced bolts. Typically, this type of flat faced, bolted joint is prone to leakage. The main countermeasure is to make the flanges very stiff (thick). Another approach was taken with a tongue and groove approach. Both the tongue and groove have draft angles—one degree on the female side, 4 degrees on the male side, with a light press fit. These housings will be CNC cut and the draft will make them assemble more easily and with the tongue filling 100% of the groove, thus providing an almost leak-proof seal. This worked per-fectly with no leakage in the prototype parts even when the unit was disassembled and reassembled many times and hot gas stand tested at 600C on numerous occasions.

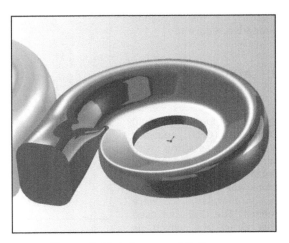

Figure 9. Volute core at full open.

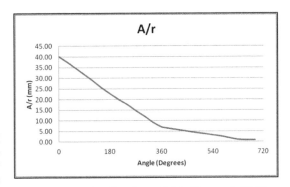

Figure 10. Distribution of A/r through 720°.

387

A machined track was inserted to guide the flexible link train and to store the excess links, and a cavity created for the actuation gear drive, Figure 11.

3   AERODYNAMIC DESIGN AND ANALYSIS

Several Computational Fluid Dynamics (CFD) analyses were done, starting with simple models, and adding detail and complexity as more was learned and the design/analyses were refined. The CFD tool used was StarCCM+ which is a general purpose CFD program and lacks specialized turbo-machinery tools. Considerable effort was put into creating equations and plots specific to turbomachinery and VGT analysis and to create templates to facilitate the import of geometry, automated meshing, and rapid change of operating points, with standardized output of parameters and plots.

The CFD was done with a computer having two 8-core Xeon processors and 128GB of RAM. All geometry manipulation was done in CAD (SolidWorks), including extracting the fluid volumes and transferring to StarCCM+ through binary Parasolid models. The model was run as adiabatic with typical diesel exhaust gas parameters. A segregated solver was used with a K-Epsilon Turbulence Model. The meshing used Polyhedral and Prism mesh.

Figure 11. Rear housing w links, rollers, and gear.

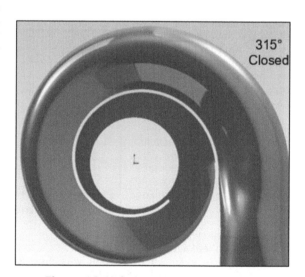

Figure 12. Volute core at 315° closed.

Since the geometry is unique circumferentially, the Frozen Rotor Method was used to eliminate the need for a mixing plane. Meshing was then done using templates to dissect the geometry into groups with assigned meshing parameters. Although it took considerable effort to create, the process became very efficient and standardized.

Design changes could be made in SolidWorks, Parasolid models transferred to StarCCM+, re-meshed and back running in a half hour. Initial models without detailed links had about 8M cells. When the link details were added, the model jumped up to about 110M cells. Extensions were added upstream and downstream. The downstream extension was one meter long to prevent reverse flow through the exit plane. The inlet temperature was 600C as is typical for testing, and typical exhaust gas properties were used.

Figure 12 shows the simplest CAD model of the housing with a cut representing the link train and at the 315° closed position.

As the analysis progressed and the link geometry took shape, additional detail was added and became a full representation of the geometry with all detailed features left in. As the link train was complex with rollers, links, tracks, passages, etc., the mesh became exceptionally large as there was an attempt to capture all the losses caused by leakage through the links and around the links, eventually exceeding 110M cells. Figure 13 shows the mesh and one can see the high density of cells in the link area and the wheel/shroud area.

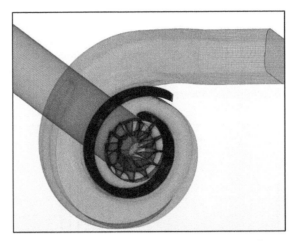

Figure 13. Mesh of fluid bodies w/link/roller detail.

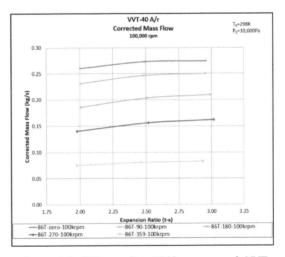

Figure 14. CFD results, 100krpm speed, VVT opening angle lines, corrected mass flow and expansion ratio.

There needed to be quite a long extension on the outlet as the high degree of swirl exiting the turbine at some conditions, resulted in reverse flow through a semi-stagnant zone in the center of the outlet. The results were not correct if the discharge boundary plane had reverse flow anywhere in the section.

389

3.1　CFD Results

The aerodynamic results of the CFD analysis indicated the unit had remarkably high efficiency, a wide and linear flow range. Gas stand test results show a limited range of the turbine speed lines, so a wider range of expansion ratio values were used in the CFD analysis. Figure 14 and Figure 15 show the corrected flow and adiabatic efficiency results for an 86mm turbine wheel run at 100,000rpm (450m/s tip speed).

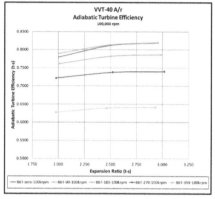

Figure 15. CFD results, 100krpm, VVT opening angle lines, adiabatic turbine efficiency vs expansion ratio.

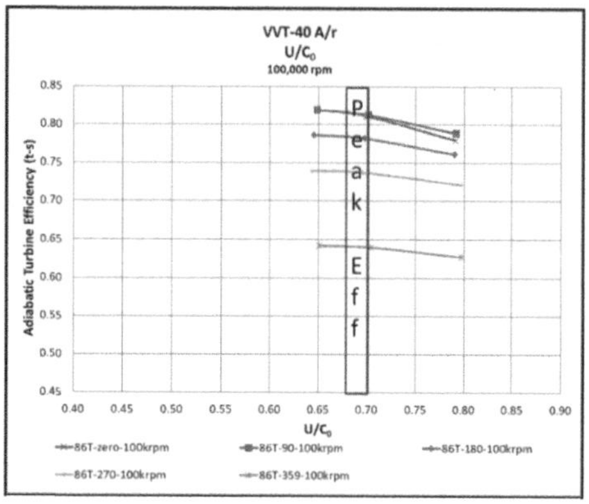

Figure 16. CFD results, 100krpm, VVT opening angle lines, corrected mass flow and expansion ratio.

To put the data in perspective, it is useful to look at the efficiency as a function of the isentropic spouting velocity (U/C_0), often referred to as the "blade speed ratio". Radial turbine's peak adiabatic efficiency happens at a blade speed ratio of about 0.7. Figure 16 indicates that at 100,000rpm, the VVT was operating near its peak efficiency.

Figure 17 and 18 depict the Corrected Flow, and Adiabatic Total-to-Static Efficiency at 60,000rpm (270m/s). Note that there is a relatively even spread of the lines from fully open to fully closed, linearity being a positive attribute for the control system.

Figure 18 is a somewhat unusual as there is a slight upward curvature in the efficiency lines rather than a slight downward curvature as one would expect.

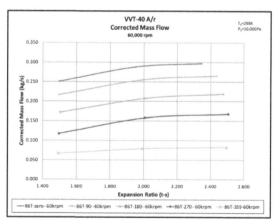

Figure 17. CFD results, 60krpm, VVT opening angle lines, corrected mass flow, expansion ratio.

390

Since it occurs in all lines, it is likely not an error in modeling or problem setup. Figure 19 shows the efficiency as a function of the blade speed ratio. We see the efficiency remains extremely high at this lower speed even though the blade speed ratio is significantly lower than the optimum 0.7. The reason for this was that the range for the expansion ratio was set too low to capture the high expansion ratio that would be seen in operation when the link-train was closed down. In hindsight, the expansion ratio should have been taken to 3.0 or 3.5 in the analysis.

The uniformity of the flow angle and mass flow circumferentially was unknown and was thought to be important to understand. The CFD model used the "frozen rotor" method and the interface between the static and rotating reference frames was 1mm larger than the turbine wheel inlet diameter. There were "bins" created circumferentially to provide the ability to integrate the flow data across the boundary into 15° wide values, Figure 20.

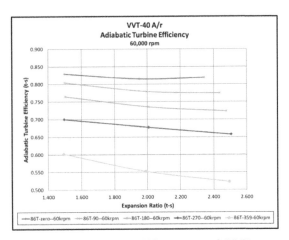

Figure 18. CFD results, 60krpm speed, VVT opening angle lines, efficiency vs blade speed ratio.

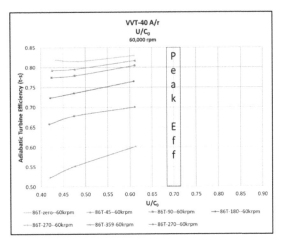

Figure 19. CFD results, 60krpm speed, VVT opening angle lines, efficiency vs blade speed ratio.

Figure 21 and 22 show the flow angle and the mass flow through each 15° sector of the turbine wheel inlet interface. Figure 21 depicts the data at the fully open position, and Figure 22 depicts the data at 315° closed. It is important to note that Figure 20-23 are taken from CFD data computed before the links and rollers were modeled and the mesh size became unwieldy.

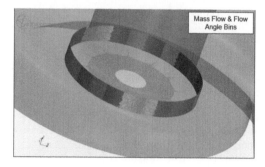

Figure 20. Integration segment bins.

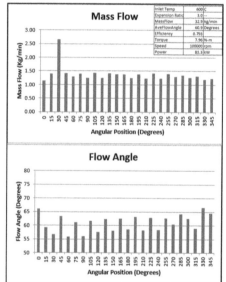

Figure 21. Mass flow and flow angle distribution for 15° peripheral segments, VVT fully open.

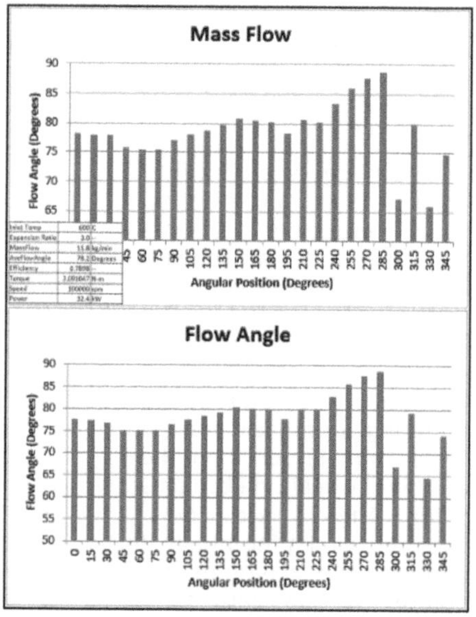

Figure 22. Mass flow and flow angle distribution for 15° peripheral segments VVT 315° closed.

When using the "frozen rotor" method, one should vary the alignment of the turbine wheel blade with the of the tongue of the housing. The spike at 30° could have been due to the alignment of "frozen rotor". Due to time constraints this step was not done.

Figure 23 summarizes the performance attributes at 100,000rpm, and ER of 3.0. At this stage of the analysis, the mechanism was designed to go to 315° as the fully closed position. After the initial analyses were done, the decision was made to increase the turndown by extending the link track to 360°.

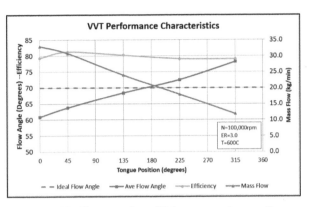

Figure 23. Flow angle, efficiency, and mass flow versus closure angle.

As pointed out previously, multi-vane VGTs have many failure modes associated with the vanes, many of which are caused by aerodynamic forces. The environment for the link based VVT is much simpler. Figure 24 shows the surface pressure of the unit at 225° closed. Note that the pressure field will apply a force to each link that is radially disposed, i.e., the link will be pressed down on its rollers. In the radial direction, the links and rollers are very stiff, so

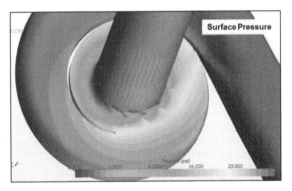

Figure 24. Surface pressure of stage, showing gradient across link boundary.

there will be minimal movement of the link as opposed to a multi-vane VGT that has significant angular movement of each vane at each cylinder firing.

The VVT was to be prototyped and tested after the design and analysis was done and comparison of the test data to the CFD data was important to understand the correlation and reconcile any differences. However, the CFD data calculates the Adiabatic Turbine Efficiency—total-to-static, whereas a hot gas test stand measures the Turbine Mechanical Efficiency—total- to-static. The hot gas test stand measures parameters that can be used to calculate enthalpy change for both the compressor and turbine and thus calculates the efficiency from the power needed to drive the compressor. The bearing power is a parasitic loss that is not included in the compressor enthalpy calculations, so it is an unknown value that reduces the turbine efficiency. Thus, the Turbine Mechanical Efficiency will be lower than the Adiabatic Turbine Efficiency. At a given turbo speed, the bearing power will be relatively constant with some small variation due to thrust loading changes across the speed line. A given speed line may have an increase of 200% in compressor power as the compressor traverses from surge to choke, so the impact of the bearing loss at compressor surge can be double the impact at compressor choke flow. These factors make it difficult to compare gas stand test data with CFD data. If the bearing power were known, then this could be added as a loss to the CFD data or subtracted from the gas stand data.

DyRoBeS "BePerf" (bearing performance) was used to calculate the bearing loss of the test turbo's journal bearings, shown in Figure 25. An equation was derived from the data using a trendline analysis and this equation was used in the post processing of the CFD analysis. The turbo has two journal bearings and one thrust bearing. A license was not available for the DyRoBeS thrust bearing design, so an approximation of the total bearing loss was calculated as 3 times the single journal bearing loss. This likely underestimates the thrust bearing loss slightly, but the overall impact on the final calculated efficiency is certainly within the measurement error of a hot gas test stand.

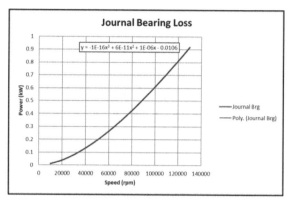

Figure 25. Journal bearing loss vs rotational speed.

Since the CFD efficiency is calculated by temperatures and pressures, not an enthalpy change, a hybrid calculation of Turbine Mechanical Efficiency was required. The turbine power was calculated and used with the total bearing power to calculate the Mechanical Efficiency of the turbo-charger, Figure 26 and 27 for 100,000rpm and 60,000rpm, respectively.

This mechanical efficiency was then multiplied with the Adiabatic Turbine Efficiency—t-s to have a calculation of the Mechanical Turbine Efficiency—t-s, which can be compared to gas stand data. This value is often called "eta-tm" or eta turbine mechanical, which is the efficiency value supplied to engine manufacturers on maps.

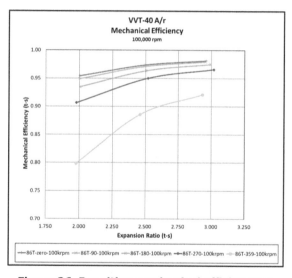

Figure 26. Resulting mechanical efficiency at 100krpm vs expansion ratio for the range of VVT closure angles.

Figure 26 and 27 also show us that although the <u>power of the bearings</u> is constant for a given speed, the impact is multi-dimensional—that is the bearing efficiency increases with expansion ratio, and as the VVT changes position, the mass flow change also increases the efficiency, as does increasing the speed of the turbo.

Figure 28 and 29 show the Turbine Mechanical Efficiency—t-s for the 100,000rpm and 60,000rpm speed line. One can see that the mechanical efficiency has a major impact on turbocharger perform-ance when the expansion ratio is low, and the vari-able turbine is closed down. This is a major contributor to "turbo lag" as the loss reduces the turbo-charger speed which must be overcome in a transient.

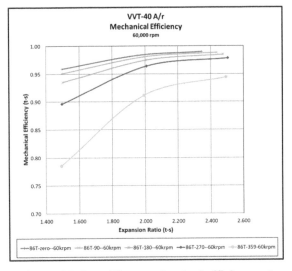

Figure 27. Resulting mechanical efficiency at 60krpm vs expansion ratio for range of VVT clos-ure angles.

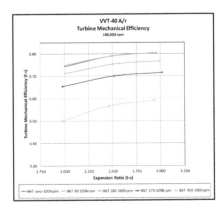

Figure 28. Turbine efficiency with bearing loss for the range of VVT closure angles at 100krpm.

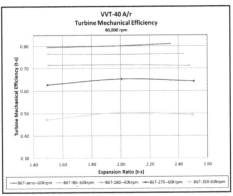

Figure 29. Turbine efficiency with bear-ing losses for the range of VVT closure angles at 100krpm.

When one designs a multi-vane VGT, there are two aerodynamic parameters that change when the VGT moves vane position; the angle of the flow directed into the wheel, and the throat area of the vane set. These two parameters are dependent on one another, so it is difficult to change them independently. As the VGT closes down, the angle goes to 90° and the area goes to zero. When the area becomes small, it is a problem as it accelerates the flow to sonic conditions. This converts nearly all the pressure to velocity to become essentially a low-efficiency impulse turbine. To improve the efficiency, the flow area must be larger while the angle approaches 90°. This is the key factor as to why the VVT is more efficient than multi-vane VGTs.

A factor to quantify this is the Degree of Reaction (RN). Most radial turbines are referred to as "50% reaction turbines". In Watson and Janota, the authors claim that it is "difficult to design a radial turbine whose reaction is far from 50 percent." [1] Perhaps this is true but being designed as a 50% reaction turbine does not mean that it operates as one. When one designs a radial turbine VGT, the Degree of Reaction can go from 0 to close to 1. Equation 3 was used to calculate the Degree of Reaction during post processing of the CFD results. RN is the Degree of Reaction, C_4 and C_5 the inlet and outlet gas velocities, and h_{01} and h_{05}, the specific inlet and outlet enthalpies.

$$RN = (C_4^2 - C_5^2)/(2 * (h_{01} - h_{05})) \qquad (5)$$

Figure 30 is very revealing as it shows all the data points at a fixed speed of 80,000rpm, where the efficiency is plotted against the RN. It becomes obvious that one would design a 100% reaction turbine if that were possible. As stated previously, an impulse turbine (0% reaction) is not an efficient turbine. To step away from the equations for a moment, this means we should minimize the conversion of velocity to pressure until the flow is in the turbine wheel. As engine BMEPs have risen through the years, the turbine housing A/r have gotten smaller and there is an acceleration (pressure → velocity) in the housing which will reduce efficiency.

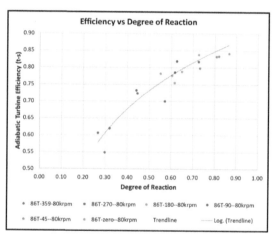

Figure 30. adiabatic turbine efficiency vs degree of reaction using a log trendline

The Degree of Reaction joins the Blade Speed Ratio (0.7), and the 70° flow angle as parameters that are associated with the highest turbine efficiency. Unfortunately, when one designs a VGT, these parameters become values that must change with the variable feature if we are to have a useful VGT.

Figure 31-33 come from other work the author was doing on a conventional multi-vane VGT. Figure 31 shows some RN data, and we can clearly see that 1) the degree of reaction went to zero (negative, but this is an aggregation error), and 2) the efficiency follows the trend shown in the VVT data, Figure 30.

We can see in Figure 32 that there is supersonic flow, and the flow angle appears to be slightly greater than 90°. Figure 33 shows the pressure has dropped from about 1.1Bar in the volute to 0.07Bar at the turbine wheel interface plane. This confirms that the low efficiency of a multi-vane VGT when closed down is due to conversion of pressure to velocity.

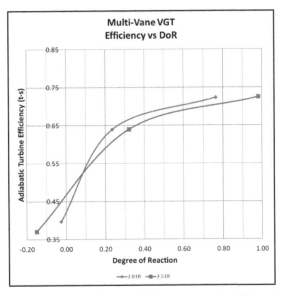

Figure 31. Data from a CFD study on traditional multi-vane VGT, efficiency vs degree of reaction.

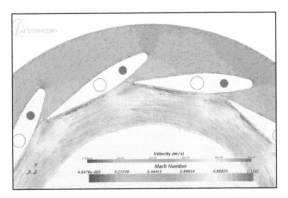

Figure 32. Velocity distribution of a multi-vane VGT with vanes closed.

4 PROTOTYPE AND TEST RESULTS

The CFD analysis showed excellent results, so the project moved to a prototype and test phase. An existing, available turbo needed to be chosen as a "mule" and the design modified to fit that turbocharger.

A production BW S300 unit from the early 1990s was supplied to ESI and the VVT design adapted to fit that design. Unfortunately, the turbine wheel was found to be an 80mm turbine wheel whereas the VVT was designed for an 86mm turbine. When reviewing the results, one must keep in mind that the VVT housing and the turbine wheel were mismatched by about one frame size. Correcting this will shift the peak performance towards the closed position.

Figure 34 shows the VVT partially disassembled and with the links moved to the 360° closed position. The author has designed many different VGT's going back to the mid-1980s. Although the designs have changed, evolved, and improved

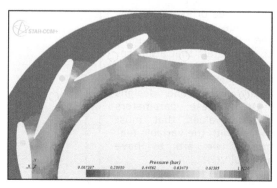

Figure 33. Pressure distribution of a multi-vane VGT with vanes closed.

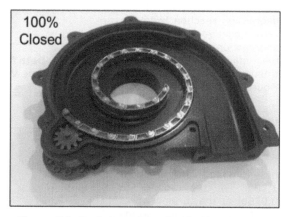

Figure 34. Prototype housing half with links, rollers, and gear.

with every successive generation, one design attribute has not changed—the reliance on net shape technologies—Metal-Injection-Molded (MIM) and Powdered Metal (PM) for low cost. The VVT follows this trend of designing the expensive parts for a net shape technology and to minimize or eliminate machining and grinding. The prototypes links were 3D metal printed out of Cobalt Chrome, which was used as there were a limited number of alloys available for use at 800C. The tolerances held by 3D printing were good, but not quite to the level of MIM. Some hand finishing required on the prototype links parts.

The rollers were screw machined from 310 stainless steel. Had this unit been slated for durability testing, the 310 rollers would have been ion nitrided.

In Figure 35, we can see the links and rollers sitting in their guide tracks along with the actuating gear drive which is engaging the gear links. The split housings were CNC machined from cast Dura Bar.

The unit was tested in the lab of a major turbocharger manufacturer and the results are shown in Figure 36 and 37. A standard temperature of 600C was used. The corrected flow was reasonably linear and had 2.5 turndown ratio at a 2.0 Expansion Ratio (ER). The efficiency is outstanding with a peak of 82% and generally stays higher than 60% except for the fully closed setting.

There was a logical question regarding the fully closed data (red line). Did it suffer from flow leakage around and through the link-train? Leakage would lower the efficiency and increase the corrected mass flow. This was one of the attributes we wanted to quantify to determine the sensitivity of the design to leakage through and around the link-train. Due to the variability of prototype parts as well as the minimum allowable tolerance of net shape technologies such as MIM, PM, and Additive Manufacturing, the clearance of the links to the side wall was increased from .25mm to .50mm total clearance to assure a fully operational unit when run on the hot gas test stand.

Figure 35. Prototype showing gear engaged with gear links.

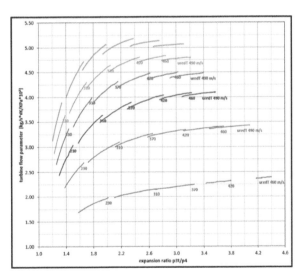

Figure 36. Test data of prototype, corrected mass flow vs expansion ratio for 6 angle positions.

The cavities that were designed into the links (as shown in Figure 38) to facilitate "net shape" manufacturing processes, increased the leakage through the barrier that the side clearances represented. Perhaps they were not as effective as hoped due their limited radial extent. To fully take advantage of any "net shape" technology requires minimization of metal mass and a uniformity of section thickness, which guided the design of the links. This will be optimized in a follow-on development program.

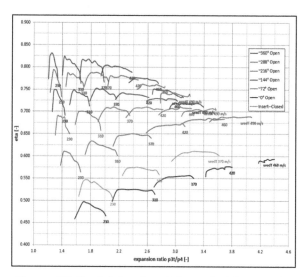

Figure 37. Test data of prototype, turbine mechanical efficiency for 6 angle positions.

The links top and bottom barriers were not fully optimized as can also be seen in Figure 38, which shows the nominal clearances of the installed links. One of the challenges facing the designer is that the passage curvature changes as one moves around the periphery of the passage, so the curvature of the links at the top and bottom and their resultant clearance to the passageway cannot be improved much. Improvement to reduce leakage will have to focus on the side clearance and increase the radial extent of the side cutouts. There are many choices to the design of the links and their geometry interconnecting with each other and forming a support bearing for the rollers. A change to the links to improve the flow resistance will be high priority for the next generation of VVT.

A fixed link-replacement that eliminated the leakage through the link-train as well as around the link-train was designed and 3D metal printed as shown in Figure 39. The test unit was rebuilt with this fixed, non-operational link and tested. The results are shown in Figure 40 and 41 and indicates that the performance when nearly closed was significantly impacted by the leakage. The turndown ratio was increased to 3.3, and the efficiency at fully closed was increased by 5 points.

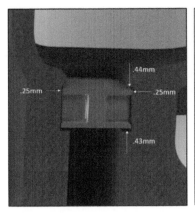

Figure 38. Clearances in the leak path around the links.

Figure 39. Special insert as exemplar to assess the leakage flow around the links.

During the gas stand testing, a rudimentary test was conducted to measure the force to move the link-train at various VVT opening angles. The torque to move the link-train was measured by use of a hand-held torque wrench. The test cell was running, but for safety concerns, the test was run cold and at 120 kg/min flow rate. Over 90% of the range, the torque required to move the link-train was below the measurement capability. When the link-train was moved to another position, it could be left free and would stay in-place. The inertia of the flow was stronger than the net force on the link-train.

Near the fully closed position torque was required to start to open the link-train, and virtually none to close. Pulling away from fully closed slowly required about 1-1.5 N-m of torque. When moving the link-train manually, it was exceptionally smooth, and one had a sense of inertia (such as trying to move a spinning gyroscope) when reconfiguring the flow path.

5   SUMMARY/CONCLUSIONS

A new type of variable geometry turbine has been developed based on a variable volute concept. The variable volute was enabled by the construction of a 720° spiral track and a series of links (or continuous flexible band). The links (or band) roll along this spiral track to create a 360° volute, which controls the A/r of the throat section.

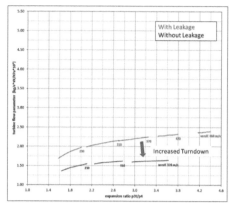

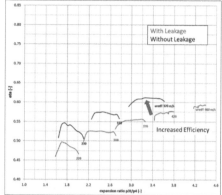

Figure 40. Test data of prototype, turbine efficiency for 6 angle positions.

Figure 41. Effect of prototype leakage at fully closed on efficiency.

The reduction of the A/r of the throat section results in a flow factor turndown more than 3.3. The links or flexible band is driven by a pinion gear that engages with teeth formed in the links or flexible band.

A prototype was designed, built, and tested and showed state-of-the-art performance with high, linear turndown and remarkably high efficiency. These results were achieved even though it was tested with a mismatched turbine wheel designed >30 years ago. As this was the first design, it is anticipated that significant improvement can be made in the unit's performance with further development as well as proper matching of the wheel with the VVT housing.

A key to the performance increase was to minimize the reduction of the Degree of Reaction of the turbine and eliminate the huge pressure drop (and subsequent conversion of pressure to velocity) of a conventional multi-vane or axial vane VGT.

Many failure modes associated with vanes have been eliminated. These include high-cycle fatigue resulting from vane/blade resonance, siren-like noise from vane/blade interaction, wear from engine firing pulsation, aero-torque reversal resulting in unstable vane control, and variation in minimum flow capacity due to vane assembly wear. The prototype has been subjected to hot performance tests, but durability and reliability have not been demonstrated.

The unit has much better control parameters than multi-vane design, which is highly non-linear when near closed, thus creating excessive gain in the transfer function. The movement of a multi-vane VGT is quite small, so electric actuation requires a large gear reduction. The VVT is much more linear and has a significantly reduced gain for the actuation system, eliminating the need for a gear reduction. In simpler terms, the multi-vane design has leverage over the actuation, whereas the VVT actuation has leverage over the link train.

Further information, including animations of the VVT can be found at: https://www.esi-inc.us/variable-volute-turbine

CONTACT INFORMATION

Steve Arnold

http://steve@esi-inc.us

http://www.esi-inc.us

+1 (310) 990-3163

DEFINITIONS/ABBREVIATIONS

A/r	Area of a section divided by the radius of its centroid
BePerf	Bearing Performance module of the rotordynamics program DyRoBeS
C_4, C_5	Turbine inlet and outlet gas velocities
CFD	Computational Fluid Dynamics
DyRoBeS	Rotordynamics analysis program
ER	Expansion Ratio
h_{01}, h_{05}	Turbine specific inlet and outlet stagnation enthalpies
MIM	Metal Injection Molding manufacturing process
PM	Powdered Metal manufacturing process
RN	Degree of Reaction
t-s	Total inlet pressure to static outlet pressure
U/C0	Tip speed divided by isentropic spouting velocity or "blade speed ratio"
VGT	Generic term for all variable geometry turbines

REFERENCES

1. Watson, N. and Janota M.S., "Turbocharging the Internal Combustion Engine," (MacMillan Press Ltd Publication, 1982), p. 152, ISBN 0 471-87082-2.
2. Arnold, S., Slupski, K., Groskreutz, M., Vrbas, G. et al., "Advanced Turbocharging Technologies for Heavy-Duty Diesel Engines," SAE Technical Paper 2001-01-3260, 2001, 10.4271/2001-01-3260.
3. Arnold, S., "Schwitzer Variable Geometry Turbo and microprocessor Control Design and Evaluation," SAE Technical Paper 870296, 1987, 10.4271/870296.
4. Cho, K., Lee, J.L., Lee, J.I., "Comparison of Loss Models for Performance Prediction of Radial Inflow Turbine", International Journal of Fluid Machinery and Systems, Vol.11, No.1, Jan-Mar 2018 http://dx.doi.org/10.5293/IJFMS.2018.11.1.097
5. Dixon, S.L., "Fluid Mechanics, Thermodynamics of Turbomachinery"

Twin scroll turbocharger turbine characterisation using a wide range multimap dual combustor gas stand

J. Thiyagarajan, N. Anton, C. Fredriksson, P.I. Larsson

Scania CV AB, Sweden

ABSTRACT

Twin scroll turbochargers are widely used in the Heavy Duty (HD) truck engines for long haulage applications. Such HD engines for pulse-turbocharging, non-EGR applications operate at higher pressure ratios compared to the automotive engines used for passenger cars and require wider operating ranges with correspondingly higher efficiencies. In order to minimize fuel consumption, the turbocharging system needs to be optimised as it is a vital component of the entire gas exchange system. This paper deals with testing of a twin scroll turbocharger turbine in a new gas stand (hot turbo test rig) facility at Scania CV AB, Sweden. Uniqueness of this gas stand is the ability to characterise a typical twin scroll turbine for a wide range of turbine loading coefficients (U/cs =0.52 to 0.83) on a dual combustor setup. In addition, it is also possible to run single, full and intermediate admission conditions (SPR=0.75 to 1.33) seamlessly on the twin scroll turbine without changing the turbocharger geometry. This is due to the capacity of the gas stand to control the compressor inlet pressure between 400 mbar to 3000 mbar, switching between open and closed loop configurations. Further, the twin scroll turbine stage is tested on a HD truck engine run at 1100 rpm full load from pressure ratios 1.3 to 3.75, to correlate the gas stand conditions and on-engine conditions. It was identified that the efficiency distribution between single admission and full admission was not linear and hence a linear interpolation must be avoided in 1D engine simulations. Further, intermediate admission maps are necessary to characterise the turbine stage and represent the on-engine conditions better. The goal of this work is to minimize the problem of interpolation & extrapolation in 1D engine simulations by characterizing a typical twin scroll turbine using gas stand measurements, by generating wider turbine maps at different admission conditions without changing the turbocharger geometry.

1   INTRODUCTION

Turbocharging the internal combustion engine has been a practice in the transportation industry to achieve downsizing and better fuel economy. With stricter emission regulations from the governmental agencies, the usage of turbocharging technology has become almost inevitable in internal combustion engines today. The concept of turbocharging the internal combustion engine was introduced by Büchi in 1905 and this had reduced the fuel consumption of such engines then. Turbocharging was later used in several aircraft engines in the conquest of higher speed and power. The passenger car industry has also been revolutionized with the use of turbocharging. The truck industry used Heavy Duty (HD) engines with turbocharging in the early 1960s for downsizing, increased power and lowered emissions.

Twin scroll turbochargers are widely used in the HD truck engines for long haulage applications. Such HD engines operate at higher pressure turbine stage ratios and require wider operating ranges with correspondingly higher efficiencies. In order to minimize fuel consumption, the turbocharging system needs to be optimised, as it is a vital component of the entire Gas exchange system. It is observed that using a twin scroll turbine volute configuration to feed a single turbine wheel is an effective method of utilising the exhaust pulse energy(3). Also, the turbine operates under varied unsteady conditions of full and partial admissions during the HD engine's operation. In order to design a turbine stage for such complex unsteady conditions for pulse turbo-charged non-EGR (Exhaust Gas Recirculation), advanced methods and test rigs are needed in order to quantify the designs. It is widely known in the industry that turbine stage operating conditions for a HD truck engine is difficult to represent on a hot gas stand. This has led to extrapolation of data from gas stand measurements which is then used in 1D simulations.

Bernhardt et. al.(1), discuss the design of a new turbine stage for passenger car engine and the difficulties in measuring relevant test data from hot gas stand. They clearly indicate the need for extrapolation due to insufficient measurement points on the gas stand. Unsteady turbine behaviour during engine's duty cycle is difficult to represent on a hot gas stand and hence needs a combination of experi-mental measurement and simulation tools to optimize the turbocharging system (2). There has been a lot of research in this area, especially in the small size engine segment, typically, passenger car applications to achieve more accurate turbine maps from test rigs. Vincenzo et al., (4), discuss the usage of steady and unsteady measurements from a hot gas stand that still needs tweaking in order to be used for 1D engine simulations. They also employ a pulse generator to measure unsteady turbine operating points. Pesiridis et. al., (5), have worked on prediction methods for turbine maps that represent the unsteady conditions better. They have discussed different types of models used, extrapolation strat-egies and effect of map width on engine performance predictions. Uhlmann T. et. al., (6), have worked on matching a double entry turbine to a 1.6 litre engine using hot gas stand measurements. It is to be noted that even though the volume of the engine is small compared to a truck engine, the turbine map could not be entirely measured in the hot gas test rig. It is required to have an extended modelling approach to replicate the turbine map to relevant engine operating conditions.

In the HD engines segment, Palenschat et al., (7), have published a work on designing radial turbines taking into account the engine load conditions and have used hot gas stand measurements for full and single admission to represent the load cycles. Also, the design and analysis of an asymmetric twin scroll turbine is presented by Müller et. al., (8), wherein experimental measurements are carried out on a hot gas stand. It is clearly mentioned that the turbine map measured is restricted due to limitations on the compressor mass flow range. The goal of this work is to minimize the problem of interpolation & extrapolation in 1D engine simulations from gas stand measurements by generating wider turbine maps without changing the turbocharger geometry to assist different loading conditions. The output from this work will provide more complete engine-matched turbochar-ger data that can be used to represent the engine conditions better in the 1D engine simulations. Using gas stand turbine maps, without consideration to 'on-engine' conditions, in 1D engine simulation can lead to severe errors in the results. Uniqueness of this gas stand is the ability to perform turbine mapping tests for a higher and lower turbine loading coefficient (U/c) on a dual combustor setup. In addition, it is also possible to run partial admission maps on the Twin Scroll Turbocharger. The wider power range (1:45) on the combustor, ability to reduce/increase compressor inlet pressure and higher mass flow rates allow wider turbine map measurements that can be further used for 1D engine simulations. A typical turbine map (blue line) run on an open loop gas stand is compared qualitatively to a map (dashed line) run on the Scania gas stand in Figure 1.

405

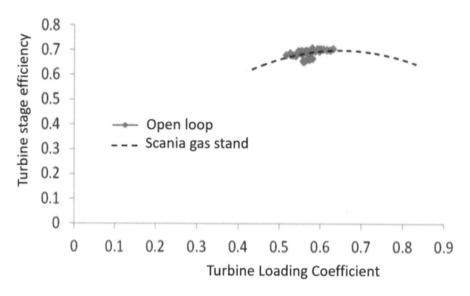

Figure 1 . Comparison of gas stand maps.

2   SCHEMATIC REPRESENTATION OF GAS STAND

The HD truck engine industry has high demands on turbocharger gas stands, especially the turbines tested for the pulse turbocharged engines. Better pulse utilization is achieved by reduction in exhaust manifold volume causing higher turbine energy variation. These higher turbine energy variations need to be mapped in an effective manner in order to better represent the engine operating conditions. A hot gas stand set up is used to generate compressor and turbine maps that represent as close as possible the on-engine quasi stationary operating conditions. The data will then be used to make crank angle resolved engine simulations accurately and predictively. Two other important usages are to validate turbine and compressor design calculations as well as to verify the component performance. A typical gas stand arrangement includes a combustor, turbocharger (test specimen), oil conditioning unit to regulate the oil lubricating the bearings in the turbocharger. Most gas stands use open loop configuration of the compressor so the turbine characteristics is based on the power change as the flow rate is altered in the compressor map with atmospheric condition at compressor inlet all the time. The gas stand facility described here uses a closed loop configuration, so the inlet pressure to the compressor can be altered and through that change the power consumption of the compressor for a given mass flow is increased or lowered depending on inlet pressure. The reason why it is called closed loop is because the outlet of the compressor is connected back to the intake via an air cooler and this closed circuit can be charged with compressed air or lowered via a vacuum pump.

Scania gas stand facility is designed to characterise turbines that are suitable for the HD engines using several unique features such as:

- Twin scroll turbine dual combustor with a wider power range of 1:45
- Compressor closed loop operation
- Seamless transition from open to closed loop operation
- Closed loop can operate both at pressure below and above atmospheric

- Turbine testing modes are single, full and partial admission testing on the twin scroll turbine without any mechanical modifications between the modes

The schematic of the gas stand facility at Scania is provided in the figure below. An external compression system is used to provide pressurised air to the dual combustors (shown in Figure 2) where a fuel is injected to generate hot gas to run the twin scroll turbines and then discharged outside the facility. The dual combustor has a maximum flow capacity of 1.2 kg/s (max 0.6 kg/s each) with a max exit temperature of 1100 degree Celsius and a wider power range (1:45). The gas stand facility can be used to test turbines sizes suitable to engine sizes of approx. 50-700kW and for twin scroll turbines approx. 90-440kW.

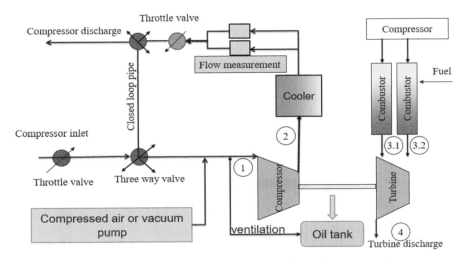

Figure 2 . Schematic representation of Scania gas stand.

3   EXPERIMENTAL SETUP AND INSTRUMENTATION

In this study, a typical twin-scroll turbocharger turbine was used for testing on the gas stand. Later, the same turbocharger was tested on a typical 13 litre, 6 cylinder engine (Figure 3). Total pressures were measured at the inlet (on each scroll separately) of the turbine stage along with static pressure and temperature, for both gas stand and engine tests. In addition, static & total pressures and temperatures have been measured at the inlet and outlet of the compressor stage. Since the temperature measurement at the outlet of the turbine is not very reliable for efficiency calculations, the work done by the compressor is calculated and combined with the transmitted losses on the bearing system in order to quantify turbine work. The data acquisition system is capable of recording steady state and transient data. All the results discussed in this work were obtained from steady state measurements in the gas stand that was recorded upon ample stabilisation. In addition, Crank Angle Resolved (CAR) pressure measurements from engine testing have also been used for analysis and discussion.

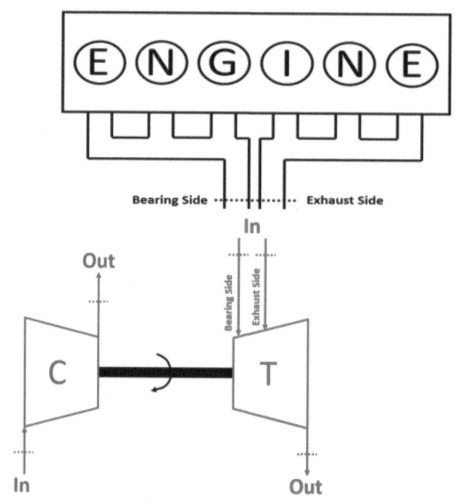

Figure 3 . A turbocharger turbine connected to a 6-cylinder engine.

4   CLOSED LOOP VS OPEN LOOP TURBINE DATA

In this section, the data obtained from gas stand tests for the twin scroll turbine is discussed in detail illustrating the differences between open loop and closed loop tests. The following equations are used to define the terminologies that are stated in the plots and will also be referred to in the text.

$$SPR = \frac{P_{t,in,T,ex}}{P_{t,in,T,be}} \tag{1}$$

$$FC = \frac{\dot{m}_T \sqrt{T_{t,in,T}}}{P_{t,in,T}} \tag{2}$$

$$PR_{TS,T,be} = \frac{P_{t,in,T,be}}{P_{s,out,T}} \tag{3}$$

$$PR_{TS,T,ex} = \frac{P_{t,in,T,ex}}{P_{s,out,T}} \tag{4}$$

$$\eta_{TS,T} = \frac{\dot{W}_T}{\dot{m}_{T,be}\Delta h_{t,is,T,be} + \dot{m}_{T,ex}\Delta h_{t,is,T,ex}} \tag{5}$$

$$\Delta h_{t,is,T,be} = c_{p,T} T_{t,in,T,be} \left(1 - \left(\frac{P_{s,out,T}}{P_{t,in,T,be}} \right)^{\frac{\gamma_T - 1}{\gamma_T}} \right) \tag{6}$$

$$\Delta h_{t,is,T,ex} = c_{p,T} T_{t,in,T,ex} \left(1 - \left(\frac{P_{s,out,T}}{P_{t,in,T,ex}} \right)^{\frac{\gamma_T - 1}{\gamma_T}} \right) \tag{7}$$

$$\frac{U}{C_S} = \frac{U}{\sqrt{2 c_{p,T} T_{t,in,T} \left(1 - \left(\frac{P_{s,out,T}}{P_{t,in,T}} \right)^{\frac{\gamma_T - 1}{\gamma_T}} \right)}} \tag{8}$$

$$\dot{W}_T = \dot{W}_C + \dot{W}_B \tag{9}$$

In the Figure 4, the turbine map obtained from the gas stand test can be seen with several operating condition such as single, full and intermediate admission. It includes the normalized stage efficiency (on the left ordinate), reduced mass flow (on the right ordinate, see Eq.2) and pressure ratio on the abscissa. The effect of closed loop is evident from the fact that the speed lines are extended for wide range of pressure ratios in all cases. For better modeling of turbine data, it is not only important to have longer speed lines mapped but also to acquire data on either side of the best efficiency point. Figure 5 shows a plot of exhaust side pressure ratio, see Eq.4, along the abscissa and the bearing side pressure ratio, see Eq.3, along the ordinate for different kinds of Scroll Pressure Ratio (SPR, see Eq.1) and admissions, namely, single, full and intermediate admissions for different turbine speeds. As seen in the figure, the pressure ratios increase with increasing turbine speed and a wider range of pressure ratios (between 1.25 to 4.75) can be mapped in the gas stand at single, full and intermediate admission conditions. This is due to the ability to control the compressor inlet pressure in the gas stand open loop or closed loop configurations. This is very effective but one has to be cautious from a mechanical strength perspective. During tests where the inlet pressure to the compressor is increased to 3 bar, the outlet pressure will be very high (5.5 bars) and this is a limitation as the typical compressor housing can easily be blown off. Also, enhancing the range by lowering the inlet pressure (using a vacuum pump in closed loop configuration) is also effective but here the issue is oil contamination in the closed loop system. At the Scania gas stand, oil discharge pressure is simultaneously reduced during such test conditions using a separate pump.

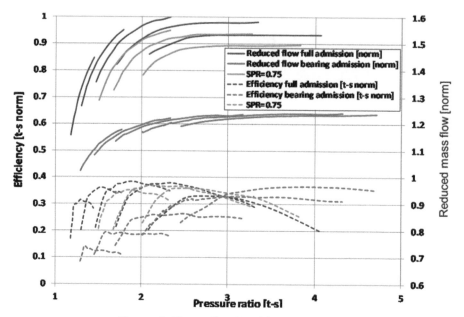

Figure 4 . Normalised turbine map data.

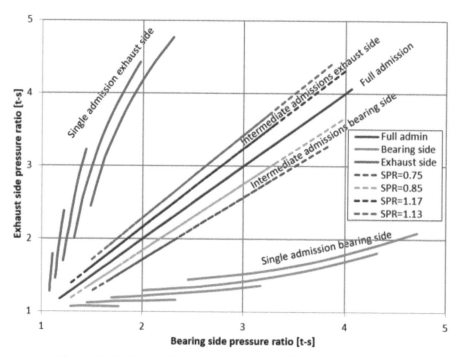

Figure 5 . Twin scroll pressure ratio for different admissions.

Mapping the turbine at peak efficiency at all speeds is important in order to create good data for 1D engine simulations at all admission conditions. This is not feasible with a gas stand that is limited to open loop configuration or limited closed loop facility. This is illustrated in Figure 6, wherein, the normalized total-static efficiencies (see Eq.5) of the twin scroll turbine at full admission condition are plotted for different speeds along with the compressor inlet pressure for a given turbine loading, see Eq.8. Colored boxes are used to indicate the different envelopes of the efficiency curves that are mapped using varied compressor inlet pressure conditions. It is interesting to note that a wider range on the efficiency curves (values to be read on the left ordinate), which also includes the peak efficiency points for the twin scroll turbine lie in the closed loop zone. This is shown by the colored box shaded in brown (1000-3000 mbar, values to be read on the right ordinate). This reinforces the necessity of a closed loop gas stand configuration in order to map the major part of the efficiency curves.

The green colored box on the plot indicates the portion of efficiency curves that can be mapped using an open loop configuration, that corresponds to 1000 mbar. It is also interesting to note that the lower part of the efficiency curves cannot be mapped using an open loop configuration as the compressor inlet pressures should be sub-atmospheric and hence need a vacuum pump to achieve 400-1000 mbar. From the above discussions, it is observed that in order to map a twin scroll turbine used for pulsed turbocharging, non-EGR applications at different turbine loading conditions (especially lower values), it is essential to have a closed loop configuration in the gas stand facility along with open loop and vacuum pump setup to seamlessly map the entire region of the efficiency curves.

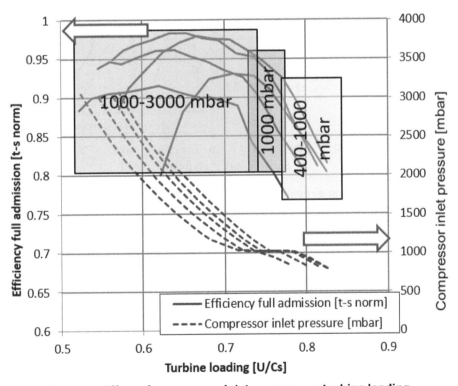

Figure 6 . Effect of compressor inlet pressure on turbine loading.

5   COMPARISON OF GAS STAND VS CRANK ANGLE RESOLVED ENGINE DATA

In this part, the matching of the gas stand turbine maps to 'on-engine' conditions will be discussed. The Crank Angle Resolved (CAR) pressure trace from the engine test at 1100 rpm engine speed and full load condition is plotted in Figure 7. The solid and dashed lines indicated the bearing side and exhaust side total to static pressure ratios respectively. It is also seen that there is an influence of pressure interference from one scroll to another indicated by the peaks at lower pressure ratios. These peaks at lower pressure ratios on one scroll lie close to the peaks at higher pressure ratios on the other scroll, phased out by small margins.

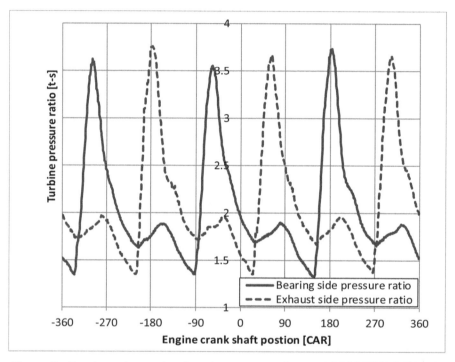

Figure 7 . Crank angle resolved engine pressure trace at 1100 RPM full load.

The engine pressure trace is overlaid on the gas stand data for the twin scroll turbine to identify the areas of interest that should be mapped with better resolution. It is seen from Figure 8 that the on-engine conditions are matched at full admission and intermediate admissions at lower pressure ratios (1.5-2). At higher pressure ratios, the engine pressure traces lie between single and intermediate admission maps. It is to be noted that available energy from the pulsating flow is higher at higher pressure ratios. It is found from the plot that this region of higher available energy lies in between the full admission and single admission conditions. In addition, it is seen that the mapping the full admission condition is not very useful as the amount of available energy from the exhaust flow is lower and also that this condition is passed quickly during on-engine operation. However, it is still necessary to include full admission mapping in order to represent the turbine stage in engine simulations. Mapping the twin scroll turbine at SPR values other than 1 seem to be more important for data interpolation and representation of on-engine conditions in a 1D engine simulation. It was also observed that in the future tests, turbine characterization should be carried out in order to trace the engine conditions (Engine op [CAR]) represented in **Figure 8** to reduce the extent of interpolation.

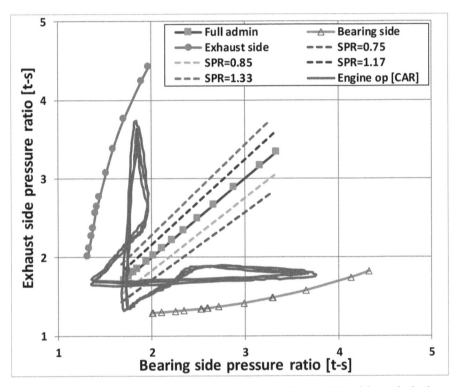

Figure 8 . Overlay of engine pressure trace on twin-scroll turbine admission at 1100 RPM full load.

Figure 9 shows an overlay of compressor inlet pressure conditions on Figure 8, representing the different gas stand running configurations (open, closed loop & sub-atmospheric) to capture the on-engine conditions in the gas stand. The higher pressure ratios encountered at 'on-engine' conditions is effectively mapped only when the compressor inlet pressure is between 100-300 kPa and this corresponds to a closed loop configuration. Simultaneously, this corresponds to the lower turbine loading coefficient zone indicated in Figure 6. The green region indicated in Figure 9 corresponds to the open loop configuration, wherein the compressor inlet pressure is atmospheric. Further below this, is the region of sub-atmospheric compressor inlet conditions driven by a vacuum pump.

Hence, in order to map a wider turbine operating range that represents the on-engine conditions, it is necessary to operate the gas stand in all the three above mentioned configurations and switch between them seamlessly.

Compressor inlet pressure
100 - 300kPa

Compressor inlet pressure
100kPa

Compressor inlet pressure
40 - 100kPa

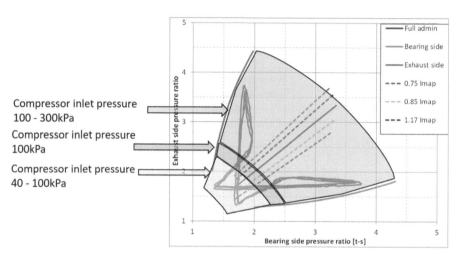

Figure 9 . Compressor inlet pressure on twin-scroll turbine engine data.

The normalized turbine stage efficiency is overlaid as iso-contours indicating islands of efficiency with the innermost red colored island indicating the maximum and the outermost green colored island indicating the minimum, see Figure 10. It shows clearly the interpolated area between full and single admissions is not linear as it is usually done in the industry and hence the intermediate admission mapping is a vital part of the testing that needs further enhancement. It is interesting to note that even the line of maximum efficiency does not occur at full admission state. It is also evident that the engine pressure trace tends to avoid the peak efficiency line which in turn explains why the turbine stage design has to be developed in close relation to the engine design in order balance the trade-offs between full admission and single admission characteristics wisely.

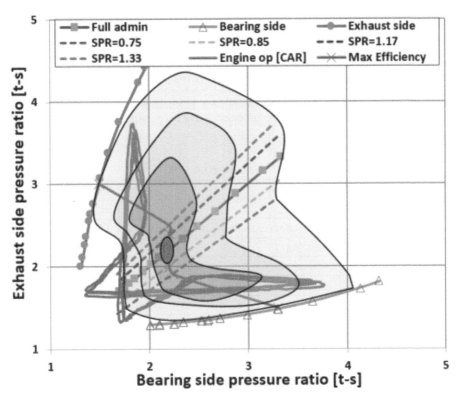

Figure 10 . Overlay of efficiency on twin scroll turbine engine data.

6   CONCLUSION

Gas stand testing is a common way of turbine stage map assessment, which is used in 1D engine simulation. For an accurate representation of the turbine stage in these engine simulations, the map must cover the 'on-engine' operating conditions. If not, there is a risk of error in the engine simulation results. In a traditional gas stand, with a single burner and open loop configuration, mapping of the turbocharger turbine is restricted to a narrow range of turbine loading coefficient. Since the pulse-turbocharged HD engine turbocharger turbine has a twin scroll volute arrangement and operates at a wide range of pressure ratios, the demands on the gas stand are high. A detailed characterization of the turbine can be carried out only with a gas stand facility equipped with sufficient burner capacity, open and closed loop arrange-ments to map higher pressure ratios along with a vacuum pump to map lower pressure ratios. These capabilities at the Scania gas stand make it a unique test facility for tur-bine mapping that can represent 'on-engine' conditions better to reduce errors from interpolation/extrapolation of maps in 1D engine simulations.

In order to characterize a typical twin scroll turbocharger turbine, a wide range of turbine loading coefficients, between 0.52 to 0.83, needs to be mapped. This necessitates a higher compressor inlet pressure range between 400-3000 mbar along with seamless transition between open and closed loop configurations. The pressure ratios from the engine test carried out at 1100 rpm full load condition range between 1.3 to 3.75. It was also observed that full admission mapping did not represent the on-engine conditions tested at 1100 engine rpm full load. It could be represented only by gas stand tests carried out under SPR values away from 1 and single admission conditions. It was also identified that in order to map different SPR values away from 1 at higher pressure ratios of 2.5 to 4.5, closed loop configuration is necessary. The plot of turbine stage normalized efficiency shows a non-linear distribution, which reinforces the necessity of full and intermediate admission mapping in order to enhance interpolation characteristics. It is also to be noted that all of the 'on-engine' conditions have been mapped seamlessly.

NOTATIONS

Notation	Description	Unit
c	Velocity	m/s
c_p	Specific heat capacity at constant pressure	J/kgK
h	Enthalpy	J/kg
FC	Flow capacity	$kg\sqrt{K}/kPa$
$\dot{m}$	Mass flow	kg/s
P	Pressure	kPa
PR	Pressure ratio	–
SPR	Scroll pressure ratio	–
T	Temperature	K
U	Blade tip speed	m/s
$\dot{W}$	Power	$Watt$
η	Isentropic efficiency	–
γ	Specific heat ratio	–

SUBSCRIPTS

Notation	Description
B	Bearing
C	Compressor
be	Bearing side
ex	Exhaust side
in	Inlet
out	Outlet
s	Static state, Spouting
T	Turbine
t	Total state
is	Isentropic
TS	Total to static

REFERENCES

[1] Bernhardt L., Dietmar F., and Jan E., (2012), International Journal of Rotating Machinery, "On Mixed Flow Turbines for Automotive Turbocharger Applications," vol. 2012, Article ID 589720, 14 pages, 2012. 10.1155/2012/589720.

[2] Piscaglia, F., Onorati, A., Marelli, S., and Capobianco, M., (2007), "Unsteady Behavior in Turbocharger Turbines: Experimental Analysis and Numerical Simulation," SAE Technical Paper 2007-24-0081, 10.4271/2007-24-0081.

[3] Björnsson H., Ottosson A., Rydquist J.-E., Späder U. and Schorn N., (2005) "Optimizing des SI-engine turbo system for maximum transient response. Methods, factors and findings," in Proceedings 10. ATK, Dresden.

[4] Vincenzo B., Silvia M., Fabio B., Massimo C., (2014), 1D simulation and experimental analysis of a turbocharger turbine for automotive engines under steady and unsteady flow conditions, Energy Procedia 45, 909–918.

[5] Pesyridis A., Salim W., Ricardo M., (2012), Turbocharger Matching Methodology for Improved Exhaust Energy Recovery, 10th International Conference on Turbocharging and Turbochargers, DOI: 10.1533/9780857096135.4a.203.

[6] Uhlmann T., Lückmann D.,Scharf J., Aymanns R., (2014), Development and Matching of Double Entry Turbines for the Next Generation of Highly Boosted Gasoline Engines, Blucher Engineering Proceedings, Número 2, Volume 1.

[7] Palenschat T., Marzolf R., Muller M., Martinez-Botas M.F., (2018), Design process of a Radial Turbine for Heavy Duty engine application taking into consideration load spectra and unsteady engine boundary conditions.

[8] Müller M., Streule T., Sumser S., Hertweck G., Nolte A., Schmid W., (2008), The Asymmetric Twin Scroll Turbine for Exhaust gas Turbochargers, Proceedings of ASME Turbo Expo 2008: Power for Land, Sea and Air, GT2008–50614.

Development of the PureFlow™ compressor inlet for added range and efficiency

S.D. Arnold

Engine Systems Innovation, Inc. , USA

ABSTRACT

Wide flow range compressors, both with and without a ported shroud, improve the surge line characteristic by increasing inducer recirculation. These compressors regurgitate hot (>125C) air when operating to the left of peak efficiency. This hot flow is re-ingested by the compressor and the result is reduced efficiency and pressure ratio. The outlet temperature can be increased up to 100C and the pressure ratio reduced by more than 20%. One result is that the turbine power needed to drive the compressor to reach the desired mass flow is increased. Another result is that the intake manifold density is reduced, and additional boost pressure is needed to achieve the desired mass air flow. This is a negative feedback loop which results in the engine performance degradation at low speed/high torque.

Two solutions to this problem are presented, neither of which require a change to the compressor design, and only require a change to the compressor inlet piping. Both result in a more favorable surge line, higher compressor efficiency and much higher boost pressure on the left side of the compressor map. The benefit of both these solutions is quantified by computational fluid dynamics (CFD) analysis and turbocharger test rig testing.

The preferred, simple solution is to divide the inlet pipe into two annular flow paths, both connected to a divided air cleaner, or two separate air cleaners. Flow in the outer flow path can flow either direction depending on the compressor operating point. The second solution has just one air cleaner, but the outer annular flow is cooled by a heat exchanger, then combined back into the core flow. The preferred PureFlow™ has potential for retrofit to existing vehicles.

1   INTRODUCTION TO THE PROBLEM

The flow range of a turbocharger compressor, defined as the ratio of choke flow to surge flow at a fixed pressure ratio, is an attribute that always needs improvement. Compared to internal combustion engines, radial compressor turbomachines are inherently a narrow range technology because they create flow into an adverse pressure gradient.

Turbochargers are used on a wide variety of engines each of which can have different characteristics and requirements. Although there is a complex set of variables that determine a turbocharger's required characteristics, the most dominant engine attribute driving the turbo's design characteristics is the speed range of the engine. At one extreme, an example would be high-performance gasoline motorcycle engines, some of which have a maximum speed of 16,000 rpm. At the other end of the spectrum an example would be gen-sets that run at one fixed speed to allow the generator to provide a fixed electrical AC frequency. Conventional radial compressors work extremely well for the gen-sets but lack flow range with good efficiency for high-speed gasoline engines. Other than constant speed engines such as a gen-set, most other engines require a broad flow range which is the weakest attribute of a radial compressor.

Diesel engines are somewhere in the middle, but still require significantly more flow range than an industrial radial compressor would provide.

The second most dominant engine attribute determining the requirements of the compressor is the Brake Mean Effective Pressure (BMEP). The higher the engine BMEP, the higher the compressor boost pressure requirement.

The maximum flow of a compressor design is determined by the throat area of the inducer. As the inducer flow area is increased relative to the maximum diameter of the compressor wheel, there is a practical limit which is often the first vibration mode of the impeller. This mode must be kept from being excited throughout the range of operation of the compressor. Therefore, to make significant improvements in flow range to a compressor of fixed diameter, the surge line must be improved, i.e., moved to lower flow as opposed to increasing choke flow.

The focus of compressor design for turbochargers has been to improve the flow range, especially to move the surge line to the left on the flow versus pressure ratio map. One can look at the historical trends to verify this. Compressors designed in the 1960s using basic principles and slide rule calculations could produce 80% efficiency. Today using computer clusters combined with auto-optimization programs that modify geometric parameters

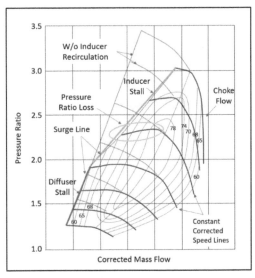

Figure 1. An example compressor map with relevant nomenclature.

then run computational fluid dynamics programs and finite element stress analysis, also result in about 80% maximum efficiency. However, the flow range of the compressors is significantly improved in today's compressors demonstrating that the engineering focus has been on off-design conditions.

An important facet of radial compressor design to note is that there are two surge/stall mechanisms. One is triggered by the compressor inducer; the other is triggered by the diffuser. This can be seen in many compressor maps as there are two slopes to the surge line. This effect can clearly be seen in Figure 1, which is a representation of a compressor circa 1980. At low pressure ratio, the diffuser stalls, then at higher pressure ratio the inducer stalls.

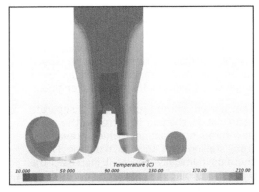

Figure 2. Longitudinal cross section temperature plot of a conventional compressor operating near surge.

Figure 1 shows several problems caused by the inducer surging or stalling. Note that the surge line at higher speeds is moved to the right compared to our expectations. Also, the constant corrected speed lines show the pressure ratio turn to a flat or downward trend near surge, and the efficiency falls significantly. The reason for these trends is that the compressor impeller is rejecting some flow back into the inlet pipe. Since the compressor has done work on this flow, it is at a higher temperature than the inlet flow stream.

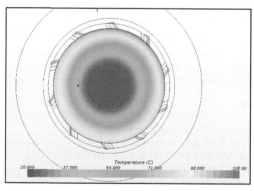

Figure 3. Axial cross section temperature plot, conventional compressor operating near surge.

A computational fluid dynamics (CFD) analysis was completed on a modern gasoline compressor design. Figure 2 shows the temperature profile on a longitudinal cross-section of the unit when it was operating near surge, and Figure 3 shows the same operating point in a section cut upstream of the compressor inlet.

These plots clearly show that 1) the inlet temperature is increased and 2) the temperature profile has a significant radial gradient when it enters the compressor impeller. The cost of increasing the flow range of a compressor through recirculation is a loss of pressure ratio and efficiency, which results in higher intake manifold temperature (thus requiring more pressure to get the desired mass flow) and more turbine work which raises the expansion ratio and puts more backpressure on the engine.

A device commonly referred to as a "ported shroud" (Figure 4) has been used since the 1980s to improve the flow range of compressors used on high BMEP engines, typically highly boosted diesel engines that operate across a relatively wide speed range (for diesel engines) such as truck engines and industrial/agricultural engines.

The ported shroud provides a separate recirculation channel to increase the mass flow through the inducer which prevents it from stalling. But this recirculation is not thermodynamically free as it increases the

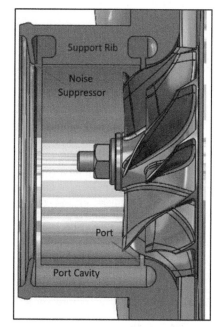

Figure 4. Cross section of a typical ported shroud compressor.

work the turbine must do to pump the extra air in this short loop. Even worse is that this recirculation heats the air that is recirculated. When this hot air is blended back into the fresh air at the inlet of the compressor, the work required of the turbine is increased as the work is directly proportional to the inlet temperature of the compressor and inversely proportional to the efficiency (equation 1).

$$\dot{W}_c = \frac{\dot{m}_a \times c_{pa} \times T_1 \times (\pi_c^{(\gamma-1)/\gamma} - 1)}{\eta_c} \qquad (1)$$

In the area of the map where the ported shroud is not necessary, it creates an additional loss and reduces the efficiency there as well. Traditional ported shroud compressors generally reduce the peak efficiency of a compressor by 1-3 points but can reduce the efficiency at surge by 10 points or more, trading efficiency for range. As noted with non-ported shroud compressors, this recirculated and heated air reduces the pressure ratio of the compressor where the recirculation happens. Figures 5, 6 and 7 show the temperature and flow fields. The reduction in pressure ratio

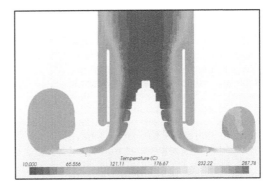

Figure 5. Longitudinal cross section temperature plot of a ported shroud compressor operating near surge.

reduces the efficiency of the compressor. This has been the trade-off to allow the engine to run at these low flows—reduced efficiency and reduced pressure ratio.

It must be noted that the ported shroud can be designed to increase choke flow by moving the port beyond the choking point of the inducer.

The recirculation is shown most clearly in Figure 7. The lack of symmetry in the flow field indicates that there is a rotating instability in the flow with recirculation penetrating further upstream, as seen on the right side of the section, alternating with less penetration on the left side. With this flow condition, the steady state CFD analysis does not converge sufficiently to meet stabilization criterion. If one looks at the results through different iterations, it becomes clear that this lack of symmetry is rotating in the lab reference frame.

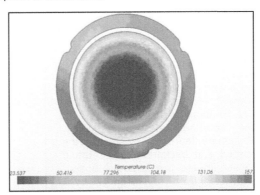

Figure 6. Axial cross section of a ported shroud operating near surge.

2   PUREFLOW DUAL INLET SOLUTION

This compressor inlet technology solves these problems. It basically splits the inlet flow into two regions as shown in Figure 8. This eliminates the losses associated with the ported shroud as well as eliminates the recirculation of hot gas into the compressor.

This technology separates the inlet into multiple, concentric regions, and the simplest version being an inner and outer portion. The feedback of the subsonic flow through the inducer blades back into the inlet flow stream allows two different amounts of swirl to be generated, which can improve the efficiency of the compressor.

As shown schematically in Figure 8, each of the concentric regions is separated at the filter end of the inlet ducting from the others. At the opposite end, the two flow streams are separated until they are in close proximity to the compressor wheel. They each have their own air filter, or the air filter is bifurcated into two separate regions. Each concentric region has the possibility of flow going in either direction. If the flow reverses in a region, it will have become hotter as the air will have gone into the compressor wheel and back out. It is important for that hotter flow not to be reingested into the compressor wheel as it reduces the efficiency of the compressor and increases the work required to drive the compressor stage.

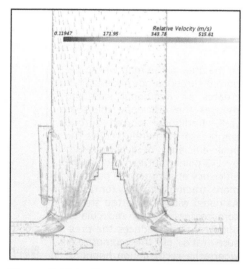

Figure 7. Velocity vectors, ported shroud unit near surge.

3   AERODYNAMIC ANALYSIS OF PUREFLOW™ CONCEPT

CAD models were constructed for a large ⌀130mm compressor with ported shroud and a smaller ⌀44mm compressor without ported shroud to assess the aerodynamic performance of the concept. Table A-1 (Appendix) lists pertinent attributes of the analysis method.

The CFD analyses were performed with the "frozen rotor" method which means that multiple frames of reference were used, and all parts within a frame of reference are stationary. The rotational velocity is applied to the "Rotating" frame of reference. Stationary surfaces such as the shroud and backplate in a rotating Region are assigned to the stationary Laboratory frame of reference.

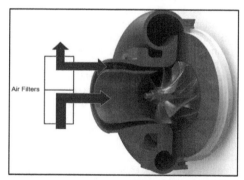

Figure 8. The basic PureFlow™ concept.

Stagnation Inlets were used on all inlet boundaries. Since turbocharger compressors have been optimized for a wider flow range compared to industrial compressors, the pressure ratio curve can go from a negative slope to zero slope or to a positive slope. This means there is not a unique flow rate associated with a given outlet pressure. Within the CFD analysis (StarCCM+) boundary options, there is not an inlet and outlet boundary condition pair that can be specified and used for all flow points. An inlet mass flow boundary can be used, but it is computationally too consumptive to use for large projects without a computing cluster. There is also the possibility of using a mass flow boundary on the outlet, however the author has not been able to be successful using that technique.

A new technique was devised to control the mass flow, mimicking the control method that is used on gas stand—a backpressure valve which discharges to atmosphere. A small cylinder was added to the outlet extension of the volute, which then discharges to atmospheric pressure, Figure 9. The diameter of the cylinder (orifice) was changed to control the flow of the compressor. This required remeshing the orifice extension, which could be done without remeshing the entire compressor system and made the process of running compressor speed lines or complete maps much simpler.

The segregated solver was used in these analyses. The author has used flow coupled solvers in the past, but they have proven to be much slower and do not appear to change the results.

With automated meshing and the power of today's computers, defeaturing or simplifying the models is no longer necessary. There is some advantage to keeping the totality of features in the model. For example, bolt holes that protrude into the volute are a loss, so there

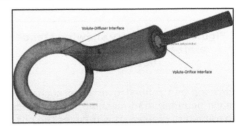

Figure 9. Mesh of volute, orifice, outlet, highlighted interfaces.

is no reason to eliminate them especially since the feature's effect will be included in any test data. Another example is the space behind the compressor wheel. It has an impact on aerodynamic efficiency as well as affecting the thrust force of the compressor wheel. This information can be passed to an engineer qualifying the turbocharger's bearing system or doing thrust bearing development.

The geometry manipulation was done in CAD (SolidWorks), output as Binary Parasolid files and imported directly into the CFD analysis. Since this analysis was adiabatic, the solid bodies provided no benefit and fluid bodies were extracted from the solid parts in CAD. Prior to extracting the fluid bodies, each Region and Interface were constructed in CAD so the geometry would need little work in CFD, Figure 10.

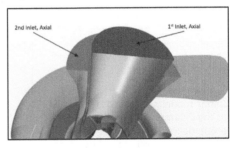

Figure 10. Geometry with 2 axial inlets.

Several permutations were designed and analyzed, with Figure 10 representing a version with both inlets axial. Figure 11 shows another configuration with the first inlet axial and the 2nd inlet radial. The results showed no appreciable change between the two.

After the binary parasolid files were imported into the CFD analysis, the "split-by-patch" function was used, and the patches were grouped in a way that facilitated meshing with pre-set mesh controls. For example, the compressor wheel groups were Hub, Blades, Leading Edges, Fillets, Shroud, and Back-disk, as can be seen in Figure 12. Templates had been

Figure 11. Geometry with one axial, one radial inlet.

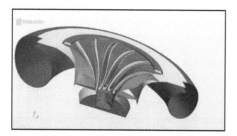

Figure 12. Geometry showing stand-ardized "grouped patches" or boundaries.

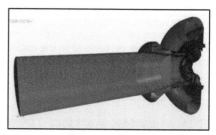

Figure 13. Mesh section, 1st inlet axial, 2nd inlet axial.

devised and refined over time such that each boundary had meshing, and physics parameters pre-set. This reduces the possibility of errors in mesh parameters and provides more consistency from project to project. Figure 13 is a section cut of the mesh of the ⌀44mm compressor with both inlets configured as axial areas. Figure 14 is a mesh of the Rotating Region. The housing shroud and diffuser walls are referenced to the stationary or Lab reference frame.

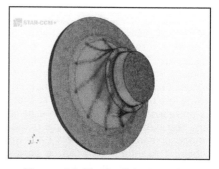

Figure 14. Mesh of the rotating region.

Table A-4 lists the Regions, Cell count, Interfaces, Solvers, Reference Frames and Stopping Criteria. The cell count varies from version to version but was typically over 10M cells.

Initial modeling and CFD analysis of the Ø44mm compressor showed dramatic improvement of the compressor on the left side of the map, with much higher-pressure ratio (Figure 15) and efficiency gain of more than 10 points (Figures 16-18). This seemed to be too good to be true, and indeed this was the case. The flow formerly being recirculated was now being rejected out of the outer inlet and was not being accounted for. This was confirmed by calculating the efficiency based on shaft work. However, there still was increased compressor pressure ratio, wider range, and decreased outlet temperature on the surge side of the map. It became somewhat difficult to assess the performance and eventually it was recognized that there

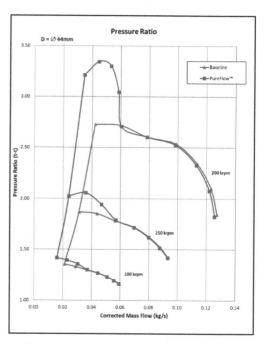

Figure 15. Comparison plot of pressure ratio vs corrected mass flow.

were two "efficiency" calculations with meaning. One is associated with the flow delivered to the engine, including the pressure, mass flow and temperature. The other is the shaft work required to drive the compressor to this point.

We will refer to the efficiency of the flow being delivered to the engine as the "aero efficiency", and the efficiency related to driving the compressor as the "shaft efficiency" or "total efficiency". Aero efficiency (η_a) is the traditional compressor efficiency calculated from inlet and outlet temperature and pressures but does not include the possibility of reverse flow in the second inlet. The shaft efficiency (η_s) calculates the efficiency based on the shaft torque and speed as well as the isentropic work. An alternative calculation, total efficiency (η_t) is like the aero efficiency but does include the enthalpy change across the second inlet. The shaft efficiency and total efficiency should be identical values depending on whether they are measured test values or calculated CFD values. Two errors that could creep in would be 1) heat transfer affecting measured values with aero or total efficiency but not shaft efficiency, or 2) compressor back-disk loss not accounted for equally in all methods. In the CFD work, both aero efficiency and shaft efficiency were calculated for a ported shroud case and were within a few tenths of a point.

The aero efficiency is equation 2, shaft efficiency is equation 3, and total efficiency is equation 4. Station 1 is the main inlet (inner inlet), Station 2 is the secondary inlet (outer inlet), and Station 3 is

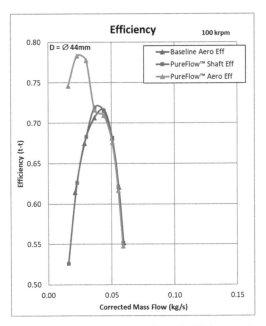

Figure 16. Comparison plot of efficiency at 100krpm.

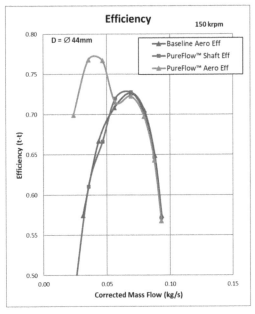

Figure 17. Comparison plot of efficiency at 150krpm.

425

the compressor outlet. These equations apply to both CFD and prototype testing.

$$\eta_a = T_1 \frac{\pi_c^{(k-1)/k} - 1}{T_3 - T_1} \qquad (2)$$

$$\eta_s = \frac{\dot{m} * cp * T_1 * (\pi_c^{(k-1)/k} - 1)}{2\pi * TQ * N} \qquad (3)$$

$$\eta_t = T_1 \frac{\pi_c^{(k-1)/k} - 1}{(T_3 - T_1) + \frac{\dot{m}_2}{\dot{m}_3}(T_2 - T_1)} \qquad (4)$$

The objective of a turbocharger is to increase the density, not the pressure, in an engine's intake manifold. The lower the temperature of air delivered to the intake manifold; the less pressure will be required. By eliminating the hot recirculated flow, the outlet temperature going to the intercooler is reduced as shown in Figure 19. This may look like a modest gain, but this plot is not an accurate comparison as the associated pressure ratio is quite different.

To better understand the real benefit, Figure 20 adds a simulated engine constant speed line going through surge of the base compressor at 200krpm. This chart shows that the boost pressure could be increased from 2.7 bar to 3.3 bar and still reduce the outlet temperature from 208C to 189C. That is an increase in air density into (not after) to the intercooler of 34%.

But the extra boost pressure and lowered temperature are not completely free. The outer compressor inlet passage has flow that it is discharging back through the inlet and that will increase the amount of turbine work required to drive the compressor to that point. Figure 21 shows that once the reverse flow starts in the compressor, the shaft power required will increase. The result will be a higher backpressure on the exhaust manifold and some

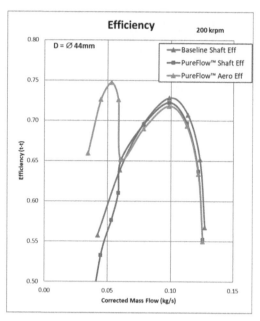

Figure 18. Comparison plot of efficiency at 200krpm.

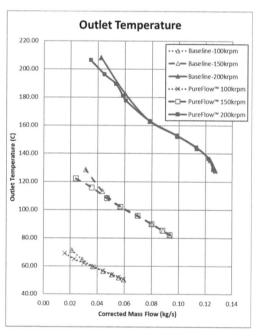

Figure 19. Comparison of compressor outlet temperature.

additional pumping work due to the increase in boost pressure.

Note that the basic concept requires no changes to the compressor wheel or the compressor housing other than to bolt a new inlet comprising two concentric inlets. All that is required is that the inlet to the compressor be divided into two concentric ducts, with the division starting immediately adjacent to the compressor wheel impeller. The ducts then extend to the air filter housing where the housing has been divided into separate filter areas taking fresh air from separate inlets.

The CVD data looked very promising, so a decision was made to move to a test of the concept. Turbo manufacturer Roto master offered to build a prototype and test it on their gas stand. A turbo from a Ford Powerstroke was chosen for convenience and the compressor housing was modified to accept a concentric split inlet. Since there was no data available on the compressor wheel or housing, it was not possible to complete CFD analysis. The concentric inlets were scaled up from the work on the Ø44mm unit. The area of the outer annulus was 23% of the total area and the core inner area was 77% of the total area. The assembled modified compressor housing is shown in Figure 22.

4   EXPERIMENTAL TEST OF PUREFLOW™ AND BASELINE

The test stand was modified with an additional inlet, and a mass flow measurement added to the outlet. The testing was accomplished smoothly with no problems. The one item of note was that the compressor was much quieter than the original in the "pre-surge" zone, and there was not a build-up of noise prior to hitting surge—it was very sudden and distinct.

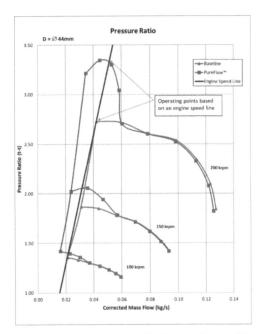

Figure 20. Pressure ratio/mass flow with constant engine speed line.

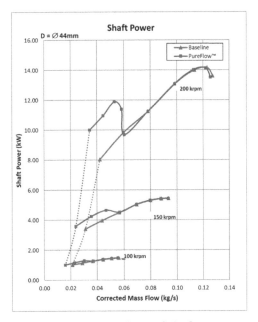

Figure 21. Comparison of shaft power.

427

The Pressure Ratio vs Corrected Mass Flow in Figure 23 shows a significant increase in pressure ratio on the left side of the map. The choke flow is slightly less, which is probably due to the blockage the concentric inlet presents along with an extra wall with boundary layer effects on both sides. The surge line was improved significantly, with the improvement increasing as the turbo speed and pressure increased. There was no optimization of the duct area ratios or distance from the wheel leading edge.

Figure 24 indicates that the shaft efficiency of the compressor was improved by about 2 points through most of the speeds. This is likely because there is typically a loss of about 2 points when a traditional ported shroud is used. The ported shroud was eliminated by the PureFlow™ inlet. This shaft efficiency improvement will reduce the backpressure required by the turbine to generate enough po power to drive the compressor to this point. And that point will be much higher boost pressure than with the ported shroud.

Figure 25 depicts the compressor aero efficiency, which is up to 20 points higher than the ported shroud. Again, the aero efficiency is associated with the pressure, mass flow, and temperature of the air delivered to the intercooler. The cold side of the engine sees the "aero efficiency" and the hot side of the engine sees the "shaft efficiency." At low engine speeds, the engine is receiving higher boost pressure at a lower temperature which results in up to 34% higher density air going into the intercooler. This could change the design point of the intercooler, resulting in lower maximum temperatures. It may also give the

Figure 22. Prototype test hardware.

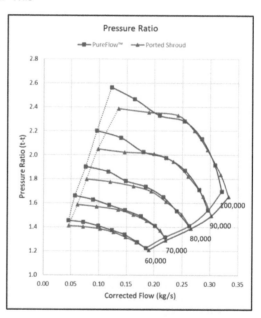

Figure 23. Comparison of pressure ratio/ mass flow test data.

vehicle manufacturer the ability to take the heat rejection savings and improve another stream of heat rejection in the vehicle by downsizing the intercooler. Figure 26 shows the temperature reduction possible in the intercooler inflow.

Figure 27 shows the flow split between the two inlets at 3 speeds. We can see that a significant portion of the mass flow at choke goes through the outer annulus.

This inlet system works with no changes to the compressor wheel—only a change to the inlet of the compressor. This can be incorporated into the design of the turbo or can be

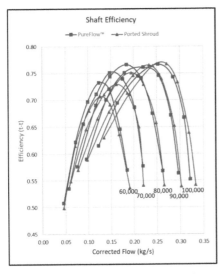

Figure 24. Comparison of shaft efficiency.

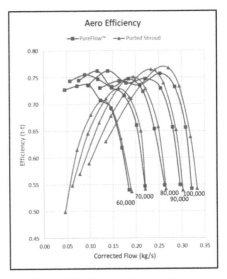

Figure 25. Comparison of aero efficiency.

incorporated into the inlet piping of the engine/vehicle. It completely transforms the performance on the left side of the compressor map—compressor pressure ratio, "aero efficiency", and compressor map width. What is the downside to this technology?

This embodiment of the technology has two inlets, one of which can have reverse flow. Many engines use closed crankcase ventilation, and the crankcase vapors are recycled into the engine in the compressor inlet. One can imagine that these vapors could be mixed into the reverse flow in one inlet. These vapors could collect on the air filter or possibly pass through the filter. Then there is the possibility of the emissions constituents being emitted to the atmosphere. If so, these emissions would need to be captured in emissions certification tests and added to the exhaust gas emissions. What are the countermeasures that could be applied? The simplest solution would be to introduce the CCV vapors into the center of the compressor inlet flow stream. These flow streams go right through the compressor without recirculating any flow. The only flow that is regurgitated is a concentric ring of flow, which proceeds from the outer boundary inwards as the compressor is moved from the middle of the map towards surge.

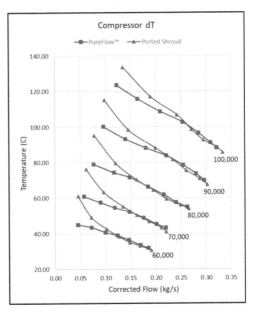

Figure 26. Comparison of compressor delta outlet temperature.

This also begs the question: do conventional compressors with or without a ported shroud, have reverse flow with CCV vapors, and do these vapors get recirculated back to the air cleaner? Figure 22 shows a dirty air filter from a diesel engine alongside a new clean one. Note that the picture is of the inside of the air filter, i.e., the "clean side".

If conventional compressors, with and without ported shroud, have recirculation extending to the air filter, then the PureFlow™ compressor could be a countermeasure. As mentioned before, constraining the CCV vapors to the central cone of air entering the compressor will assure that it does not become part of the air that is regurgitated and sent out the second air filter.

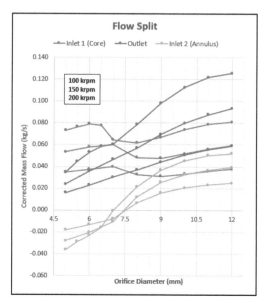

Figure 27. Flow split between two inlets, outlet flow.

Another aspect of the this compressor technology is that you will now have two inlet air streams, one of which can reverse flow. If the engine control system is using a mass air flow sensor upstream of the compressor in the fueling control, then there will have to be 2 sensors and a temperature sensor to measure the flow. The temperature sensor is a clear indicator of reverse flow. A much simpler method would be to simply move the mass air flow sensor to after the intercooler. This will be accurate, relatively cool, and will measure the flow going into the engine with virtually no lag. If a mass air flow sensor is used upstream of the compressor, then there is substantial lag before the flow reaches the intake manifold, as the flow must go through the piping and the intercooler.

Figure 28. Comparison engine side of dirty and clean air filter.

There are valid objections to the use of two inlets however, there appear to be relatively straightforward solutions. For those who want a single inlet solution, there is PureFlow-HX™.

5   PUREFLOW-HX™, A SINGLE INLET SOLUTION

The essence of this technology is to prevent the waste heat of a recirculating compressor from being recycled causing further performance deterioration from the loss of fluid density. The alternative to dumping the heat flow overboard is to cool the recirculated flow and return it to the compressor inlet. This design eliminates the inconvenience attributes of a two-inlet design. However, for this to work, there must be a low temperature coolant source, either liquid or air.

A new design was completed with only one inlet and incorporated a heat exchanger in the space of the inlet. Figure 29 is a cutaway of this design. An annular "shell and tube" heat exchanger was designed, and the end "tanks" were connected to the inlet and the outer annular passage leading to the compressor inlet. A CFD model was constructed and an analysis was done.

The analysis revealed that the heat exchanger had to have an extremely low pressure drop. The reason for this is that in normal operation with positive flow, 20% of the air flow goes through the outer annulus leading to the compressor, and thus through the heat exchanger. If the heat exchanger imposed a significant pressure drop, the performance of the compressor from the middle of the map to choke flow would suffer a flow loss with attendant efficiency

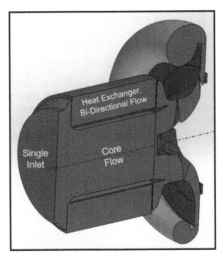

Figure 29. Initial model of Pure-Flow-HX™.

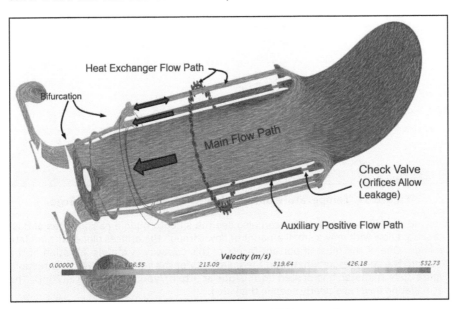

Figure 30. Velocity vectors of PureFlow-HX™ operating near surge.

loss. The heat exchanger was enlarged to reduce the pressure drop through it, but the unit became unwieldy and was deemed to not be commercially viable. Another solution had to be found.

Since the reverse flow rate near surge is so much smaller than the positive flow rate near choke, it was decided to incorporate two concentric passages. The inner passage was smooth and sized to accommodate the positive flow. The outer passage was a concentric heat exchanger sized for the reverse flow. A reverse flow valve was added to the smooth passage to prevent reverse flow. The valve would be like a reed valve used on HD EGR systems or two stroke engines. For CFD modeling, the passage was simply closed off and two small orifices were added to simulate the effect of leakage through the valves. The analysis was done on another turbo, slightly larger at ⌀56mm OD and also without a ported shroud.

Figure 30 is a cutaway section of the CFD analysis identifying the components and passages as well as showing the velocity vector field with the compressor operating near surge. One can clearly see the flow going through the center core passage, and the outer portion of the flow reversing at the bifurcation of the passage. The flow passes through the heat exchanger then reverses course again and merges smoothly

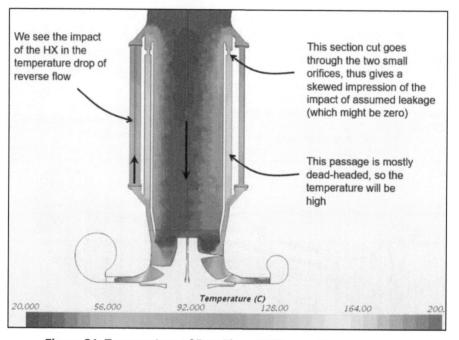

We see the impact of the HX in the temperature drop of reverse flow

This section cut goes through the two small orifices, thus gives a skewed impression of the impact of assumed leakage (which might be zero)

This passage is mostly dead-headed, so the temperature will be high

Temperature (C)					
20.000	56.000	92.000	128.00	164.00	200.

Figure 31. Temperature of PureFlow-HX™ operating near surge.

into the core flow passage. One can also see the smooth middle passage has almost stagnated flow with only a small amount of flow through the orifices placed to simulate leakage. One of the most important facets of this cutaway is the visualization of the flow into the compressor wheel and the uniformity of the velocity as well as direction. Figure 30 is animated and posted on YouTube at https://youtu.be/iuWHKHI7BhU. The recirculating flows are clearly seen in the video.

One of the objectives of the design is to provide low temperature air to the compressor wheel inlet as well as temperature uniformity. Figure 31 shows a cutaway of the temperature profile in a point near surge. Examining this profile, we can see that the temperature of the positive flow in the core passage is much more uniform.

Since we are near surge, we can see a larger 20C area on the right side compared to the left side. This is indicative of a rotating stall starting to emerge just before the full surge point is reached.

There is a consistent temperature drop through the annular heat exchanger. The flow exiting the heat exchanger is visible as it is slightly hotter than the core flow, and we can follow the flow down the core passage to the compressor inlet and the passage primary bifurcation. The smooth outer passage has elevated temperature since it is fed from the reverse flow, but since this is just leakage flow from the one-way valve, it is of little consequence.

Two speed lines were analyzed, 80,000rpm and 160,000rpm, for both the baseline compressor as well as the PureFlow-HX™ configuration. We see similar behavior as the basic Pure-Flow™ in Figure 32, where the elimination of the heat from the recirculated flow restores the pressure ratio on the left side of the map. In addition, the surge line is moved to the left as well.

Again, we see that there are two efficiency calculations that are germane to the analysis of the benefit of this technology—the "aero efficiency" which represents the conditions of the flow going to the intercooler, and "shaft efficiency" which represents the impact on the back pressure of the engine from the increased turbine power requirement.

Figure 33 depicts the aero efficiency and shaft efficiency for the PureFlow-HX™ and the efficiency of the baseline compressor for both 80,000rpm and 160,000rpm. We see that the Pure-Flow-HX™ raises the peak efficiency at both speeds by 2-3 points. The reason for this improvement is that the diffuser width was increased slightly. Normally, diffusers are narrowed to trade-off efficiency for range. With the PureFlow™ and PureFlow-HX™, there is no need to give this efficiency away for a better surge line.

The aero efficiency of the PureFlow-HX™ is more than 15 points better at .045kg/s but gives up nearly an equivalent amount in shaft efficiency. Of course, the pressure ratio is dramatically higher at 3.4 vs the baseline at 2.75 which will drive more mass flow to the engine.

The inlet and outlet temperatures of the compressor are dramatically different for the baseline vs the PureFlow-

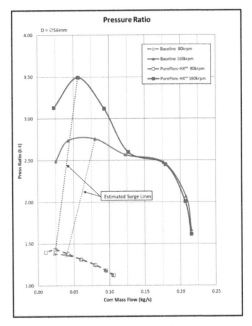

Figure 32. Comparison of pressure ratio/mass flow for PureFlow-HX™.

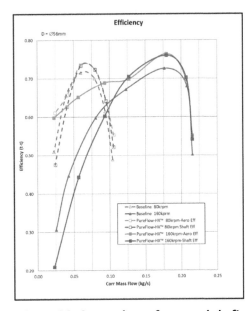

Figure 33. Comparison of aero and shaft efficiencies for Baseline and PureFlow-HX™.

HX™ near the surge line as is shown by Figure 34 and Figure 35. Surprisingly, the Baseline unit shows the inlet temperature rising for both speed lines clear back to the middle of the map. The pinched diffuser of the Baseline compressor is most likely the culprit and that inlet temperature rise is undoubtedly the reason the peak efficiency of the PureFlow-HX™ is higher than the Baseline. This can also be seen in Figure 35 in the choke flow temperatures. The Baseline temperature is 18C higher than the PureFlow-HX™. The outlet temperature is not reduced as much as would be expected from the inlet temperature reduction.

But the reduced inlet temperature of the PureFlow-HX™ should have a positive impact on the aluminum compressor wheel low cycle fatigue life. With the PureFlow-HX™, the cool inlet temperature flowing through the inducer blading will provide exceptional cooling for the compressor wheel, reducing the temperature at the core high-stress area, the compressor wheel hub.

Figure 36 shows that nothing is free. The additional boost pressure provided near surge drives up the required turbine power as one would expect.

It should be pointed out that if the engine manufacturer does not want to provide additional boost/mass flow, then the turbo speed can be reduced, the driving power requirement reduced commensurately. Depending on what type of turbine (fixed, wastegate, variable geometry) is used would determine how that would be implemented.

The PureFlow-HX™ will need to reject the heat pulled out of the compressor recirculation flow. Figure 37 shows the heat rejection required for this ⌀56mm compres-

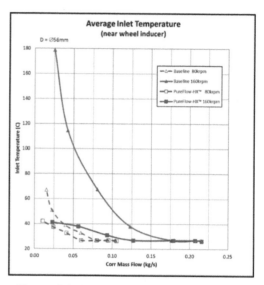

Figure 34. Comparison of inlet temperature of Baseline and PureFlow-HX™.

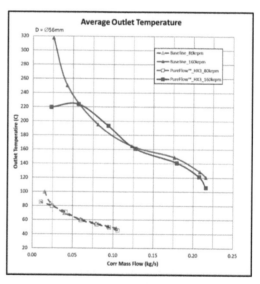

Figure 35. Comparison of outlet temperature of Baseline and PureFlow-HX™.

sor at 80krpm and 160krpm. The engine will probably not require the full heat rejection shown as it is doubtful an engine would be driven clear to surge. Also note that his is a steady state number and depending on the type of application the engine is designed for will affect the duty cycle. For many engine types, operation in this region will only be a few seconds. There would be a significant thermal capacitance in the cooler which would absorb much of the transient load.

Also note that if that heat were not removed by the heat exchanger, that thermal load would have to be removed by the charge air cooler, and it would be dramatically increased by the inefficiency of the compressor. The outlet temperature could be 320C, which is beyond the capabilities of most aluminum charge air coolers.

6   SUMMARY/CONCLUSIONS

The subject technology is a concentrically bifurcated compressor inlet that allows hot recirculated flow to be expelled from one inlet to improve the performance of the compressor and its effect on the engine. The PureFlow-HX compressor is a single inlet system with a heat exchanger to take the excess heat out of flow being recirculated. It also has a one-way valve to open a passage to bypass the heat exchanger when all flow is going to the compressor. These technologies are somewhat difficult to assess since they have two "efficiencies" depending on whether one is assessing the impact on the cold side or the hot side of the engine.

What is clear is that they both remove the heat of a compressor loss mechanism that occurs at high torque, low engine speed. This has a significant impact on the intake manifold pressure and temperature, and thus elimination of this loss will increase intake manifold density, the primary function of the turbocharger. Both technologies also reduce the thermal load on the charge air cooler at low engine speed, high load by way of a lower temperature at the highest pressure the intercooler will experience.

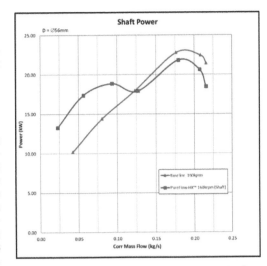

Figure 36. Comparison of baseline shaft power and PureFlow-HX™.

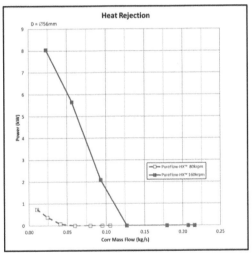

Figure 37. Heat rejection of PureFlow-HX™ at two speeds.

The increase of boost pressure near the surge line results in higher turbine power as physics would predict, thus the extra mass flow through the engine is not entirely "free". It does broaden the operating envelope of the turbo and redefines what is possible when calibrating an engine at low engine speed and high torque.

The patented technology may be implemented by the engine manufacturer or the turbocharger manufacturer depending on the preferences of the engine manufacturer. This technology could also be used to retrofit existing vehicles to improve torque at low speed, improve BSFC at low engine speed, and improve reliability of the intercooler.

Further information, including animations of flow within the PureFlow-HX™ compressor can be found at https://www.esi-inc.us/PureFlow-compressor.

CONTACT INFORMATION

Steve Arnold

http://steve@esi-inc.us

http://www.esi-inc.us

+1 (310) 990-3163

ACKNOWLEDGMENTS

The author would like to acknowledge the assistance of Garret Wiebe and Chris Scremin of Rotomaster for preparing and testing the prototype unit.

The author would like to acknowledge the assistance of Leon Hu, James Yi, and Eric Curtis in correcting the initial aero efficiency calculations and Dr. Apostolos Pesyridis for reviewing the paper.

DEFINITIONS/ABBREVIATIONS

BMEP	Brake Mean Effective Pressure
CAD	Computer Aided Design
CCV	Closed Crankcase Ventilation
CFD	Computational Fluid Dynamics
c_p	Constant Pressure Specific Heat
EGR	Exhaust Gas Recirculation
HD	Heavy Duty
HX	Heat Exchanger
k	Ratio of specific Heats
	Mass flow
N	Speed
T	Temperature
TQ	Torque
η_a	Aero Efficiency (calculated from inlet and outlet temperatures and pressures
η_s	Shaft Efficiency (calculated from torque, speed, isentropic work)
η_t	Total Efficiency (calculated from enthalpies of both inlets and outlet)
π	Pressure Ratio
subscript **1**	Station 1—Main Inlet
subscript **2**	Station 2—Second (outer) inlet
subscript **3**	Station 3–Outlet

REFERENCES

[1] Watson, N. and Janota M.S., "Turbocharging the Internal Combustion Engine," (MacMillan Press Ltd Publication, 1982), p. 152, ISBN 0 471-87082-2.
[2] Fisher, F., Langdon, P, 1985, *Improvements in and relating to compressors*, USP 0229519B2.
[3] Hu, L, Sun, H., Krivizhy, E., Larosiliere, L., Zhang, J., Lai, M., "Experimental and Computational Analysis of Impact of Self Recirculation Casing Treatment on Turbocharger Compressor", SAE Technical Paper 2010-01-1224, 4/12/2010. 10.4271/2010-01-1224.
[4] Dehner, R., Selamet, A., Steiger, M., Miazgowicz, K., Karim, A., "The Effect of Ported Shroud Recirculating Casing Treatment on Turbocharger Centrifugal Compressor Acoustics", SAE Journal Article 2017-01-1796, June 5, 2017. 10.4271/2017-01-1796.
[5] Arnold, S., Calta D., 2013, *Ported Shroud with filtered external ventilation* USP 8,511,083.
[6] Arnold, S., 2020, *Inlet system for a radial compressor with a wide flow range requirement*, USP 10,662,949.

Appendix

Table A-1

Analysis Framework	
Tool	StarCCM+ 12.04.01
Method	Frozen Rotor
Inlet	Stagnation Inlet
Outlet	Simulated Gas Stand–atmosphere
Flow Control	Orifice Simulating Gas Stand Valve
Geometry	imported fluid solids–binary parasolid
Mesh Process	Split-by-Patch, group into Boundaries
Mesh Control	Customize per Boundary
Regions/Boundary	Recirculation Inlet
	Interface-recirculation
	Plenum
	Ribs
	Struts
	Rotating
	Backdisk
	Backplate
	Blades
	Wheel Shroud

(Continued)

Table A-1. (*Continued*)

	Diffuser
	Wheel Fillets
	Hub
	Inlet Interface
	Recirculation Inlet Interface
	Volute Interface
	Leading Edges
	Housing Shroud
	Volute
	Scroll
	Tongue
	Diffuser Interface
	Orifice Interface
	Orifice
	Cylinder
	Volute Interface
	Outlet

Table A-2

Regions	cells
Inlet_Main	1,217,026
Inlet_recirc	3,133,063
Orifice	39,749
Rotating	6,250,831
Volute	775,938
Total	11,416,607
Interfaces	
Inlet_main	Internal Interface
Inlet_Recirc	Internal Interface
Interface_volute	Internal Interface
Orifice	Internal Interface

(*Continued*)

Table A-2. (*Continued*)

Solvers	
Partitioning	
Wall Distance	
Segregarted Flow	
Velocity	0.6
Pressure	0.25
Segregated Energy	
Fuild Under-Relaxation Factor	0.9
AMG Linear Solver	
Max Cycles	30
K-Epsilon Turbulence	
Under-Relaxation Factor	0.8
K-Epsilon Turbulent Viscosity	
Under-Relaxation Factor	1
Reference Frames	
Lab Reference Frame	
Rotating	
Axis Direction	[0.0, 0.0, 1.0]
Axis Origin	[0.0, 0.0, 0.0] m
Rotation Rate	100000.0 rpm
Stopping Criteria	
00-Efficiency – tt Monitor Criterion	
Std Deviation	5.00E-04
Number of Samples	300
01- phy_Mass_Flow Monitor Criterion	
Std Deviation	0.005 kg/s
Number of Samples	300
Maximum Steps	1000

Optimization of a centrifugal turbine rotor for robustness and reliability against manufacturing tolerances

M.H. Aissa[1], T. Verstraete[1], A. Racca[2], C.E. Williams[3]

[1]Von Karman Institute for Fluid Dynamics – Sint-Genesius-Rode, Belgium
[2]PUNCH Torino S.p.A. – Torino, Italy
[3]General Motors Global Propulsion Systems – Warren (MI), USA

ABSTRACT

The classic approach in turbomachinery design optimization considers only nominal geometries while manufacturing tolerances are evaluated in post-processing. Without any knowledge about such deviations, the optimizer chooses the solution corresponding to the highest attainable performance. However, such a shape may require tight and expensive tolerances to maintain the expected performance in large-scale populations or even necessitate tolerances not realizable with the available manufacturing process. Therefore, the entire optimization activity would need to be re-run to look for a more robust solution.

In contrast, Uncertainty Quantification (UQ) methods take into account the tolerances applicable to every design parameter and propagate those uncertainties to the output. When applied in combination with an optimizer, performance data along with information related to the robustness and reliability of every geometry become available, thus allowing to find optimal solutions that are less sensitive to the design parameters deviations.

In this work, we present the development of an efficient evaluation framework through the application of the Probabilistic Collocation method, a sampling-based UQ technique generating high-fidelity statistical data of the uncertainties impact on the output solution, coupled with the Smolyak algorithm to reduce the sampling size. A metamodel-assisted approach steers the optimization using surrogate modeling to rapidly approximate the statistical data using the Monte-Carlo method. The UQ model is invoked only when the optimizer approaches convergence in order to confirm the robustness of the identified optimum or to lead the optimization to a more robust region of the design space.

The methodology is applied to the robust optimization of a radial turbine rotor with 32 design variables, nine of which were affected by uncertainties. The proposed approach provides a more robust and reliable design against geometrical uncertainties, in addition to the sensitivity analysis, with an estimated overhead below 30% of the cost of a traditional design optimization.

Keywords: Multidisciplinary Optimization, Uncertainty Quantification, Robust Design, Manufacturing Tolerances, Radial Turbine

1   INTRODUCTION

The design optimization of turbocharger components is a demanding task requiring the adoption of suitable multidisciplinary optimization methods [1]. The standard approach to identify an optimal solution is through the analysis and evolution of the nominal geometry of the component of interest, till saturating the exploration of the design space. Once the final shape is available, it is common practice to perform rigorous verifications of the geometry in order to account for the impact of manufacturing deviations typically occurring in large scale populations.

The uncertainties introduced by the manufacturing processes might alter the original shape returned by the optimization and, therefore, possibly infringe some project constraints. In those cases, the unfeasible solution must be discarded, and a new optimization process, accounting for more severe margins, should take place in the search for a new optimal design.

It is evident that such a procedure may result in an increased lead-time. Therefore, the aim of the following paper is to present a newly developed methodology that accounts for manufacturing deviations since the early stages of the optimization process, in order to return an optimal yet robust solution in a unique execution. Moreover, large attention is spent in the identification of a technique capable of fulfilling pre-determined limits in terms of computational budget, aiming at attenuating a well-known pain-point in the evaluations of uncertainties by means of classical methods, and guaranteeing the suitability of the process in an industrial framework.

The test case of a radial turbine impeller for turbocharger applications is adopted in this paper with the scope of proving the methodology within a multidisciplinary and multipoint optimization problem presented in Section 4. In particular, a demanding rotor architecture was considered in this study, with a high number of blades potentially inducing risks of infringement of the structural constraints by manufacturing deviations because of the limited space available in the hub region.

In order to thoroughly discuss the proposed methodology, an introduction to the problem of uncertainties evaluation is herein offered.

1.1   Uncertainty Quantification methods

In Uncertainty Quantification (UQ), all process variables might be considered uncertain variables with an associated Probability Density Function (PDF). If the PDF of the input variables are known, then UQ methodologies allow to compute the PDF of the output variables. For example, consider the process that computes the outflow angle for a given blade metal angle distribution. If the input, in this case the outlet blade metal angle, would be an uncertain input variable with a known PDF, then UQ is the process that computes the PDF of the outlet flow angle. This allows, for instance, to compute the mean and variance of the outflow angle, and to verify if, under the given uncertainty, the design is acceptable.

There are two major challenges in the application of UQ methods in design optimization. The first one is about the identification of the input uncertain parameters and their PDF function. In an ideal case, a database with a large set of raw data should be available a priori in order to generate an approximation of the PDF. In most cases, the uncertainty on the input variable follows from the manufacturing process and requires detailed measurements of a large set of manufactured products. Hence, this data is very costly to acquire and quite often not available. In that case, some assumptions can be taken, given the variation interval (min/max) and the mean value. In general, the input PDF is uniform, normal, and sometimes following a Beta distribution [2].

The second challenge is about propagating the input uncertainty to the output quantity of interest (QoI). For instance, how would a modification of the inlet blade angle influence the efficiency? The Monte-Carlo simulation [3], a sampling-based intuitive method, evaluates a large number of uncertain input parameters chosen according to the input PDF. The histogram of the output QoI describes the output PDF. Once an output PDF is available, the designer could check if the interval of +/- 3σ fits within the initial tolerances. The effect of the output PDF's head and tail is known as a rare event (black swan) and requires a high-quality PDF involving the skewness and kurtosis in addition to the mean and variance [4]. The amplitude of an output PDF can be a measure of the influence of the input variations. A larger deviation correlates with a higher influence.

1.2   Optimizations with uncertainties

Traditional design optimization methods are fully deterministic and do not consider the effect of manufacturing tolerances nor the fluctuations of the operating conditions. One of the more commonly used deterministic methods, namely the Metamodel-assisted multidisciplinary design optimization, builds a Design of Experiment (DOE) at first, which consists of a cloud of designs around a baseline sample. The DOE serves as a database to train a fast-response model, which will be used in a second moment to explore the design space, replacing the time-consuming high-fidelity simulations. Afterwards, a high-fidelity simulation evaluates the most prominent design provided by the metamodel and adds it to the database [5].

Robust optimization, on the other hand, considers the effect of a variation on the input parameters by using an uncertainty propagation method. The latter should evaluate the effect of the inlet uncertainty on the output, such as the effect of a slight deviation on the blade thickness (caused by the manufacturing tolerances) on the efficiency of a turbine.

A body of methods is available in the literature to propagate the uncertainty [3, 2]. Some of them are "non-intrusive", such as Monte-Carlo simulations (MC), while others are "intrusive". An intrusive method, such as the perturbation method, expands the random variables via Taylor series around their mean and truncates them at a certain order changing the governing Partial Differential Equations [3]. The issue with those methods is that they involve potentially error-prone modifications to the source code of the solver. Moreover, in most of the cases, the source code might not be even available to the user.

In the family of "non-intrusive" methods, some techniques are based on the MC method, such as the multi-level MC. Other algorithms, instead, are surrogate-based involving a low-fidelity model for the determination of the statistical quantities of interest (mean, deviation, skewness, etc.). In this case, the surrogate models may be represented by the Collocation method or the Polynomial Chaos [3,6]. These techniques, in their original formulation, require a large number of samples to operate statistical evaluations, consequently driving to a prohibitive cost in case of time-consuming evaluations (such as CFD simulations). Moreover, for the sampling-based non-intrusive methods, the computational cost is also proportional to the accuracy of the method. For example, the accurate evaluation of the variance is more sampling-intensive than the estimation of the mean. Similarly, the accurate evaluation of a PDF would require the computation of skewness and kurtosis, which require additional samples. Seshadri et al. [7] used 3000 CFD evaluations for a robust optimization of a highly loaded, transonic compressor rotor blade. Wang et al. [8] did a MC of more than 4000 evaluations to quantify the impact of geometric variations on turbine blade performance. Such a computational budget might be applicable for standalone design purposes, but it would not be suitable for iterative routines, like within the framework of gradient-free optimizations, which are known to require a large number of geometrical evaluations [5].

The present work studies the effect of the manufacturing variability on the multi-point design optimization problem of a radial turbine impeller for turbocharger applications. While standard optimizations rely on deterministic simulations (e.g. CFD, CSM) for the evaluation of different designs, robust optimizations add an uncertainty measure to each computation, such as the mean and the variance.

Some optimization algorithms search for a design with an optimal mean value as a first objective while minimizing the variance as a second objective of the optimization. The purpose is to maximize the system performance while minimizing the part-to-part dispersion. In this work, instead, we optimize a combination of the mean and the variance (defined as mean +/- 3*standard deviation). This formulation does not discard designs with large variance unless the mean is not optimal; instead, the method penalizes those geometries that have a certain probability in the tail of their PDF's of not satisfying the constraints.

The literature provides indications for many different types of uncertainties: the operational uncertainties and manufacturing uncertainties belong to the most addressed ones. In this work, the operational uncertainties, for instance represented by a variance on the boundary conditions, is not considered as it is included in the definition of the multi-point optimization strategy as described in Section 4. In fact, if a design is not robust to small deviations around the nominal condition of an operating point, such assessment will be indirectly provided by the expected performance degradation in the other off-design key points. Such evaluation is possible without invoking operational uncertainties because the strategy herein adopted for the execution of the multipoint optimization includes a large span in the operating range, with points covering the whole extension of the turbine map. For this reason, the present work will focus only on manufacturing uncertainties, resulting from the deviations w.r.t. the nominal shape induced by the expected manufacturing tolerances.

1.3   Computational cost

The inclusion of the uncertainties in the evaluation of the design problem increases the computational cost compared to a deterministic simulation. Using a very efficient uncertainty quantification method for 10 uncertain parameters could be 20 times more time-consuming than the standard evaluation process [9]. In an optimization framework driven by gradient-free methods, an overhead increase of 20x makes the direct integration of UQ methods not affordable.

Limiting the number of uncertain parameters would reduce the computational cost, but not to a sufficient extent to run a Design of Experiment (DOE) within a limited budget, in case of metamodel-assisted optimizations [10]. In order to maintain a low number of uncertain parameters, some authors use statistical analysis (e.g. Spearman rank correlation [11]) to filter out highly correlated parameters and keep the most independent ones. A correlation between the parameters can also be considered. For instance, Principal Component Analysis (PCA) [8, 12] can analyze a random field reflecting a diaspora of geometries, and only important modes would be used as uncertain parameters. In the present work, nor SA or PCA could be adopted because all the selected parameters demonstrate to be uncorrelated, and therefore varying independently from each other.

As this work is driven by the requirement of keeping the overhead of the robust optimization below 1.5x w.r.t. a traditional optimization, the developed method involves a standard fractional factorial DOE, as used in deterministic optimizations.

443

During the optimization, the variance will be estimated by a Monte-Carlo simulation, which uses a Kriging model [13,14]. Kriging is a fast-to-train accurate interpolation method used to approximate computer models [15]. The process runs similarly to a deterministic optimization until it stagnates by converging to the optimal solution. Afterwards, a Probabilistic Collocation method is applied to the selected geometry in order to evaluate its mean and variance. If this high fidelity UQ method confirms the previous Kriging estimation, the optimization finishes and exits the workflow. Otherwise, the performance data of the designs evaluated by the Collocation method is added to the database, and as such, drives the optimization to explore different parts of the design space.

The Collocation method, herein considered as a high-fidelity source of mean and variance estimates, could be computationally prohibitive. In this respect, the Smolyak algorithm is adopted to reduce its computational cost [16]. Even though the overhead requirement of 1.5x led to a very simplistic approach, the results show that the proposed method is able to reduce the variance, consequently finding an optimal design.

In light of this objective, the remaining portion of this document is structured as follows. First, the methodology developed in this work is presented. Then the design problem is discussed, along with the augmented optimization strategy and the selection of the design variables affected by manufacturing uncertainties. Finally, the analysis of the results concerning the application of the Uncertainty Quantification method to the optimization is presented, followed by an overview of additional sensitivity evaluations available in the post-processing phase.

2   METHODOLOGY

The robust optimization follows the algorithm shown in Figure 1. First, a standard DOE explores the design space and builds the initial database. Then, a Kriging model, called "Kriging 1" in Figure 1, is trained to approximate the deterministic performance parameters, such as efficiency, choking mass flow rate, etc. The algorithm then uses the Kriging model to run a Monte-Carlo (MC) aimed at calculating the mean and the variance of the relevant performance parameters for each sample in the DOE. Hence, the database initially generated is invoked for the training of two additional models, used to predict the mean and the variance of the new designs. Those two models, named "Kriging 2" and "Kriging 3" in Figure 1, are built on top of a metamodel. Therefore, the quality of their prediction depends on the maturity of the surrogate model, with expected low accuracy at the beginning of the optimization process. Since the system is self-learning [17], the quality of all Kriging models increases with the expansion of the database along the optimization. As a next step in the workflow, the optimizer has access to low-fidelity values of mean and variance for all the individuals in the design space, and a traditional gradient-free optimization loop can take place, with a modified objective and constraints formulation as indicated in Eq. 4 and Eq. 5.

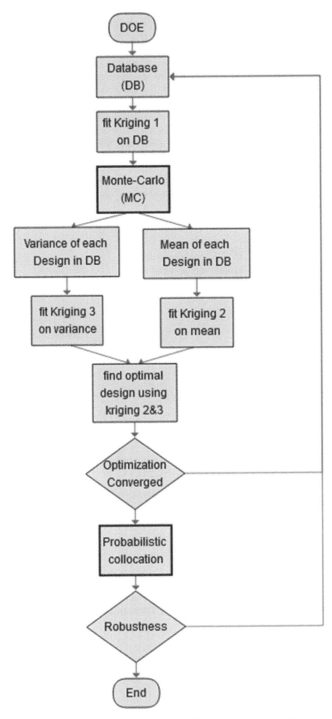

Figure 1. Proposed robust optimization algorithm.

To further enhance the accuracy of the robust optimization method, an additional feature is implemented, which ensures that the UQ results at convergence are reliable. The optimization progresses first, with the low-fidelity statistical data until it converges to the optimum. Afterwards, a higher fidelity UQ module (such as the Probabilistic Collocation method indicated in Figure 1) verifies the mean and the variance values estimated by the Kriging model for the optimal geometry. If the UQ module confirms the Kriging-based statistical data, the robust optimization is successfully concluded. Otherwise, the samples used for the high-fidelity UQ validation are added to the database, and the simulation restarts from the first step in the workflow. In this respect, Figure 2 qualitatively shows the continuation of the optimization after the first call to the UQ module. Thanks to the additional information just injected into the database in a relevant region of the design space, the accuracy of the Kriging model prediction increases. Therefore, the second call to the optimizer is deemed more expedite in the identification of the robust optimum, satisfying the maximum computational budget previously indicated.

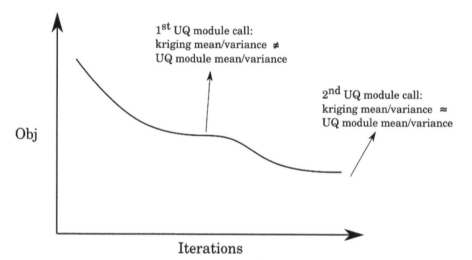

Figure 2. Illustrative example of the use of the UQ module within the robust optimization. Convergence occurs after 2nd UQ module call.

The algorithm of the proposed robust optimization, shown in Figure 1, is close to a traditional deterministic optimization. There are only two added functionalities, namely the Monte-Carlo simulation and the Probabilistic Collocation (PC). The latter defines a random process as follows:

$$u(x,t,\omega) = \sum_{i=1}^{Np} u_i(x,t)h_i(\xi(\omega)),$$

(1)

with $u_i(x,t)$ indicating the deterministic solution at the collocation point, ω as a function of space and time, and h_i as the corresponding Lagrange interpolation polynomial. The collocation points and their weights are set by the Golub-Welsch algorithm

[18,19]. The PDF influences the position of the selected points for the Probabilistic Collocation.

The mean and the variance are then calculated as follows:

$$\mu = \sum_{i=1}^{N_p} u_i(x,t)\omega_i, \tag{2}$$

$$\sigma = \sum_{i=1}^{N_p} u_i(x,t)^2\omega_i - \mu, \tag{3}$$

While for one random variable, a PC of level 1 needs three samples and a PC of level 2 requires five samples, in case of a problem exhibiting five random variables, the method would require 243 samples for the level 1 and 3125 samples for the level 2 [9]. Therefore, it is evident that PC would be less suitable for robust optimizations because of the significant increase in the computational overhead unless it is coupled with some sparse sampling methods.

The Smolyak algorithm is one of those methods. While for each dimension we conserve the number of evaluations (3 for level 1 and 5 for level 2), in the case of five random variables the method reduces the number of samples to 11 for level 1 and 71 for level 2. The combined cases are indeed not all covered, depending on the level adopted by the Smolyak algorithm. The higher the level, the more combinations are covered, the closer is the prediction to the full sampling PC one, but the higher the computational budget. For the present study, a level 1 is chosen, which is the lightest option covering no combinations.

3   BASELINE DESIGN PROCESS

Figure 3 shows an example of a Variable Geometry Turbine layout along with the baseline design of a periodic sector of the radial inflow turbine impeller considered in the present study. This preliminary geometry was generated using analysis tools developed at the von Karman Institute for Fluid Dynamics (VKI) with a special focus on turbomachinery applications [20, 21].

The extent of the baseline design investigation was limited to a minimum level, in order to gain sufficient confidence with the classes of design variables demonstrating to be more relevant for the optimization process. The resulting preliminary loading coefficient is about $\psi = 1.16$ at design point, with an exit flow coefficient $\phi = 0.55$. Although the operative conditions for this application lead to a highly loaded impeller blade, the initial results indicate the need for an optimization step in order to increase the total-to-static efficiency.

The choice of this specific test case is justified by the goal of proving the optimization procedure against a highly constrained problem, in which stretched performance targets are demanded while complying with manufacturing and mechanical constraints in the limited space available at the hub surface of the rotor. With regard to this, Figure 3 shows the proximity of the blade hub fillet to the periodic surfaces, highlighting the tough problem expected in satisfying the targets of maximum von Mises stresses and blade first eigenfrequency, as well as guaranteeing the fulfilment of such constraints with respect to manufacturing uncertainties in large scale populations. Therefore, this paper aims at describing a novel approach conceiving the application of the in-house developed multidisciplinary optimization framework to the baseline rotor geometry.

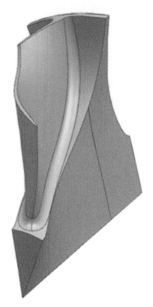

Figure 3. Exemplary Variable Geometry Turbine layout (left) and periodic sector of the radial inflow turbine wheel (right).

In total, 32 degrees of freedom were identified for the optimization process, nine of which are affected by uncertainties. In particular, the design features relevant for the blade thickness distributions at hub and shroud, and its shape in proximity of the trailing edge are referenced, among the others. Only symmetric tolerances were considered for such optimization variables. Additional details about the optimization problem are reported in section 4.

4   OPTIMIZATION PROBLEM

The design optimization aims to maximize the total-to-static isentropic efficiency of the Variable Geometry Turbine rotor over a wide range of operative conditions while minimizing the moment of inertia. Additionally, a set of constraints are imposed, including manufacturing deviations, to achieve a robust optimal design.

A multipoint approach accounts for three key points (namely OP1, OP2 and OP3, whose characteristics are reported in Table 1) relevant for the engine operations. Two points show closed vanes positions, as a demonstration of the intent of optimizing the rotor for high inlet swirl. The third operative condition is chosen in order to set the characteristics of the machine in the peak power region of the engine map.

Table 1. Turbine operative conditions.

	VANES POSITION	TURBINE SPEED	MASS FLOW RATE	EXPANSION RATIO
OP1	2deg from maximum closure	low	low	low
OP2	6deg from maximum closure	high	mid	high
OP3	mid-high opening	high	high	high

All the points operate at different rotational speeds, with imposed inlet mass flow rate in OP1 and OP2 and varying turbine back-pressure. The third point is exposed to a fixed expansion ratio in order to keep the choking mass flow rate above a given threshold which preserves the operating range of the baseline. Additionally, at each specific operative condition, the impeller should deliver a power output above the pre-defined minimum requirements. Finally, in order to maintain the structural integrity of the wheel, the maximum von Mises stresses and the first eigenfrequency are constrained.

The resulting multidisciplinary optimization problem can be written as follows:

Maximize

$$
Obj \equiv \omega_{Obj} \underbrace{\left(\omega_1 \frac{\mu_{\eta_{ts,OP1}} - 3\sigma_{\eta_{ts,OP1}}}{\eta_{ts,OP1,ref}} + \omega_2 \frac{\mu_{\eta_{ts,OP2}} - 3\sigma_{\eta_{ts,OP2}}}{\eta_{ts,OP2,ref}} \right)}_{Obj_{Aero}} + (1 - \omega_{Obj}) \underbrace{\left(1 - \frac{\mu_{I_{xx}} + 3\sigma_{I_{xx}}}{I_{xx,ref}} \right)}_{Obj_{Inertia}}, \quad (4)
$$

Subjected to

$$
Constr_1 \equiv 1 - \frac{\mu_{P_{OP1}} - 3\sigma_{P_{OP1}}}{P_{OP1,ref}} \leq 0, \quad Constr_2 \equiv 1 - \frac{\mu_{P_{OP2}} - 3\sigma_{P_{OP2}}}{P_{OP2,ref}} \leq 0,
$$

$$
Constr_3 \equiv 1 - \frac{\mu_{P_{OP3}} - 3\sigma_{P_{OP3}}}{P_{OP3,ref}} \leq 0, \quad Constr_4 \equiv 1 - \frac{\mu_{\dot{m}_{OP3}} - 3\sigma_{\dot{m}_{OP3}}}{\dot{m}_{OP3,ref}} \leq 0, \quad (5)
$$

$$
Constr_5 \equiv \frac{\mu_{\sigma_{vM,max}} + 3\sigma_{\sigma_{vM,max}}}{\sigma_{vM,max,ref}} - 1 \leq 0, \quad Constr_6 \equiv 1 - \frac{\mu_{f_{1st}} - 3\sigma_{f_{1st}}}{f_{1st,ref}} \leq 0.
$$

In these expressions, μ is the mean, σ is the standard deviation, η_{ts} is the total-to-static isentropic efficiency, I_{xx} the moment of inertia with respect to the axis of rotation of the wheel, $\dot{m}$ the choking mass flow rate, $\sigma_{vM,max}$ the maximum von Mises stress, and f_{1st} the first eigenfrequency. The weighting coefficient $\omega_{obj} \in [0, 1]$ in the objective function (Eq. 4) is a user-defined blending factor between the efficiency and moment of inertia terms. Additionally, the user-defined coefficients ω_1 and ω_2 determine the weights given to the efficiencies respectively at operating point OP1 and OP2.

5 DESIGN AND RANDOM VARIABLES

In turbomachinery, it is a common practice to use Bézier control points as design variables for continuous functions. For instance, Bézier curves define the meridional contours and store the thickness distribution of a radial turbine impeller. In this work, the thickness has a tolerance of +/- T mm. A PDF characterizes the random control points for the thickness distribution at hub and shroud, as shown in Figure 4. The Monte-Carlo procedure can evaluate different values for these control points following the given PDF. Therefore, it is crucial to find a suitable range of variation aimed at generating the required tolerance of +/- T mm.

The Bezier curve is represented by the following equation:

$$C(u) = \sum_{i=0}^{n} B_{n,i}(u) P_i,$$

(6)

with B_n as the Bernstein polynomials, and P_i the Bezier control points.

The deviations in the control points positions represent the required tolerance on the real curve, divided by the Bernstein polynomial evaluated at that position. If all the control points of a Bezier curve are defined by random variables and $\sum_{i=0}^{n} B_{n,i} = 1$, because of the Bernstein's property, then the change of each Bezier control point would lead to $\delta P = T$. In other words, when all the control points are moved, each deviation should not exceed +/- T in order to limit the accumulated curve deviation to the desired tolerance.

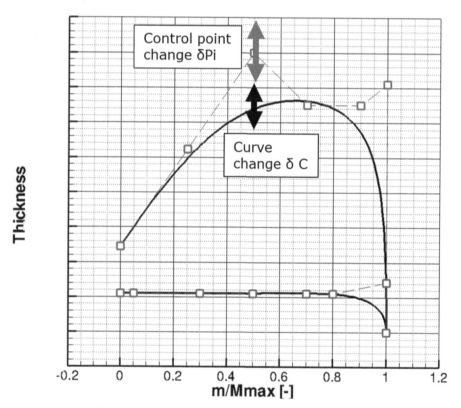

Figure 4. Effect of a change on one control point on the Bezier curve.

Table 2 presents the nine random variables considered in this study: one concerns the backplate and eight of them are related to the impeller. For the blade, six random variables refer to the thickness at hub and shroud, while two variables are related to the trailing edge (TE), as shown in Figure 5. All the random variables are associated with a normal probability distribution and symmetric tolerances of +/- T (trimmed to the values corresponding to +/- 3 σ).

Table 2. List of the nine random variables of the robust optimization.

	Random variable	tolerances	PDF
1	Shroud TE radius	Symmetric (+/- T)	Normal
2	Trailing Edge Sweep	Symmetric (+/- T)	Normal
3	Hub Meridional Point 3 Y	Symmetric (+/- T)	Normal
4	Blade Thickness Control Point1 (at Hub)	Symmetric (+/- T)	Normal
5	Blade Thickness Control Point3 (at Hub)	Symmetric (+/- T)	Normal
6	Blade Thickness Control Point4 (at Hub)	Symmetric (+/- T)	Normal
7	Blade Thickness Control Point5 (at Hub)	Symmetric (+/- T)	Normal
8	Blade Thickness Control Point1 (at Shroud)	Symmetric (+/- T)	Normal
9	Backplate Thickness	Symmetric (+/- T)	Normal

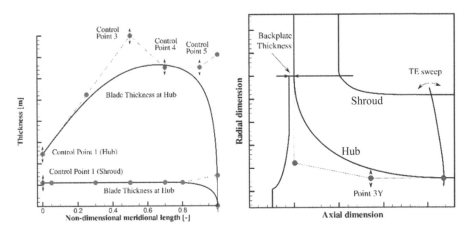

Figure 5. Parametrization of the radial turbine showing the random variables from Table 2.

6   RESULTS

6.1   Importance of the UQ

To assess the importance of considering the Uncertainty Quantification for design optimization purposes, we analyze the effect on the von Mises stresses induced by the variation of the hub fillet radius within the manufacturing tolerances. This design variable describes the blade hub contour, which is exposed to a peak in the structural stresses, as shown in Figure 6.

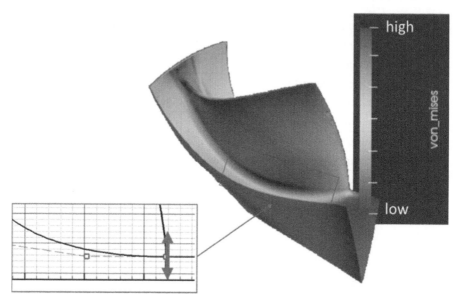

Figure 6. Relation of the hub contour radius with the peak structural stress at blade roots.

Table 3. Von Mises stresses variation at the change of the hub fillet radius.

Hub radius	Peak of von Mises stresses
R_{Ref}	σ_{Ref}
$R_{Ref} + 3\sigma$	$\sigma_{Ref} - 0.8\%$
$R_{Ref} - 3\sigma$	$\sigma_{Ref} +2.1\%$

Table 3 shows that a decrease in the hub fillet radius by 3 times the standard deviation σ would correspond to an increase of the maximum von Mises stresses by 2.1%. Such a high value may turn an originally optimal design into an unfeasible one because not fulfilling the structural constraint. Therefore, this finding confirms the need of possibly surpassing the concept of deterministic optimizations, in favor of a new framework acknowledged of the manufacturing uncertainties.

6.2 Convergence of the robust optimization

The robust optimization following the algorithm shown in Figure 1 is executed for 184 iterations by using only the Monte-Carlo process to compute the mean and the variance. The convergence plot in Figure 7 shows the evolution of the von Mises stresses along the first 72 iterations, tracking both the low-fidelity metamodel and the high-fidelity solutions for the best design identified at each optimization loop. In particular, a standard behavior is shown in the initial learning phase until iteration 35, with a closing gap between the predicted and the simulated values. Starting from iteration 45, the optimizer is capable of reducing the von Mises stresses below the desired threshold. Moreover, the decrease continues further, such that the stresses are lowered at a sufficient level to satisfy also the margin of -3σ w.r.t. the threshold (as mentioned in Section 6.1). Thus, the resulting failure probability is attested around 0.17% assuming a normal probability distribution.

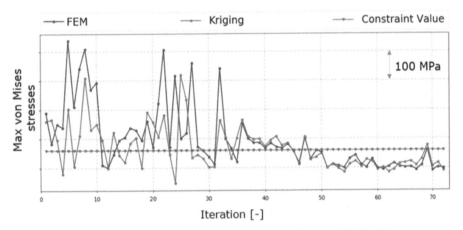

Figure 7. Evolution of von Mises stresses along optimization.

Figure 8 shows a snapshot from the Monte Carlo process on the evolution of the database of the turbine along the optimization. The cluster contains the DOE samples, presenting high standard deviation, and the samples generated by the optimization, following a descending trend. Therefore, it is proven that the optimizer successfully reduces the variance during the optimization compared to the original DOE designs.

453

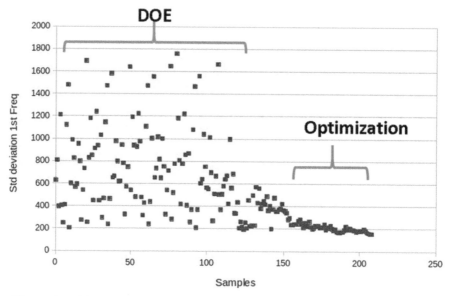

**Figure 8. MC based evaluation of the standard deviation for samples origin-
ated by a DOE and by the optimization iterations.**

Starting from Iteration 184, the Probabilistic Collocation module is called in 4 subsequent iterations, as shown in Figure 9 in the case of the total-to-static efficiency and in
Figure 10 for the stresses. At each call to the high-fidelity UQ module, the PC evaluates
19 designs around the optimal solution in order to assess the mean, the variance and
the sensitivities.

Before the first call, the Kriging prediction of the efficiency reaches already a high
accuracy of about 0.1% at Iteration 184. After the first call to the UQ module and the
enrichment of the database with the 19 designs generated by such verification process, the prediction accuracy for the efficiency is further improved, such that almost
all the samples from the Kriging and the high-fidelity simulations overlap. This outcome means that the mean and variance originated by the surrogate model almost
coincides with those generated by the higher-fidelity UQ module.

Additionally, Figure 11 shows the evolution of the gap between the variance prediction
emanating from the Kriging model based on Monte-Carlo against the high fidelity
results by the Probabilistic Collocation method at each activation of the UQ module.
The progressive drop in such curves demonstrates the increased accuracy of the
metamodel w.r.t. the expensive 3D simulations, confirming its suitability for the identification of an optimal robust design solution. In particular, the self-learning Kriging
reaches an optimal level of accuracy at iteration 185. The error in the efficiency estimation concerning the mean value for the following iterations is below 0.1%. However,
the optimization continues until iteration 187, as two more steps (with their relative
calls to the UQ module) are necessary to confirm also the robustness of the variance.

Figure 10 shows that the von Mises stresses follow the same convergence pattern as
the efficiency in Figure 9, but the Kriging and PC results start to fully overlap only at
iteration 186, one iteration later than the efficiency, as confirmed by Figure 11.

In total, the robust optimization evaluated 187 designs and called four times the PC module, which evaluates every time 19 additional geometries. Since the robust optimization is considered converged already at iteration 186, the overhead is only 30% compared to a deterministic optimization. The cost of all the Monte-Carlo simulations is negligible compared to the high-fidelity evaluation of a sample, and therefore does not contribute to the overhead.

The robust optimization was able to improve the objective of the baseline by 2.9% while reducing the variance by one order of magnitude.

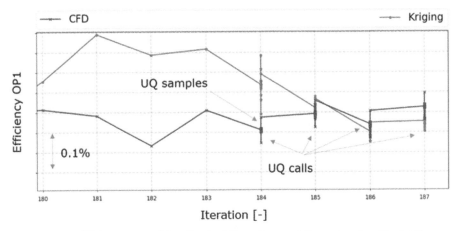

Figure 9. Efficiency trend for the last iterations of the robust optimization, including four activations of the UQ module.

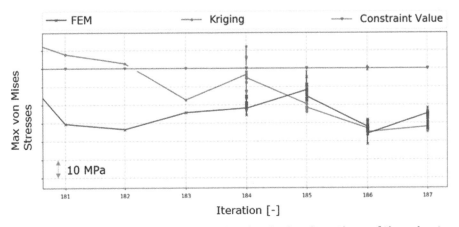

Figure 10. Von Mises stresses evolution for the last iterations of the robust optimization, including four activations of the UQ module.

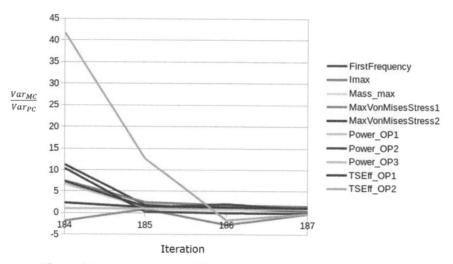

$$\frac{Var_{MC}}{Var_{PC}}$$

Iteration

Figure 11. Improvement of the accuracy of variance generated by Monte-Carlo on Kriging against the Probabilistic Collocation method.

6.3   Sensitivity analysis

A first method to evaluate how influential a certain random variable is on a performance parameter consists in checking how the interval of 6 standard deviations σ compares with the mean value μ $(6\sigma/\mu)$. Figure 12 shows the ratio $6\sigma/\mu$ generated by the low-fidelity Kriging-based Monte-Carlo and the higher-fidelity PC UQ module. The rather similar results from Kriging and PC reflect the level of accuracy reached by the metamodel after several optimization iterations. Moreover, the results show that the impact of the standard deviation on the von Mises stresses is more significant than on the other performance parameters, followed by the inertia (Imax), the first frequency mode, and the power at the highest mass flow operating point (OP3).

A second method for the evaluation of the impact of the uncertainties consists in using the results of the samples evaluated by the PC module to compute the sensitivity analysis of all the performance parameters for the random variables [9]. This analysis provides the pie charts shown in Figure 13 for the inertia, the peak stresses and the choking mass flow rate. It turns out the stresses are mainly influenced by the shroud thickness in proximity of a first location of high stress concentration, while the shroud radius at the trailing edge is responsible for the centrifugal forces determining a second peak in the impeller. Additionally, the inertia is influenced by the shroud thickness as well, since it increases the impeller mass, and by the DX_Flat variable controlling the volume of the wheel. Finally, the choking mass flow rate is influenced mainly by the blade trailing edge radius, with direct consequences on the power in OP3.

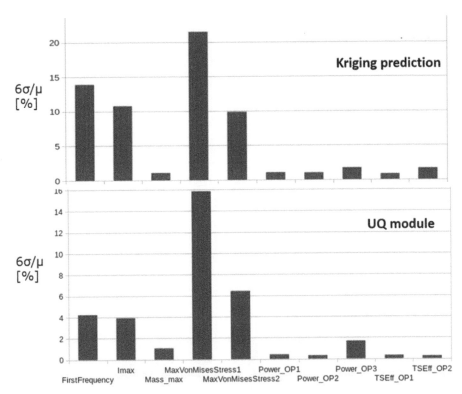

Figure 12. Monte-Carlo on Kriging predictions, compared to UQ module for all performance parameters.

7   CONCLUSIONS

The problem of the robust optimization of turbocharger components, with focus on a radial turbine impeller, is undertaken by the present work. The paper presents a novel optimization framework accounting for manufacturing uncertainties along the search for the optimal solution, yet preserving the computational budget within defined bounds.

The low-cost robust optimization methodology discussed herein accounts for surrogate models aimed at evaluating the mean and the variance of the selected design variables using the Monte Carlo method. A high-fidelity Uncertainty.

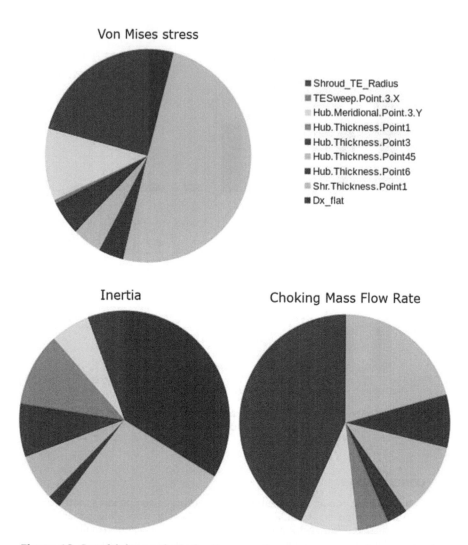

Figure 13. Sensitivity analysis for three performance parameters w.r.t. nine uncertain design variables.

Quantification technique based on the Probabilistic Collocation method is invoked in the last stages of the optimization process in order to validate the accuracy of the robustness predicted by the metamodel and to expand the database in regions of the design space requiring further exploration.

The application of the framework to the case of a multidisciplinary and multipoint optimization of a turbine impeller confirms the possibility of identifying an optimal yet robust solution within a limited computational budget, exceeding by 30% the duration of an equivalent deterministic optimization. The iterations executed by the optimizer to improve the robustness of the solution are discussed, along with the evaluation of the performance parameters mostly impacted by the uncertainties, and the identification of the key random variables controlling them.

In conclusion, it is shown that robust optimizations are a viable means for the design of turbomachinery components, ruling out the risk of generating unfeasible designs infringing any constraint due to the impact of manufacturing tolerances.

REFERENCES

[1] Mueller, L., Prinsier, T., Verstraete, T., Racca A. (2019) CAD Based Multidisciplinary Adjoint Optimization of a Radial Compressor Impeller, 24, Aufladetechnische Konferenz, Dresden, Germany, pp. 415–434

[2] Hirsch, C., Wunsch, D., Szumbarski, J., Pons-Prats, J. (2019) Uncertainty Management for Robust Industrial Design in Aeronautics.

[3] Montomoli, F., Carnevale, M., D'Ammaro, A., Massini, M., Salvadori, S. (2015) Uncertainty quantification in computational fluid dynamics and aircraft engines (pp. 33–34). Berlin: Springer.

[4] Ghisu, T., Shahpar, S. (2018) Affordable uncertainty quantification for industrial problems: application to aero-engine fans. Journal of Turbomachinery, 140(6).

[5] Verstraete, T. (2010) CADO: a computer aided design and optimization tool for turbomachinery applications. In Proceedings of the 2nd international conference on engineering optimization, Lisbon, Portugal (Vol. 69, p. 69).

[6] Najm, H. N. (2009) Uncertainty quantification and polynomial chaos techniques in computational fluid dynamics. Annual review of fluid mechanics, 41, 35–52.

[7] Seshadri, P., Shahpar, S., Parks, G. T. (2014) Robust compressor blades for desensitizing operational tip clearance variations. In Turbo Expo: Power for Land, Sea, and Air (Vol. 45608, p. V02AT37A043). American Society of Mechanical Engineers.

[8] Wang, X., Zou, Z. (2019) Uncertainty analysis of impact of geometric variations on turbine blade performance. Energy, 176, 67–80.

[9] Wunsch, D., Nigro, R., Coussement, G., Hirsch, C. (2019) Non-intrusive Probabilistic Collocation Method for Operational, Geometrical, and Manufacturing Uncertainties in Engineering Practice. In Uncertainty Management for Robust Industrial Design in Aeronautics (pp. 143–167). Springer, Cham.

[10] Lönn, D., Fyllingen, Ø., Nilssona, L. (2010) An approach to robust optimization of impact problems using random samples and meta-modelling. International Journal of Impact Engineering, 37(6),723–734.

[11] Lange, A., Voigt, M., Vogeler, K., Schrapp, H., Johann, E., Gümmer, V. (2010) Probabilistic CFD simulation of a high-pressure compressor stage taking manufacturing variability into account. In Turbo Expo: Power for Land, Sea, and Air (Vol. 44014, pp. 617–628).

[12] Garzon, V. E., Darmofal, D. L. (2003). Impact of geometric variability on axial compressor performance. J. Turbomach., 125(4),692–703.

[13] Javed, A., Pecnik, R., Van Buijtenen, J. P. (2016) Optimization of a centrifugal compressor impeller for robustness to manufacturing uncertainties. Journal of Engineering for Gas Turbines and Power, 138(11).

[14] Dellino, G., Kleijnen, J. P., Meloni, C. (2012). Robust optimization in simulation: Taguchi and Krige combined. INFORMS Journal on Computing, 24(3),471–484.

[15] Martin, J.D.; Simpson, T.W. (2005) Use of Kriging Models to Approximate Deterministic Computer Models. AIAA J., 43 (4), 853–863

[16] Smolyak S. (1963) Quadrature and Interpolation formulas for tensor products of certain classes of functions, Dokl. Adad. Nauk USSR B, 240-243.

[17] Aissa, M.H.; Verstraete, T. (2019) Metamodel-Assisted Multidisciplinary Design Optimization of a Radial Compressor. Int. J. Turbomach. Propuls. Power, 4, 35

[18] Loeven, G. J. A., Witteveen, J. A. S., Bijl, H. (2007) Probabilistic collocation: an efficient non-intrusive approach for arbitrarily distributed parametric uncertainties. In 45th AIAA Aerospace Sciences Meeting and Exhibit (p. 317).

[19] Golub, G. H., Welsch, J. H. (1969) Calculation of Gauss quadrature rules. Mathematics of computation, 23(106),221–230.

[20] Verstraete, T. (2008) Multidisciplinary Turbomachinery Component Optimization Considering Performance, Stress, and Internal Heat Transfer, PhD thesis, Von Karman Institute for Fluid Dynamics - University of Gent, Belgium.

[21] Mueller, L. (2019) Adjoint-Based Optimization of Turbomachinery with Applications to Axial and Radial Turbines. Ph.D. Thesis, Université libre de Bruxelles & von Karman Institute for Fluid Dynamics, Brussels, Belgium.

14th International Conference on Turbochargers and Turbocharging
Institution of Mechanical Engineers, ISBN: 978-0-367-67645-2

Multidisciplinary adjoint optimization of radial turbine rotors

L. Mueller[1], T. Verstraete[1], A. Racca[2]

[1]Von Karman Institute for Fluid Dynamics, Waterloosesteenweg Sint-Genesius-Rode, Belgium
[2]General Motors Global Propulsion Systems – Torino S.r.l., Corso Cas telfidardo Turin, Italy

ABSTRACT

Multidisciplinary optimization methods are routinely applied in product development processes in the automotive industry. The wide popularity gained by gradient-free methods – such as Evolutionary Algorithms – is due to their generality and relative simplicity in implementation. Nevertheless, the new performance goals demanded to the turbocharging business require more detailed geometrical investigations and comprehensive optimizations capable of capturing the interactions of the different involved physical phenomena. Such holistic view is not easily achievable by gradient-free methods which suffer from the so-called "curse of dimensionality" and high computational cost. A more feasible approach to richer design spaces is therefore offered by gradient-based methods, and in particular by the adjoint methods, that prove to have the highest potential in this field.

This paper discusses the development of a multidisciplinary gradient-based methodology, which includes the CAD model for the shape parametrization, hence allowing the involvement of manufacturing considerations within the optimization. In addition to the aerodynamic performance, mechanical and vibrational predictions are invoked as constraints in order to prevent the search beyond mechanically feasible shapes.

The method is applied to the design optimization of a turbocharger radial turbine rotor for automotive applications. Multiple operative conditions within the engine map are included in the definition of the objective function, allowing for optimal performance over a large range of regimes. The results demonstrate that with limited computational resources a competitive rotor design can be efficiently achieved through a holistic design approach.

Keywords: Multidisciplinary, Optimization, Adjoint, CAD, Radial Turbine

1   INTRODUCTION

The development of modern internal combustion engines is progressively growing in complexity because of the combination of the latest regulations, the need of providing excellent fuel economy, and the request of delivering the best driving experience. These essential elements have to be satisfied by the identification of cost effective design solutions while shortening the product development cycle. In order to cope with such competitive targets, the engineering community recognized an enabler in the concurrent engineering approach [1], offering a holistic view over the problem in object through the exploration of the mutual interactions of the engine subsystems at multidisciplinary level. This design philosophy proved to be effective in disclosing the optimal synergies among the different technologies, explored from system perspectives up to details level.

Nowadays, the challenge is in the identification of robust optimization methods supporting the experts' decision making process while dealing with such extended domain spaces and the consequent high number of design variables. Moreover, the successful exploitation of such techniques can be achieved only through a comprehensive multi-disciplinary outlook during the product optimization process, posing even more severe challenges to the adopted design workflow.

The implementation of gradient-free optimization methods, like Evolutionary Algorithms–boosted design tools [2], is an established practice in the automotive business. These techniques are proven to be robust in delivering high-quality design solutions and relatively simple in their application to industrial development processes. Moreover, the continuous improvement in reliability of the computational analysis methods allowed a progressive reduction in prototyping and validation testing. However, the advent of more stringent requirements and the need of evaluating larger domain spaces are exposing some intrinsic limitations of gradient-free methods, mainly related to the so-called "curse of dimensionality" [3]. Additionally, the integration of more physical disciplines in the design process and their mutual interactions increase significantly the computational cost, with non-negligible impacts even in the framework of HPC environments.

An opportunity is offered by the adoption of gradient-based methods and in particular adjoint methods [4, 5, 6], whose cost in the computation of the sensitivity derivatives of the objective function is essentially independent of the number of design variables.

This paper addresses the development of a novel multidisciplinary approach to the adjoint optimization of a Variable Geometry Turbine (VGT) impeller for automotive applications. Turbochargers indeed represent an outstanding example of technologies whose performance is sensitive to the interactions with the surrounding subsystems and the achievement of high operative efficiencies is well regarded all over the engine map.

The adjoint optimization of a radial compressor impeller was previously reported in [7]. The present paper is devoted to the development of an adjoint optimization architecture suitable for the design of the hot side of the turbomachine with focus on the complex multidisciplinary interactions determining the final geometry of the turbine rotor. A highly constrained test case is selected in order to demonstrate the challenges in the achievement of stretched performance targets while satisfying the component mechanical integrity and manufacturability aspects. In particular, the proposed optimization addresses the improvement of the total-to-static isentropic efficiency of the turbine rotor assessed over three relevant operative conditions in the engine map, in compliancy of robustness requirements evaluated by means of mechanical stresses and blade vibrational modes.

In light of this objective, the remaining portion of this document is structured as follows. First, the baseline design is presented. Then, the adjoint optimization framework is described outlining the numerical methods and tools involved, followed by the definition of the optimization problem. Finally, the results of the optimization process are discussed with a performance comparison with the original geometry and the conclusions are drawn from this study.

2 BASELINE DESIGN PROCESS

Figure 1 shows an exemplary representation of a Variable Geometry Turbine layout along with the baseline design of a periodic sector of the radial inflow turbine impeller considered in the present study. This preliminary geometry was generated using analysis tools developed at the Von Karman Institute for Fluid Dynamics (VKI) with special focus on turbomachinery applications [8, 9].

The extent of the baseline design investigation was limited to a minimum level, in order to gain sufficient confidence with the classes of design variables demonstrating to be more relevant for the optimization process. The resulting preliminary loading coefficient is about $\psi = 1.16$ at design point, with an exit flow coefficient $\phi = 0.55$. Although the operative conditions for this application lead to a highly loaded impeller blade, the initial results indicate the need of an optimization step in order to increase the total-to-static efficiency. Therefore, this paper is aimed at describing a novel approach conceiving the application of the in-house developed multidisciplinary adjoint optimization framework [10] to the baseline rotor geometry.

The choice of this specific test case is justified by the goal of proving the optimization procedure against a highly constrained problem, in which stretched performance targets are demanded while complying with manufacturing and mechanical constraints in the limited space available at the hub surface of the rotor. On this regard, Figure 1 shows the proximity of the blade hub fillet to the periodic surfaces, highlighting the tough problem expected in satisfying the targets of maximum von Mises stresses and blade first eigenfrequency.

Moreover, high values for the second and third vibrational modes are desired in order to extend the operative range of such Variable Geometry Turbine to high rotational speeds avoiding critical resonance conditions.

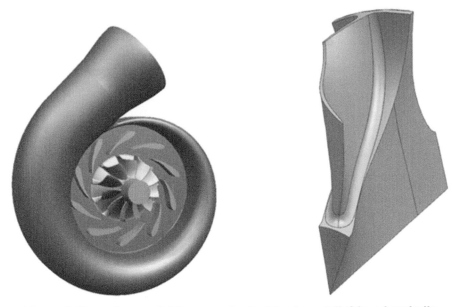

Figure 1. Exemplary variable geometry turbine layout (left) and periodic sector of the radial inflow turbine wheel (right).

In total, 83 degrees of freedom were identified for the optimization process. While in this work we opted for a rather low number of design variables in the meridional plane (22), 61 additional parameters were selected to control the shape modifications of the blade at different sections in the span-wise direction. Additional details are reported in section 3.

3   OPTIMIZATION FRAMEWORK

A brief summary of the optimization framework is presented here. A more detailed description of the system can be found in Refs. [7, 10]. The main components of this system are the gradient-based optimizer, the geometry and mesh generation, the analysis tools, and the gradient evaluation.

In this work, we use a Sequential Quadratic Programming (SQP) algorithm [11] with a line-search procedure to solve the optimization problem. The remaining components of the optimization framework evaluate the objective and the constraints as well as their gradients with respect to all design variables, which are required by the optimizer in each optimization step. They are briefly explained below.

3.1   Geometry parameterization

The geometry parameterization of the radial turbine is accomplished using B-Spline representations, where the position of the B-Spline control points are the design variables. In general, the three-dimensional model of the wheel is defined by 1) the meridional flow channel, 2) the blade camber surface, 3) a blade thickness distribution on several span-wise positions, and 4) the number of blades. In this work, the blade camber surface is constructed by integrating a blade-angle distribution at hub and shroud to allow for a non-radial fibered blading. Additionally, to obtain more complex bowed blade shapes, two mid-sections are used where the camber surface can be displaced in circumferential direction. Additional optimization parameters are a rake angle at the blade leading edge and a trailing edge cut-back to control the vibrational response of the wheel. Furthermore, several dependent parameters are used to account for manufacturing requirements, e.g., geometric tangency across patches or axisymmetric end-walls. However, some parameters such as the number of blades remain constant due to external constraints. As anticipated, in total, 83 free design parameters are used and are subject to optimization.

3.2   Mesh generation & update

The B-Spline surface model of the radial turbine wheel is used to derive the computational domain for the aerodynamic and structural analyses. In particular, two different approaches are used to discretize both domains as shown in Figure 2.

The fluid domain is discretized by a multi-block structured grid of about 1.3 Mio. cells with a boundary layer refinement at viscous walls, resulting in a y^+ value around unity. A constant tip clearance gap is included in the fluid model to account for its large impact on the loss generation inside the machine.

The solid domain is discretized by an unstructured grid of about 200k tetrahedral elements. Here, the hub fillet is modelled to prevent a large stress concentration in this region (Figure 2).

During the optimization process the modified B-Spline surface model acts as an intermediate layer between both domains, thus avoiding possible overlaps or voids between the two different meshes. For the unstructured grid, the mesh-deformation tool developed by Verstraete et al. [12] is used in which the volume grid is updated after updating the surface grid to comply with the new modified geometry. The computational mesh of the fluid domain is re-meshed in every iteration using elliptic grid generation where source terms [13, 14] are introduced to ensure common grid quality standards such as cell thickness, skewness, or expansion ratio.

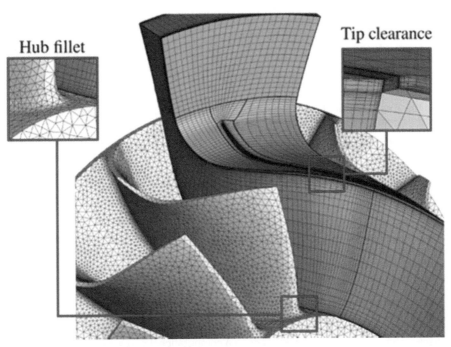

Hub fillet

Tip clearance

Figure 2. Computational meshes of the solid domain (full wheel in grey) and fluid domain (orange). Coarse meshes for improved visualization.

3.3 Analysis methods

3.3.1 *Computational fluid dynamics solver*

The governing equations are the compressible Reynolds-Averaged Navier-Stokes equations discretized in space with a cell-centered finite volume formulation on multi-block structured grids. The convective fluxes are computed by Roe's upwind scheme [15] with a MUSCL-type extrapolation [16] for 2nd-order spatial accuracy. Viscous fluxes are calculated with a central discretization. Turbulent effects are modelled with the negative Spalart-Allmaras turbulence model [17] assuming fully turbulent flow from the inlet (Re_{Inlet} = 450K based on the wheel diameter). For the temporal discretization, we use an implicit multi-stage Runge-Kutta scheme which has been proposed to stabilize discrete adjoint solvers for mildly unsteady flows [18]. Convergence to a steady-state solution is accelerated by local time-stepping and geometric multigrid.

The same time-marching method is used to solve the corresponding discrete adjoint equations. Both the flow governing equations and the turbulence model have been manually differentiated with selective use of algorithmic differentiation to reduce the development time [9].

3.3.2 *Computational structural mechanics solver*

In this work, the maximum von Mises stresses in the material as well as the natural eigenfrequencies are computed by the structural solver. Therefore, two different structural analyses are performed. For the maximum von Mises stresses, the linear elastic equations are discretized using quadratic mesh elements. The resulting linear system is solved with an iterative preconditioned conjugate gradient solver. To preserve continuity for the gradient calculation, the maximum von Mises stresses are approximated using the p-norm as follows:

$$\sigma_{vM,max} \approx \sqrt[p]{\frac{\int_V \sigma_{vM}^p dV}{\int_V dV}} \quad \text{with } p = 75, \tag{1}$$

For the frequency analysis, a generalized eigenvalue problem is iteratively solved, which is described in Ref. [19].

The algorithmic differentiation tool CoDiPack [21] has been used to derive a discrete adjoint model of the solver. The structural solver is also used to deform the unstructured mesh using a linear elastic analogy and its adjoint enables the gradient calculation of the structural mesh with respect to the design variables. More details on the structural solver and its adjoint are given in Refs. [19, 20].

3.3.3 *Moment of inertia computation*
The moment of inertia of the turbine wheel is computed at very low computational cost during the meshing of the solid domain. For this purpose, the algorithm proposed by Tonon [22] is used, which employs explicit formulas using the coordinates of the tetrahedral elements of the solid unstructured mesh.

3.4   Gradient evaluation
The computational cost of gradient-based optimization methods is significantly determined by the gradient calculation. Using inefficient methods to approximate these gradients, such as finite differences, leads to a computational cost which is proportional to the number of design variables. For example, in the present work, the computation of the aerodynamic gradients for a single operating point would require 83 + 1 CFD evaluations, which increases the optimization run-time rapidly beyond feasible limits. On the contrary, with the adjoint method, the computational cost is nearly independent of the number of design variables and scales only with number of objectives and constraints involved. Since for most of the engineering problems the number of objectives and constraints is much lower than the number of design variables, the adjoint method is the key enabling technology to solve large scale optimization problems.

In this work, the gradient of the cost function J with respect to the design variables α is evaluated by a two-step approach (Figure 3). Essentially, it decouples the adjoint solver from the geometry and grid generation and can be summarized as follows:

- Compute the sensitivity of the cost function with respect to the grid point coordinates dJ/dX.
- Compute the sensitivity of the grid point coordinates to the design variables dX/dα.

The final gradient is then calculated as scalar product of these two sensitivities as follows:

$$\frac{dJ}{d\alpha} = \frac{dJ}{dX}\frac{dX}{d\alpha} \tag{2}$$

The sensitivity of the cost function with respect to the grid point coordinates (dJ/dX) is computed by the CFD and CSM adjoint solvers for each objective and constraint. The complementary sensitivity information of the grid to the design variables (dX/dα) is computed in parallel to the adjoint solvers with the complex-step method.

466

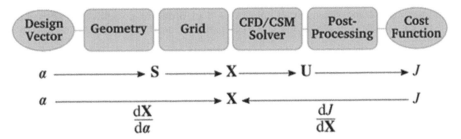

$$\alpha \longrightarrow S \longrightarrow X \longrightarrow U \longrightarrow J$$

$$\alpha \xrightarrow{\quad\quad\quad} \underset{\dfrac{dX}{d\alpha}}{} X \longleftarrow \underset{\dfrac{dJ}{dX}}{} J$$

Figure 3. Schematic representation of the gradient evaluation.

4   PROBLEM STATEMENT

The aim of this design optimization is to maximize the total-to-static isentropic efficiency of the Variable Geometry Turbine rotor over a wide range of operative conditions while minimizing the moment of inertia.

A multipoint approach accounts for three key points (namely OP1, OP2 and OP3, whose characteristics are reported in Table 1) relevant for the engine operations. Two points show closed vanes positions, as a demonstration of the intent of optimizing the rotor for high inlet swirl. The third operative condition is chosen in order to set the characteristics of the machine in the peak power region of the engine map.

Table 1. Turbine operative conditions.

	VANES POSITION	TURBINE SPEED	MASS FLOW RATE	EXPANSION RATIO
OP1	2deg from maximum closure	low	low	low
OP2	6deg from maximum closure	high	mid	high
OP3	mid-high opening	high	high	high

All points operate at different rotational speeds, with imposed inlet mass flow rate in OP1 and OP2 and varying turbine back-pressure. The third point is exposed to a fixed expansion ratio in order to maximize the choking mass flow rate. Additionally, at each specific operative condition the impeller should deliver a power output above the predefined minimum requirements. Finally, in order to maintain the structural integrity of the wheel, the maximum von Mises stresses and the first and second eigenfrequencies are constrained. In particular, the stresses are evaluated through the p-norm function reported in Eq.1 and integrated over the entire rotor sector represented in Figure 1 in order to control the mechanical response of both the blade and the back-plate region, till the connection with the shaft.

The resulting multidisciplinary optimization problem can be written as follows:

Maximize

$$Obj \equiv \omega_{Obj} \underbrace{\left(\omega_1 \frac{\eta_{ts,OP1}}{\eta_{ts,OP1,ref}} + \omega_2 \frac{\eta_{ts,OP2}}{\eta_{ts,OP2,ref}} \right)}_{Obj_{Aero}} + (1 - \omega_{Obj}) \underbrace{\left(1 - \frac{I_{xx}}{I_{xx,ref}} \right)}_{Obj_{Inertia}} \qquad (3)$$

Subjected to

$$Constr_1 \equiv 1 - \frac{P_{OP1}}{P_{OP1,ref}} \leq 0, \qquad\qquad Constr_2 \equiv 1 - \frac{P_{OP2}}{P_{OP2,ref}} \leq 0$$

$$Constr_3 \equiv 1 - \frac{\dot{m}_{OP3}}{\dot{m}_{OP3,ref}} \leq 0, \qquad\qquad Constr_4 \equiv \frac{\sigma_{vM,max}}{\sigma_{vM,max,ref}} - 1 \leq 0, \qquad (4)$$

$$Constr_5 \equiv 1 - \frac{f_{1st}}{f_{1st,ref}} \leq 0, \qquad\qquad Constr_6 \equiv 1 - \frac{f_{2nd}}{f_{2nd,ref}} \leq 0,$$

$$Constr_7 \equiv \frac{I_{xx}}{I_{xx,max}} - 1 \leq 0.$$

In these expressions, η_{ts} is the total-to-static isentropic efficiency, I_{xx} the moment of inertia with respect to the axis of rotation, $\dot{m}$ the mass flow, $\sigma_{vM,max}$ the maximum von Mises stresses, and f_{1st} and f_{2nd} the first and second eigenfrequency, respectively. The weighting coefficient $\omega_{obj} \in [0,1]$ in the objective function (Eq. 3) is a user-defined blending factor between the efficiency and moment of inertia terms. Additionally, the coefficients ω_1 and ω_2 determine the weights given to the efficiencies respectively at operating point OP1 and OP2.

One of the crucial points in adjoint optimization is the convergence rate of the CFD solver. In particular, at off-design operating conditions with strong secondary flow motion, standard iterative schemes often fail to converge to a steady-state solution, entering in so-called limit cycle oscillations around the stationary point. Applying the discrete adjoint approach in such situation, with the adjoint equations linearized around this point, may lead to divergence of the adjoint solver because the system matrix is not contractive, with some eigenvalues larger than unity [23]. Therefore, one solution to this problem, which has been adopted in this work, is to use an implicit time-marching scheme [18] with strong temporal damping able to stabilize the flow solver and the adjoint counterpart over a wide range of flow regimes. This is illustrated in Figure 4, which shows the convergence histories of both solvers at operating point OP1, characterized by high inlet swirl due to closed vanes position. As can be seen from this plot, both solvers feature identical convergence characteristics till machine level accuracy. Since such operative conditions must be considered in realistic design problems, the proposed method is a pivotal aspect for adjoint-based optimization in automotive applications.

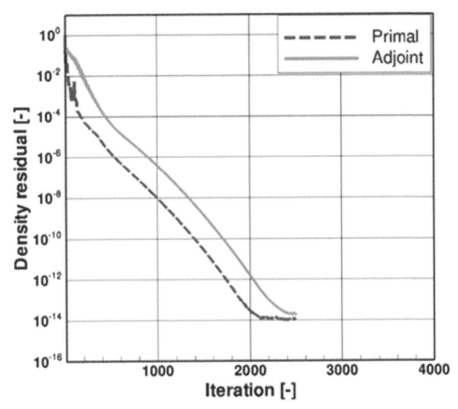

Figure 4. Convergence history of the CFD primal and adjoint solvers at OP1.

5   RESULTS

The convergence history of the optimization process is shown in Figure 5, including the evolution of the weighted objective function (top), the mechanical constraints (mid), and the individual aerodynamic constraints (bottom). The target regions highlighted in grey represent the areas where all the constraints of this design problem are satisfied, in accordance with Eq. 4. Geometries outside such target regions are considered as "infeasible" because not delivering the expected minimum performance on a pointwise perspective or not mechanically robust to safely operate within the boundaries of the engine map.

The optimization algorithm performed in total eighteen iterations till convergence of the design problem, in which the combined efficiency objective could be improved by about 3% relative to the datum geometry, at the expense of the rotor moment of inertia which experienced a deterioration of about 3% (Figure 5, top). Such scenario is considered acceptable based on the project goals, since the inertia increase still resides within the target region, as reported in the Mechanical Constraints chart (Figure 5, mid). The trend in efficiency improvement is almost specular to the inertia deterioration, as indication of the highly constrained problem described in the current paper.

Most of the advancement in the efficiency objective can be attributed to the evolution of the rotor meridional contour reported in Figure 6, where a reduction of blade trim and increase in hub contour radius in regions beyond the 25% of the blade chord is shown. Additionally, a significant increase in blade chord length is reported, aimed at decreasing the aerodynamic loading of the baseline geometry. Here the modifications address the efficiency in operative conditions at mid-low mass flow rates, fulfilling the constraints of turbine power in OP1 and OP2 while reducing the rotor choking mass flow rate in OP3, but still within the acceptable boundaries of the project (Figure 5, bottom).

The inertia deterioration is related to the longer rotor and locally increased blade thickness, whose countermeasures invoked by the optimization algorithm are represented by the trim reduction, a refined back-plate thickness, and a smaller hub contour radius in the area corresponding to the first 25% of the chord (Figure 6).

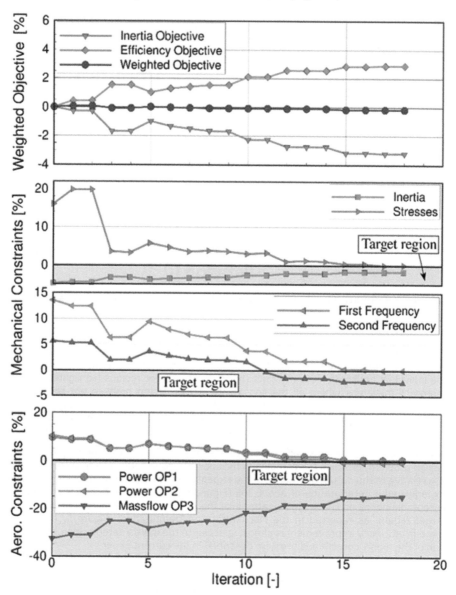

Figure 5. History of the weighted objective function (top), mechanical constraints (mid), and aerodynamic constraints (bottom). Constraint target regions indicated in grey.

As shown in Figure 5 (mid), the baseline design does not provide the required robustness to the problem of High Cycle Fatigue. The flattering issue at the blade exducer is addressed by optimization of the first eigenfrequency, demonstrating an improvement in excess of 15% respect to the datum.

The problem of fatigue related to the stator-rotor interaction in the variable nozzle turbine is indirectly tackled through optimization of the second eigenfrequency. With reference to standard analyses on the Campbell diagram, here the goal is to maximize the second eigenfrequency in order to avoid any critical intersection of the engine order characteristic of the nozzle vanes count with the second mode of the rotor within the operative range of the turbine. The optimization algorithm could increase such value of about 8%, improving the robustness of the rotor to the interaction with the stator vanes.

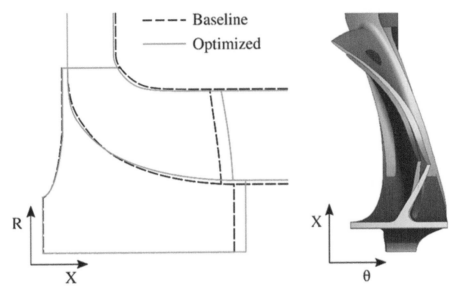

Figure 6. Comparison of the baseline turbine wheel with the optimized geometry. Meridional contours (left) and 3D blade shape (right).

Furthermore, the baseline impeller does not pass the structural requirements (Figure 5, mid). More specifically, the maximum von Mises stresses are about 15% too high. After the first iteration, the aerodynamic performance of the turbine mildly improved, at the expense of a further penalization in stresses of about 2% and a slight increase in moment of inertia. Such consideration, in combination with the reduced space available at the hub as shown in Figure 1, is the demonstration that the multidisciplinary optimization problem presented in this work is particularly difficult to solve.

On the other hand, starting from the fourth iteration the optimization algorithm could recover the mechanical constraints till the achievement of the target regions, while the aerodynamic performance continued evolving towards a progressive improvement. The distribution of maximum von Mises stresses is reported in Figure 7, comparing the baseline design with the optimized layout. Indeed, it is proven a lower value for the peak of stress is experienced at the hub of the blade on the pressure side.

A comprehensive summary of all performance and constraint values for the optimized geometry relative to the datum layout is reported in Table 2.

Table 2. Optimized Vs. baseline turbine rotor performance and constraints comparison.

Performance, Constraint	Optimized Vs. Baseline
Efficiency OP1	+1.9%
Efficiency OP2	+3.9%
Power OP1	+10.1%
Power OP2	+12.5%
Mass flow OP3	-13.4%
Inertia	+3.2%
Max. Stress	-15.2%
1st Frequency	+15.6%
2nd Frequency	+8.5%

Finally, the aerodynamic performance of the optimized blade shape is analysed in terms of isentropic Mach number distribution and compared with the datum layout, considering the operative condition OP2 at medium mass flow rate and high expansion ratio. Figure 8 refers to a section at 80% of the blade span. As can be seen, the optimized geometry is less loaded, as a result of the longer blade. Moreover, concerning the baseline geometry, the flow on the suction side surface decelerates after the 20% of the chord. This phenomenon is not visible on the optimized shape, presenting a sustained flow acceleration till the trailing edge, resulting in a thinner boundary layer and thus lower profile losses.

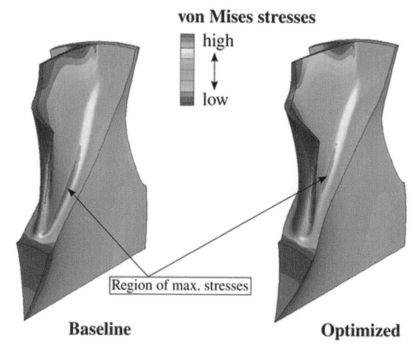

Figure 7. Comparison of von Mises stresses.

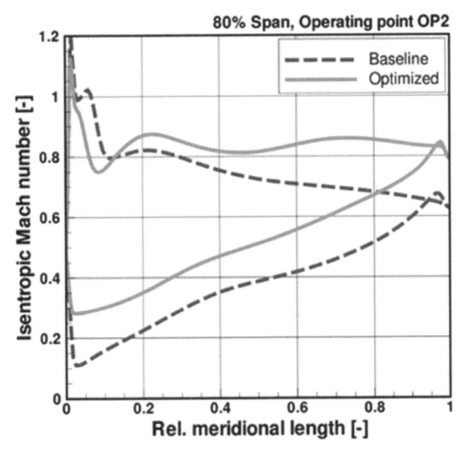

Figure 8. Blade isentropic Mach number distribution at 80% span at operating point OP2.

6   CONCLUSIONS

This paper addresses the problem of the multidisciplinary and multipoint adjoint optimization of a variable geometry radial turbine impeller for automotive applications. A highly constrained test case is selected in order to verify the performance of the novel gradient-based optimization framework developed at the von Karman Institute. The aim of the study is to demonstrate the possibility of increasing the total-to-static isentropic efficiency of a highly loaded turbine blade and decreasing its moment of inertia while complying with the manufacturability requirements and the mechanical constraints in terms of stresses and first and second eigenfrequencies.

The CAD-based parametrization approach, the mesh and analysis procedures, and the gradients evaluation methods are introduced. Moreover, a Sequential Quadratic Programming algorithm is adopted for the evaluation of the adjoint sensitivities and the evolution of the rotor geometry.

The optimization results demonstrate that the direct application of a gradient-based method to a manually-designed baseline geometry is effective in producing a high-quality solution satisfying the project targets in only eighteen design iterations. In particular, the combined total-to-static isentropic efficiency objective improved by about 3% respect to the original value, with an increase in moment of inertia controlled within the project admissible limit, while maintaining a non-radial fibered blade configuration, which has a higher number of degrees of freedom respect to a traditional radial-fibered layout.

Additionally, all mechanical constraints were satisfied, leading to a design solution compatible with the limit of maximum von Mises stresses within the range of operative speed and temperature of the turbine, and compliant with the requirements for improved robustness to High Cycle Fatigue.

The work reported in this paper is a second step towards large-scale optimizations of variable geometry turbochargers by gradient-based methods where in the future additional degrees of freedom and more physical models will be included.

REFERENCES

[1] Quan, W., Jianmin, H., "A study on collaborative mechanism for product design in distributed concurrent engineering", in 2006 7th International Conference on Computer-Aided Industrial Design and Conceptual Design, IEEE, 2006.

[2] Abbas, H.A., Sarker, R., and Newton, C., "PDE: A Pareto-Frontier Differential Evolution Approach for Multi-objective Optimization Problems", in Proceedings of the Congress on Evolutionary Computation, 2, pp.971–978, Piscataway, New Jersey, USA, 2001

[3] Trunk, G.V., "A Problem of Dimensionality: A Simple Example", IEEE Transactions on Pattern Analysis and Machine Intelligence, PAMI-1, 3, 1979

[4] Pironneau, O., "On Optimum Design in Fluid Mechanics", Journal of Fluid Mechanics, 64, pp.97–110, 1974

[5] Jameson, A., "Aerodynamic Design via Control Theory", Journal of Scientific Computing, 3(3),pp.233–260, 1988

[6] Giles, M.B., Duta, M.C., Mueller, J.D., and Pierce, N.A., "Algorithmic Developments for Discrete Adjoint Methods", AIAA Journal, 41(2), pp.198–205, 2003

[7] Mueller, L., Prinsier, T., Verstraete, T., Racca A., CAD Based Multidisciplinary Adjoint Optimization of a Radial Compressor Impeller, 24, Aufladetechnische Konferenz, Dresden, Germany, pp. 415–434, 2019

[8] Verstraete, T., 2008, Multidisciplinary Turbomachinery Component Optimization Considering Performance, Stress, and Internal Heat Transfer, PhD thesis, Von Karman Institute for Fluid Dynamics - University of Gent, Belgium.

[9] Mueller, L., 2019, Adjoint-Based Optimization of Turbomachinery with Applications to Axial and Radial Turbines. Ph.D. Thesis, Université libre de Bruxelles & von Karman Institute for Fluid Dynamics, Brussels, Belgium.

[10] Mueller, L., Verstraete, T., 2017, CAD Integrated Adjoint-Based Optimization of a Turbocharger Radial, Turbine, Int. J. Turbomach. Propuls. Power, 2(3), 14.

[11] Nocedal J.; Wright S.J. Numerical Optimization, volume 2. Springer Science + Business Media, LLC, 2006.

[12] Verstraete, T., Mueller, L., and Mueller, J.-D., 2017, CAD-Based Ad-joint Optimization of the Stresses in a Radial Turbine, In Proceedings of ASME Turbo Expo 2017, GT2017-65005.

[13] Steger, J., and Sorenson, R., 1979, Automatic Mesh-Point Clustering Near a Boundary in Grid Generation with Elliptic Partial Differential Equations, J. Comput. Phys., 33, pp. 403–410.

[14] Thomas, P.D. and Middlecoff, J.F., 1980, Direct Control of the Grid Point Distribution in Meshes Generated by Elliptic Equations, AIAA Journal, 18(6), pp. 652–656.

[15] Roe, P., 1981, Approximate Riemann Solvers, Parameter Vectors, and Difference Schemes, J. Comput. Phys., 43, pp. 357–372.

[16] van Leer B. Towards the Ultimate Conservative Difference Scheme, V. A Second Order Sequel to Godunov's Method. Journal of Computational Physics, 32(1):101–136, 1979.

[17] Allmaras, S.R., Johnson, F.T., Spalart, P.R., 2012, Modifications and Clarifications for the Implementation of the Spalart-Allmaras Turbulence Model. In Proceedings of the 7th International Conference on Computational Fluid Dynamics (ICCFD7-1902), Big Island, HI, USA.

[18] Xu, S., Radford, D., Meyer, M., Müller, J.-D., 2015, Stabilisation of Discrete Steady Adjoint Solvers, J. Comput. Phys., 299, pp. 175–196.

[19] Schwalbach, M., Verstraete, T., Gauger, N.R., 2019, Discrete Adjoint Gradient Evaluations for Linear Stress and Vibration Analysis, Computing and Visualization in Science, Springer.

[20] Schwalbach, M., Verstraete, T., Müller, J.-D., Gauger, N.R., 2019, A Comparative Study of Two Different CAD-Based Mesh Deformation Methods for Structural Shape Optimization, In Evolutionary and Deterministic Methods for Design Optimization and Control With Applications to Industrial and Societal Problems, Springer, pp. 47–60.

[21] Sagebaum, M., Albring, T., Gauger, N.R., 2017, High-Performace Derivative Computation using CoDi-Pack, arXiv preprint arXiv: 1709.07229.

[22] Tonon, F., 2004, Explicit Exact Formulas for the 3-D Tetrahedron Inertia Tensor in Terms of its Vertex Coordinates. J. Math. Stat., 1, pp. 8–11.

[23] Campobasso, S.G., and Giles, M.B., 2003, Effects of Flow Instabilities on the Linear Analysis of Turbomachinery, J. Propuls. Power, 19(2), pp. 250–259.

An electric-potential turbocharger speed sensor

P.M. Fussey, C. Bennett, A. Kermani

Department of Engineering, University of Sussex, Brighton, UK

ABSTRACT

An electric-potential (EP) sensor detects changes in an electric field. The authors have applied this principle to produce a low-cost turbocharger speed sensor which generates a signal with a high signal to noise ratio across the elevated speed ranges found in automotive applications. A range of packaging and signal processing approaches have been investigated for application in the hostile automotive environment. The sensor has also been tested with a damaged compressor wheel to simulate a turbocharger failure. The paper concludes with details of how the sensor can improve engine control and diagnostic algorithms.

1   INTRODUCTION

Over the last decade, the drive for reduced emissions and fuel consumption has increased the proportion of turbocharged engines [1]. Improvements to the control of the turbocharger can be achieved by measuring the turbocharger speed which allows the turbocharger to operate closer to its limits whilst also providing information to diagnose faults with the turbocharger and other engine components.

There are several existing methods used to measure turbocharger speed, reviewed in [2]. They include; passive and active eddy current sensors, variable reluctance sensors, Hall effect sensors and acoustic emissions. The sensors need to operate at temperatures above 160°C, at vibration levels of up to 50g whilst being robust, low cost and suitable for high volume production.

This paper presents a new method for measuring the speed of automotive turbochargers to address the challenges with existing methods, providing a cost effective solution that is both robust and generates a signal with a high signal to noise ratio at the elevated speeds found in a turbocharger. The new sensor uses a measurement of the electric potential (EP) to detect individual compressor blades passing the sensor, from which, the instantaneous speed of the turbocharger can be calculated. There are several potential applications for the turbocharger speed; turbocharger control, engine control and diagnostics.

Turbocharger control seeks to control the state of the turbocharger. Since the operation of the turbocharger is defined by its speed, the speed can be considered as the state of the turbocharger. Turbocharger control typically targets an intake manifold pressure and measures manifold pressure signal for feedback, providing an indirect control of the turbocharger state. Speed feedback from the turbocharger improves the control of the turbocharger by directly controlling the state of the turbocharger [3]. This is beneficial when adapting for operating the vehicle at different altitudes [2].

In addition to boost pressure control, the turbocharger speed can be used to estimate engine operating conditions. In [4] the authors estimate time varying exhaust pressure pulsations to gain insight into the engine combustion. Whilst [5] uses the turbocharger speed fluctuations to diagnose variations in injector performance.

The turbocharger is an integral component in automotive combustion engines, and it is essential to detect potential failures as soon as possible. Currently, turbocharger failures may be detected through a loss in boosting performance. This paper also investigates whether the EP sensor can provide advance warning of an imminent failure of the turbocharger itself, to allow a limp-home strategy to be adopted and avoid leaving the vehicle stranded.

This paper is structured as follows; Section 1 reviews existing turbocharger speed sensors and introduces the EP sensor, Section 2 presents the experimental investigation, Sections 3 and 4 discuss the application to engine control and diagnostic algorithms with conclusions presented in Section 5.

1.1 Turbocharger speed sensing technologies

Eddy current sensors [6] have been used in production for over 10 years and are being continuously updated, for example with active sensors [2] and low temperature co-fired ceramic sensors [7], [8]. However, the signals from an eddy current sensor typically require a dedicated and relatively expensive, ASIC to process the signal.

Variable reluctance sensors have been in production for over 15 years [2]. They are low cost but require a ferrous target wheel, which limits their use in many current turbochargers that have aluminium compressor blades. In addition, the signal magnitude varies with speed, again requiring expensive signal processing. Due to their low cost and robustness, research is continuing in exploiting their usage [9].

Hall effect sensors [3] also need a ferrous target wheel and since the electronics are situated near the sensor tip, the temperature range of the sensor is limited. These sensors are used on research engines, with little or no examples in mass production.

Finally, acoustic emissions measurements [10] have challenges with both signal processing to isolate the turbocharger signal from the combustion noise and secondly to establish speed fluctuations from the mean turbocharger speed. For these reasons, there are few examples adopting this approach.

1.2 Electric potential sensor

The EP sensor has been developed at the University of Sussex over many years, with applications ranging from electric cardiological measurements through to detecting fingerprints [11].

The EP sensor can be modelled as a perfect voltmeter/voltage amplifier with an input impedance approaching infinity and where no conduction current is drawn from the sample under test. The sensor utilizes a positive feedback scheme to achieve an input resistance of 2 GΩ and an input capacitance less than 1 pF. The sensor bandwidth spans from 100 Hz up to 2 MHz and is DC stable which eliminates the need for zeroing circuitry to stop the sensor from saturating. This simplification also reduces potential sources of noise.

The measurement principle is illustrated in Figure 1. The tri-axial sensing electrode of the EP sensor is mounted in the compressor housing. The inner core of the tri-axial electrode acts as the sensing electrode while the middle conductor carries an active guard signal and the outer conductor is grounded. Every time a grounded object (individual blades of the turbo in this case) approaches the sensing electrode, a voltage is induced on the sensing electrode. The ultra-high input impedance EP sensor enables the measurement of this induced voltage signal. The sensing electrode is only capacitively coupled to the blades making the measurement both mechanically and electrically non-invasive (i.e. There is no resistive contact between the electrode and blades).

The sensing electrode is completely passive and can be fabricated from variety of alloys with low or high electrical conductivity. This allows for flexibility in mounting the sensing electronics, they can be located remotely to reduce circuit complexity and thermal noise effects.

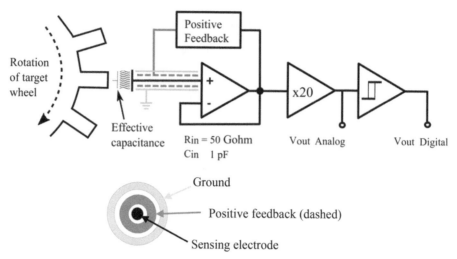

Figure 1. EP sensor operating principle, showing the change in capacitance as the grounded target wheel passes the end of the electrode. Also showing simplified conditioning circuit and cross-section through the tri-axial electrode.

1.3   Automotive turbocharging

An automotive turbocharger has a radial compressor attached to a radial turbine via a shaft. The compressor is typically made from aluminium and the turbine from steel. The performance of the turbocharger is constrained by several limits. The surge limit when the compressor blades stall, a choke limit and a turbocharger speed limit.

The work in this paper describes the application of the EP sensor to measure the speed of an automotive turbocharger to allow the turbocharger speed to be controlled within these limits. Whilst in current engine controllers, turbocharger speed can be estimated from pressure ratios and operating conditions, there is an error band that requires the control to be conservative to avoid exceeding the constraints. A turbo- charger speed signal offers the potential to run the turbocharger closer to the limits and therefore improve the overall performance. A second, potential benefit of the direct feedback on the turbocharger speed is the capability to diagnose faults and adapt to changing con- ditions. For example, local ambient conditions or gradual degradation over the lifetime of the turbocharger.

Automotive sensors are typically used in mass production and therefore are sensitive to piece cost. The piece cost is made up from the sensing element, mounting and signal conditioning. In the case of the EP sensor, the costs are minimised through a low requirement for signal processing, a cost-effective sensing element and location of electronics remote from turbocharger allowing packaging flexibility and keeping away from high temperatures.

Potential locations for a turbo speed sensor are the compressor, shaft or turbine. The turbine is exposed to exhaust gases and will reach temperatures too high to allow a cost-effective sensor element to be used. The shaft is often used for turbo- speed measurements on large engines where a feature is added to the shaft to allow the speed to be detected. These sensors typically use hall or variable reluctance principles to detect the speed.

The compressor was selected for this application since it did not require an additional feature to be added to the shaft. Secondly, with the multiple blades, it offers a higher resolution than a single pulse per revolution. And additionally, the sensor is further away from the high temperatures in the turbine.

2   EXPERIMENTAL INVESTIGATION

2.1   Objectives
The objectives of the experimental investigation were to assess the potential of an EP sensor for measuring turbocharger speed and detecting out of balance forces within the turbocharger, simulating a potential fault.

2.2   Test configuration
The EP sensor was fitted to a typical turbocharger, Figure 2. The turbocharger was mounted in a test rig with a lubrication circuit. A compressed air supply was connected to the turbine to spin up the turbocharger to assess the sensor performance without using an engine. Two compressed air systems were used, a low pressure system for initial assessment followed by a larger system for achieving speeds over 200,000 rev/min.

Figure 2. Photographs of sensor tip location, looking at the compressor housing without the compressor wheel fitted, and of the prototype sensor, probe and associated amplifier circuit.

2.3   Sensing element
The sensing element is a tri-axial electrode without any electronics. The electronics are mounted at the remote end of the electrode, allowing their location away from the turbocharger. For these tests, the electrode used in these measurements is approximately 100mm long, see Figure 2.

The location of the sensor tip is shown in Figure 2. The location was selected to target the longer of the compressor blades (the compressor has interleaved, long and short blades in an alternating pattern).

2.4   Assessment of raw signal

The experimental development of the sensor started with an assessment of the raw signal from the EP sensor. The signal was captured in a digital acquisition system at 2.0 MHz. The signal quality versus turbocharger speed is examined by inspecting of the time history, Figure 3, shows a clean signal suitable for turbocharger speed calculations. There is a smooth, cyclic signal with a peak for each blade passing. The compressor wheel has 12 blades.

The frequency analysis in Figure 4. shows the main blade passing frequency (red), peaking at approximately 14,000 Hz (corresponding to 70,000 rev/min). This can also be expressed as the 12th order fundamental frequency. There is a harmonic of this frequency at double the blade passing frequency (light blue) and some lower order content including at the rotational frequency (first order).

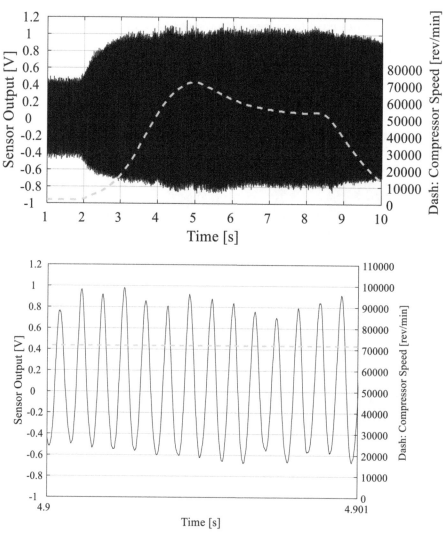

Figure 3. Raw sensor output for spin up to approx. 70,000 rev/min with the standard compressor wheel. Zoom on lower plot.

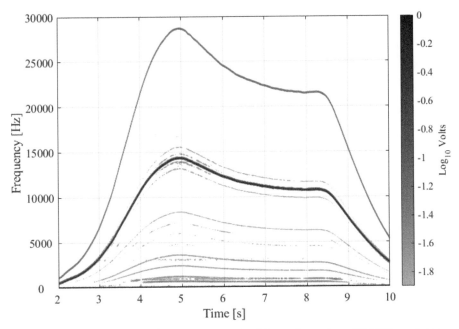

Figure 4. Sensor output frequency content for spin up to ~70,000 rev/min. This figure also highlights the good performance at low speeds, down to ~3,000 rev/min. Main blade passing frequency can be seen as dominant frequency with harmonics present.

2.5   High speed testing

The test rig was connected to a high-pressure air supply and spun up to speeds in excess of 200,000 rev/min. Similar results were observed in the time and frequency domain, Figure 5. Above 200,000 rev/min, there is an increase in noise, however the main blade passing frequency is still clearly dominant and suitable for speed measurement.

481

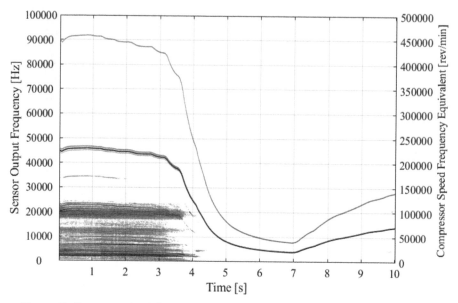

Figure 5. Sensor output frequency content for spin up to approx. 200,000 rev/min. Main harmonics can be seen clearly, above the increase in low frequency noise.

2.6   Protrusion

A key design parameter is the distance of the electrode tip from the compressor wheel. Closer improves signal but increases risk of hitting the compressor. The sensitivity of the induced current to the separation between the target and sensor was measured. As expected, the amplitude of the signal reduces with increased distance, Figure 6.

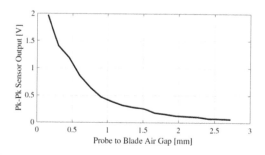

Figure 6. Sensor peak to peak output as a function of air gap.

2.7   Imbalance

An imbalance was introduced on the compressor wheel to assess the feasibility of the EP sensor to diagnose a turbocharger that is failing. The imbalance was created by machining material from one side of the compressor, see Figure 7.

482

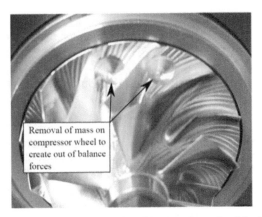

Figure 7. Modified compressor wheel to create out-of-balance force.

From Figure 8 the sensor signal has a 12th order fundamental frequency modulated by first order.

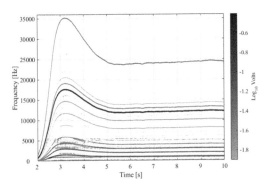

Figure 8. Sensor output frequency content for spin up to approx 90,000 rev/ min with the unbalanced compressor wheel. Additional harmonics visible compared to Figure 4.

3   INTEGRATION INTO AUTOMOTIVE CONTROL SYSTEMS

3.1   Signal processing

The EP sensor only requires a simple signal conditioning circuit, as shown in Figure 1. For this application, it was found that the addition of a differentiator circuit improved the digital output by removing the DC offset. Combination with a Schmitt trigger generates a series of pulses suitable for direct input into an Engine Control Unit (ECU). Electronics have been designed for a compact form factor using cost effective components with differential and Schmitt trigger.

A Schmitt trigger was designed to convert the analogue signal into a digital waveform. The clean signal from the analogue electronics enabled the use of an off-the-shelf

Schmidt trigger gate IC to convert the analogue waveform to a square wave. The design was iterated to set the voltage thresholds for the desired operating speed range, but otherwise the implementation was straightforward.

Whilst local processing of the pulse train could be envisaged, this would add cost to the sensor, and it is more cost effective to use additional input channels of the main ECU processor.

3.2   Robustness

When considering the development of sensors, it is important to quantify their robustness to disturbances. For the measurement of the electric field potential, the sensor should be independent of variations in temperature and humidity.

Changes in temperature will vary the DC offset of the sensor. The impact of this drift can be eliminated by using the differential of the original signal, thus removing the DC offset. Changes in humidity have been found to have no significant impact on the EP sensor, as shown in [12].

Interference of the signal could occur from the sensing tip to the sensor electronics and from the sensor to the ECU, however, the raw sensor signal is over 1V and shielded so has a good signal to noise level. The signal being transmitted from the sensor to the ECU is the output from a Schmitt trigger, giving a standard pulse train which will also have a robust signal to noise level.

3.3   Cost

The sensor has been developed with cost in mind. Minimal electronic components are required, and the probe is fabricated from low cost materials. Mass production costs are difficult to estimate at this stage, but the target would be to achieve a cost of the order of one US dollar for the sensor electronics, probe and housing.

4   APPLICATIONS FOR TURBOCHARGER SPEED SIGNAL

4.1   Engine control

The principal application for the turbocharger speed sensor is to provide a speed feedback signal to engine control algorithms. The turbocharger speed defines the turbocharger operating point.

The turbocharger performance is often illustrated with a compressor performance map with contours of compressor efficiency at different turbocharger speeds, pressure ratios and mass flows. The quantities are typically corrected to reference pressures and temperatures [13]. The performance map shows the operational area of the compressor to the right of the surge line, which is the boundary between stable and unstable operation. The surge line represents the maximum pressure ratio for a given flow rate. If the flow rate is reduced at a constant pressure ratio and crosses the surge line, local flow reversal occurs at the boundary layer followed by full flow reversal, resulting in a drop in pressure ratio. With the reduced pressure ratio, the flow can now re-establish and a periodic instability is established that can damage the turbocharger and reduce performance.

A challenge when controlling the turbocharger is to operate at maximum efficiency whilst avoiding the compressor surge. Turbochargers are controlled at the turbine side with a bypass valve (wastegate) or a variable geometry turbine that has vanes to throttle or direct the flow. These actuators control the flow of exhaust gases through the turbine and hence control the energy extracted from the exhaust gases. At a given pressure ratio, the turbocharger speed is a function of the mass flow and pressure ratio.

The direct speed feedback allows the controller to extend the operation of the turbocharger, closer to the boundaries of the compressor map, allowing increased performance.

The turbocharger control must also compensate for reduced pressure at altitude, typically taking a conservative approach to ensure component protection. In these situations, the turbocharger speed can improve the performance by allowing operation closer to the limits through increased confidence in the operating point of the compressor.

4.2   Turbocharger diagnostic

The failure of a turbocharger can be catastrophic, leaving the vehicle stranded and causing expensive damage to the engine. Turbochargers can fail for many reasons, see [14] for an analysis of typical causes of failures. For these reasons it is important to diagnose a potential turbocharger fault before the complete failure of the boosting system. In this section, a diagnostic algorithm is proposed using the speed signal from the unbalanced turbocharger speed signal.

The results with the unbalanced compressor rotor demonstrate the potential for the sensor to detect a faulty turbocharger. The signals from the sensors have been analysed further to assess a potential diagnostic algorithm.

The raw signals were analysed by calculating the speed based on one full revolution, by taking the time for 12 blades to pass. In addition, the speed can also be calculated by considering the speed for one blade to pass. Since the speed fluctuates, this signal appears as noisy for the unbalanced rotor whereas it shows significantly less variation for the balanced rotor.

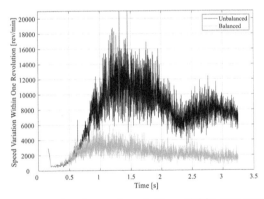

Figure 9. Compressor wheel speed variation within one revolution for the unbalanced (black) and balanced (grey) rotors.

485

These results led to the analysis of the speed variation within one revolution. The speed variations are shown in Figure 9 and demonstrate a clear separation between the balanced and unbalanced rotors. This metric can therefore be used to identify a damaged rotor and trigger the appropriate failure management within the ECU.

4.3 Engine diagnostic

In addition to engine control improvements and turbocharger diagnostics, the instantaneous turbocharger speed can be used to enhance the diagnostics of various engine components. In [5], the authors propose using turbocharger speed fluctuations to monitor injector performance, correcting for drift and detecting misfire events. Two approaches have been studied, based on the time varying acceleration of the turbo- charger which shows good correlation to exhaust manifold pressure, or examination of the frequency content in the turbocharger speed signal. In addition to injector monitoring, a broad range of diagnostics are presented in [2]; detecting clogged air filters, improve EGR diagnostics, improve air leak detection and compression efficiency monitoring.

5 CONCLUSIONS

This paper presents a novel and cost-effective sensing method for turbocharger rotational speed measurements using the Electric Potential (EP) sensor. The method has been successfully tested on an automotive turbocharger for the first time and outputs a clean signal at high rotational speeds. The signal has several uses from improving engine control to reduce emissions and fuel consumption, to engine diagnostic algorithms. This work also examined the signal from a damaged turbocharger and presents a diagnostic algorithm to detect the damaged component.

Future work will take this sensing approach and apply to a firing engine, where the ability of the sensor to operate in a more hostile environment will be assessed. In addition, since the sensor has a relatively high resolution for turbocharger speed, the ability of the sensor to detect cylinder pulsations will also be assessed. Production activities would include cost optimisation and durability development.

ACKNOWLEDGMENT

The authors would like to thank the University of Sussex for funding this work as part of the Research Development Fund.

REFERENCES

[1] A. J. Feneleya, et al, "Variable geometry turbocharger technologies for exhaust energy recovery and boosting a review" Renewable and Sust. Energy Rev, 2017.
[2] J. Tigelaar, et al, "Utilization of turbocharger speed data to increase engine power and improve air path control strategy and diagnostics," SAE Technical Paper 2017-01-1068, 2017.
[3] R. Holmbom, et al, "Implications of using turbocharger speed sensor for boost pressure control," IFAC-Papers On Line, vol. 50, no. 1, pp. 11 040–11 045, 2017, 20th IFAC World Congress.

[4] V. Macian, et al, "Exhaust pressure pulsation observation from turbocharger instantaneous speed- measurement," Meas. Sci. Technol, vol. 15, 2004.

[5] M. Becciani, et al, "Innovative control strategies for the diagnosis of injector performance in an internal combustion engine via turbo speed," Energies, 2019.

[6] J. D. Rickman, "Eddy current turbocharger blade speed detection," IEEE Transactions on Magnetics, vol. Mag-18, no. 5, pp. 1014–1021, Sept 1982.

[7] M. Ihle, et al, "Low temperature co-fired ceramics technology for active eddy current turbocharger speed sensors," Microelectronics International, 2018.

[8] N. Jamiaa, et al, "Modelling and experimental validation of active and passive eddy current sensors for blade tip timing," Sensors and Actuators A: Physical, 2019.

[9] T. Addabbo, et al, "Instantaneous rotation speed measurement system based on variable reluctance sensors for torsional vibration monitoring," IEEE Transactions on Instrumentation and Measurement, vol. 68, no. 7, pp. 2363–2373, 2019.

[10] V. Ravagliolia, et al, "Automotive turbochargers power estimation based on speed fluctuation analysis," Energy Procedia, no. 82, pp. 103–110, 2015.

[11] P. Watson, et al, "Imaging electrostatic fingerprints with implications for a forensic timeline," Forensic Science International, vol. 209, pp. e41–e45, 2011.

[12] L. H. Ford, "The effect of humidity on the calibration of precision air capacitors," J of the Inst. of Elec. Eng - Part II: Power Eng. vol. 95, no. 48, pp. 709–712, 1948.

[13] J. B. Heywood, Internal Combustion Engine Fundamentals. McGraw Hill, 2018.

[14] J. Filipczyk, "Causes of automotive turbocharger faults," Transport Problems, vol. 8, no. 2, 2013.

Author Index